T0141831

Advances in Intelligent Systems and Computing

Volume 1055

The series "Advances in Intelligent Systems and Computing" contains publications on theory, applications, and design methods of Intelligent Systems and Intelligent Computing. Virtually all disciplines such as engineering, natural sciences, computer and information science, ICT, economics, business, e-commerce, environment, healthcare, life science are covered. The list of topics spans all the areas of modern intelligent systems and computing such as: computational intelligence, soft computing including neural networks, fuzzy systems, evolutionary computing and the fusion of these paradigms, social intelligence, ambient intelligence, computational neuroscience, artificial life, virtual worlds and society, cognitive science and systems, Perception and Vision, DNA and immune based systems, self-organizing and adaptive systems, e-Learning and teaching, human-centered and human-centric computing, recommender systems, intelligent control, robotics and mechatronics including human-machine teaming, knowledge-based paradigms, learning paradigms, machine ethics, intelligent data analysis, knowledge management, intelligent agents, intelligent decision making and support, intelligent network security, trust management, interactive entertainment, Web intelligence and multimedia.

The publications within "Advances in Intelligent Systems and Computing" are primarily proceedings of important conferences, symposia and congresses. They cover significant recent developments in the field, both of a foundational and applicable character. An important characteristic feature of the series is the short publication time and world-wide distribution. This permits a rapid and broad dissemination of research results.

** Indexing: The books of this series are submitted to ISI Proceedings, EI-Compendex, DBLP, SCOPUS, Google Scholar and Springerlink **

More information about this series at http://www.springer.com/series/11156

Kim-Kwang Raymond Choo ·
Thomas H. Morris · Gilbert L. Peterson
Editors

National Cyber Summit (NCS) Research Track

 Springer

Editors
Kim-Kwang Raymond Choo 🆔
The University of Texas at San Antonio
San Antonio, TX, USA

Thomas H. Morris
University of Alabama in Huntsville
Huntsville, AL, USA

Gilbert L. Peterson
Air Force Institute of Technology
Wright-Patterson AFB, OH, USA

ISSN 2194-5357 ISSN 2194-5365 (electronic)
Advances in Intelligent Systems and Computing
ISBN 978-3-030-31238-1 ISBN 978-3-030-31239-8 (eBook)
https://doi.org/10.1007/978-3-030-31239-8

This Springer imprint is published by the registered company Springer Nature Switzerland AG
The registered company address is: Gewerbestrasse 11, 6330 Cham, Switzerland

Preface

"The cyber threat is one of the most serious economic and national security challenges our nation faces. Cybersecurity affects nearly every aspect of our lives – critical infrastructure, banking, healthcare, transportation, and more. Our nation's economic prosperity in the 21st century will depend on how well we manage cybersecurity," said National Cyber Summit Chair Rodney Robertson.

This book contains 22 papers from the 2019 National Cyber Summit, held in Huntsville, Alabama, from June 4 to 6, 2019. The papers were selected from submissions from universities, national laboratories and the private sector from across the USA (AL, CA, FL, GA, IL, MD, NJ, OH, TX, VA, and WV), the UK, and Saudi Arabia. All of the papers went through an extensive review process by internationally recognized experts in cybersecurity.

The research track at the 2019 National Cyber Summit has been made possible by the joint effort of a large number of individuals and organizations worldwide. There is a long list of people who volunteered their time and energy to put together the conference and deserved special thanks. First and foremost, we would like to offer our gratitude to the entire Organizing Committee for guiding the entire process of the conference. We are also deeply grateful to all the Program Committee members for their time and efforts in reading, commenting, debating, and finally selecting the papers. We also thank all the external reviewers for assisting the Program Committee in their particular areas of expertise as well as all the authors, participants, and session chairs for their valuable contributions.

Last but not least, we would like to acknowledge the support of the National Initiative for Cybersecurity Education (NICE), a program of the National Institute of Standards and Technology in the U.S. Department of Commerce, under Grant # 70NANB16H016.

Thomas H. Morris
General Chair

Kim-Kwang Raymond Choo
Gilbert L. Peterson
Program Committee Chairs

Eric Imsand
Program Committee Vice-chair

Organization

Organizing Committee

General Chair

Thomas H. Morris   The University of Alabama in Huntsville, USA

Program Committee Chairs

Kim-Kwang Raymond Choo   The University of Texas at San Antonio, USA
Gilbert L. Peterson   Air Force Institute of Technology, USA

Program Committee Vice-chair

Eric Imsand   The University of Alabama in Huntsville, USA

Program Committee and External Reviewers

Program Committee Members

Ali Dehghantanha   University of Guelph, USA
Bhupendra Singh   Defence Institute of Advanced Technology, India
Huijun Wu   Arizona State University, USA
Jaewoo Lee   University of Georgia, USA
Junggab Son   Kennesaw State University, USA
Junghee Lee   Korea University, South Korea
Junyuan Zeng   The University of Texas at Dallas, USA
Ke Zeng   Microsoft, USA
Kewei Sha   University of Houston-Clear Lake, USA
Reza M. Parizi   Kennesaw State University, USA
Robin Verma   The University of Texas at San Antonio, USA

Rongxing Lu   University of New Brunswick, Canada
Vijayan Sugumaran   Oakland University, USA
Wei Zhang   Virginia Commonwealth University, USA
William Glisson   Sam Houston State University, USA
Xiaolu Zhang   The University of Texas at San Antonio, USA
Yifei Wang   Alipay, USA

External Reviewers

Aaron Werth   Lee Mauldlin
Amanda Fernandez   Nannan Xie
Amin Azmodeh   Petr Matousek
Amit Ahlawat   Raj Boppana
Cary Pool   Ron Sikes
Chen Chen   Sanaz Nakhodchi
David Coe   Shengyi Pan
Duo Lu   Sueanne Griffith
Feng Wang   Timothy Yuen
Hamed Haddadpajouh   Umara Noor
Jingyan Jiang   Vahid Heydari
Jiuming Chen   Wei Gao
John Springer   Yao Li
Kailiang Ying   Zahid Anwar

Contents

Cyber Security Education

Towards Secure Password Protection
in Portable Applications

Hossain Shahriar[1]([⊠]), Atef Shalan[2], and Khaled Tarmissi[3]

[1] Kennesaw State University, Kennesaw, GA 30144, USA
hshahria@kennesaw.edu
[2] Alderson Broaddus University, Philippi, WV 26416, USA
shalanm@ab.edu
[3] Umm Al-Qura University, Mecca, Kingdom of Saudi Arabia
kstarmissi@uqu.edu.sa

Abstract. The security dangers to mobile and web applications are developing explosively. Programming security has now turned into a more extensive security idea. Secure software development is fundamental and vital for Confidentiality, Integrity, and Availability of all product applications. Most vulnerabilities should be resolved in the versatile programming implementation stage. However, most application developers frequently spend little or no effort for security bugs remediation, as they are generally entrusted for undertaking due dates. Moreover, most developers are unaware of secure coding or cryptography from the degree programs they have obtained from schools. Software security for secure portable and web application advancement is of enormous interests in the Information Technology fields. In this paper, we proposed and developed an innovative learning module based on several real world scenarios to broaden and promote for secure software development.

Keywords: Password security · Cryptography · Secure coding · Active learning

1 Introduction

Secure software is essential and crucial for Confidentiality, Integrity, and Availability. Most malicious attacks are being launched by exploiting the vulnerabilities in applications [8]. Potential vulnerabilities in an application should be identified and mitigated before its uses. The best approach should be to address them during the software development phase – identify them when vulnerable codes are being written. However, most application developers have little to no time for reviewing code for security to emphasize on the project deadlines [8]. Worst, professionals lack awareness of the importance of security vulnerabilities and the knowledge and skills to mitigate them during application development stage [1, 2, 9].

With more schools developing teaching materials on mobile and web application development, more educational activities are needed to promote security education especially for secure software development. However, secure software development is a relatively weak area, and it is not well represented in most schools' computing curriculum. It is essential to integrate software security in computing and security education.

K.-K. R. Choo et al. (Eds.): NCS 2019, AISC 1055, pp. 3–13, 2020.
https://doi.org/10.1007/978-3-030-31239-8_1

OWASP Top 10 emphasizes the following measures to be enforced towards secure software development, which include (i) Define Security Requirements, (ii) Leverage Security Frameworks and Libraries, (iii) Secure Database Access, (iv) Encode and Escape Data, (v) Validate All Inputs, (vi) Implement Digital Identity, (vii) Enforce Access Controls, (viii) Protect Data Everywhere, (ix) Implement Security Logging and Monitoring, and (x) Handle All Errors and Exceptions [2]. This paper presents hands-on lab examples for implementation of digital identity (i.e., control - vi). Our module focuses on cryptography approach for the web applications.

The paper is organized as follows. Section 2 describes the learning module design principles. Section 3 discusses a module on digital identity implementation, demonstrating insecure and secure practices of password storing securely with cryptography techniques. Section 4 presents our evaluation approach performed on classroom students and corresponding results. Finally, Sect. 5 concludes the paper.

2 Learning Module Design

We propose to build the capacity on ProActive Control for Software Security through two venues: (1) curriculum development and enhancement with a collection of ten transferrable learning modules with companions hands-on labs on mobile and web software development which can be integrated into existing undergraduate and graduate computing classes that will be mapped to ISA KAs proposed in CS curricula 2013 to enhance the student's secure mobile software development ability; [5] (2) The mobile and web hands-on learning modules and labs are designed based on the OWASP 2018 Top Ten Proactive Controls open source project with 10 most important security techniques that should be applied proactively at the early stages of software development to ensure maximum effectiveness. The project provides ten transferable learning modules including Security Requirement Specification Control, Data Store and Database Security Control, Data Communication Control, Input validation Control, Output Decoding Control, Access control, Logging Monitoring and Exception Handling Control, Framework API Control, File Inclusion Control, Session Control and Digital Identity Implementation [2–4].

The learning modules are designed to map to Information Security Area (ISA) knowledge areas (KAs) of CS curricula 2013 [5] so that they can be easily "plugged" into existing CS/IA courses. Hands-on labs will be incorporated into these modules to challenge and engage students with real-world problems and build skills in developing secure mobile applications. All the hands-on labs are real-world based to support authentic teaching and learning.

3 Password Protection and Digital Identity

In this learning module, we provide resources of securing data within application, during processing and storage. The module consists of pre-lab, hands-on lab, and post-lab. The following are the learning objectives

- To explain the ways passwords are stored insecurely
- To apply the basic defensive practice skills against password leakage attacks.

3.1   Prelab

The pre-lab provides students a concept overview on the vulnerability of storing passwords in plain text and not using up-to-date encryption mechanism.

Storing a plaintext password either in source code and database need to be avoided as part of secure programming practices. An application need to verify a password instead of checking the content of the password. Hash function is a one-way operation, which can be applied for verification without revealing the content of password. Choosing a higher length password is more effective than lower length password. An attacker can perform brute force attack to uncover the plaintext if the length is too short.

Hashing algorithms such as MD5 or SHA1 are currently available as APIs in development environment to verify contents against a known hash value. An application developer should store only the hash value and avoid saving plaintext password in source code. Given their relatively limited output space it was easy to build a database with known passwords and their respective hash outputs, the rainbow tables.

Adding a salt to the password before hashing is a good practice to make a rainbow table based brute force attack useless. It is also suggested to use hash algorithms that generate larger size hashes (*e.g.*, 256 bits). The two popular choices of hash algorithms are PBKDF2 (Password Based Key Derivation Function v2) and bcrypt (Blowfish).

3.2   Hands on Lab Activity

In this part, we show example codes and their outcomes demonstrating insecure and secure practices on password security.

(1)   Insecure Vulnerable Misuse Case

The misused case lab is designed based on a real world scenario. Students will see code examples where username and passwords are saved in the source code. See Fig. 1, an example of code, where Bob and Tim are two users and their passwords are saved in *test.php*.

```php
1
2
3  <?php
4
5      $fileuser[1] = "Bob"; // Password 123
6      $fileuser[2] = "Tim"; // Password 321
7
8      $filepassword[1] = "123"; // User Bob
9      $filepassword[2] = "321"; // User Tim
10
11
12  ?>
```

Fig. 1. Plain text username and password

Figure 2 shows an example code where hash value of the password is stored instead of the plain text. Though the hash does not reveal the content, it may be possible for an attacker to use the hash against a set of dictionary words to recover the plaintext password. Figure 3 shows an example of code (*login.php*) processing the password input in a web form and matching with the hashes stored in the source file password.php.

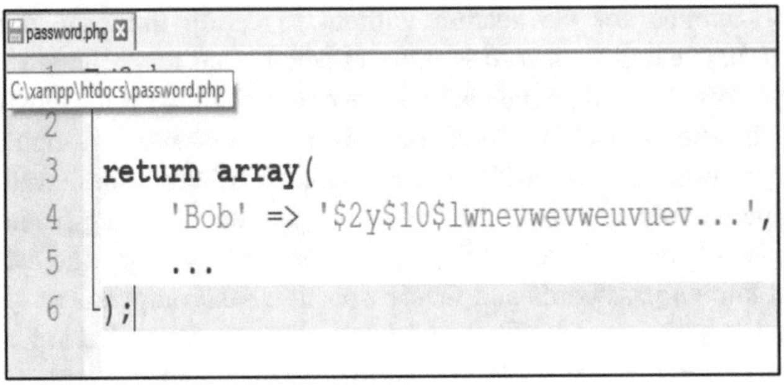

```
password.php ⊠
C:\xampp\htdocs\password.php
   2
   3    return array(
   4        'Bob' => '$2y$10$lwnevwevweuvuev...',
   5        ...
   6    );
```

Fig. 2. password.php file storing hash

```php
<?php
    require_once 'lib/password.php';
    $passwords = require 'passwords.php';
    if (isset($_POST['username'], $_POST['password'])) {
        if (!isset($passwords[$_POST['username']])) {
            die('Invalid username');
        }
        if (!password_verify($_POST['password'], $passwords [$_POST
['username']])) {
            die('Invalid password');
        }
        echo 'Hi there!';
    }
?>
```

Fig. 3. Password validation using webform (*insecure-login.php*)

The source code can be applied in this hands-on lab-activity to show the weakness of password storing in source files.

(2) Secure Use Case

A good practice is to use the supported API library functions that can help in encrypting texts and storing the encrypted texts such as bcrypt(). Before encrypting, it is important to use a salt. The crypt() method supports including a unique salt for a given password. The database table need to then store the hashed value instead of the plaintext password in source code or database tables.

A Saltisa random string that is added during hashing, and used to make that hash even more obscure. A developer should choose a large random string for salt value where the letters within the salt are 22 characters from the collection of: [/a-zA-Z0-9]. A random function be used either a microtime of the moment, Below is an example of salt:

$$\$2a\$15\$Ku2hb./9aA71tPo/E015h.\$$$

The example above has three parts. The $2a$ section is an identifier to let the PHP method know using the BlowFish hash algorithm. The second part, the 15 section is the cost parameter (to be chosen in the range of 4–31), and it is the iteration count for the algorithm. The higher the number the more time it takes for brute force attacks. With the salt, the hashing of a password should look like below:

$$crypt('password',\ '\$2a\$15\$Ku2hb./9aA71tPo/E015h.\$');$$

```php
<?php
$checkPassword = $_POST['password'];
  $passwords = db.query("select password from user where name =
'$POST['username']'");
   if (isset($_POST['username'], $_POST['password'])) {
   if (!isset($passwords[$_POST['username']])) {
         die('Invalid username');
   }
   }
   if(crypt($checkPassword,$currentPassword)=== $currentPassword){
     echo 'You are in!';
   }else{
     echo 'You entered the wrong password';
   }
?>
```

Fig. 4. Secured password validation (secure-login.php)

To check that a password is valid, the secured code is shown in Fig. 4, where crypt() method being used to match the generated hashes with stored hashes from databases.

3.3 Postlab

In this part, we provide more guidelines on practices for restoring passwords. For example, the NIST guidelines explain three levels of a authentication assurance called a authentication assurance level (AAL) [14]. AAL level 1 is reserved for lower-risk applications that do not contain PII or other private data.

AAL level 2 is for higher-risk applications that contain personal information and where level 2 multi-factor authentication is required. Multi-factor authentication ensures that users are who they claim to be by requiring them to identify themselves with a combination of: (i) something you know – password or PIN; (ii) something you own – token or phone; (iii) something you are – biometrics, such as a fingerprint.

Using passwords as a sole factor provides weak security. Multi-factor solutions provide a more robust solution by requiring an attacker to acquire more than one element to authenticate with the service.

It also includes some examples of best practices for login with password, reset, etc. For example, locking down a user's account after a set number of invalid login attempts, requiring users to wait a set period before continuing further, emailing users links to unlock accounts after failed attempts.

4   Evaluation

We conducted pre and post lab survey in one grade course section (Ethical Hacking and Networking Security – IT6843) having total 35 students (with response rate 85%). The course covers fundamental concepts of using various tools of performing enumeration, probing of computers, while assessing threats imposed by deployed software operating on the hosts. Specific modules covered on web and mobile security threats, where the labware was included.

The prelab survey was conducted before releasing the hands-on lab materials and the postlab was conducted after completing the lab. Students were asked to complete each labware module in two weeks. The surveys were designed to assess the gain in understanding and apply the security concepts. We have total 7 prelab questions as shown below (Fig. 5).

Q1: Have you been ever working on proactive control security based software development?
Q2: Have you been ever educated on secure software development?
Q3: I learn better by hands-on lab work
Q4: I learn better by listening to lectures.
Q5: I learn better by personally doing or working through examples.
Q6: I learn better by reading the material on my own.
Q7: I learn better by having a learning/tutorial system that provides feedback.

Fig. 5. Prelab questionnaires

Figures 6(Q1–Q2) and 7(Q3–Q7) show the response of the classes. We had 18 responses in both class sections. Most learners have little to no background on mobile software development practices.

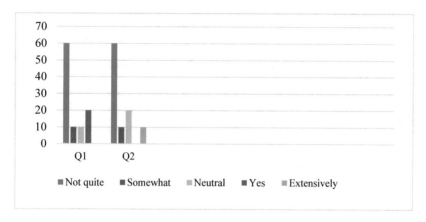

Fig. 6. Prelab survey response – Q1–Q2 Scale: [Not quite], [Somewhat], [Neutral], [Yes], [Extensively];

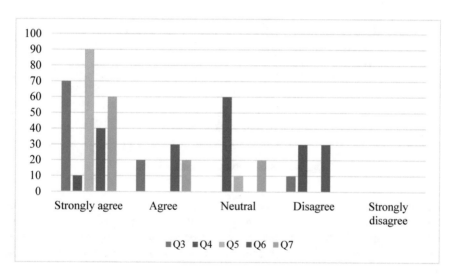

Fig. 7. Prelab survey response – Q3–Q7 Scale: [Not quite], [Somewhat], [Neutral], [Yes], [Extensively];

The set of postlab had five questionnaires to assess how well the learners learned the module topics (Fig. 8).

Q1: I like being able to work with this hands-on labware

Q2: The real world security threat and attacks in the labs help me understand better on the importance of proactive security control based learning.

Q3: The hands-on labs help me gain authentic learning and working experience on proactive security control

Q4: The online lab tutorials help me work on student add-on labs/assignments

Q5: The project helps me apply learned proactive security control to develop secure applications

Fig. 8. Postlab questionnaires

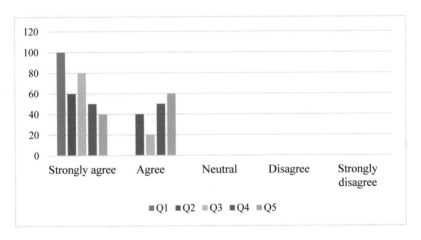

Fig. 9. Postlab survey response (Q1–Q5)

Figure 9 shows the survey response of the class. Most students agreed that our developed resources enabled them gaining authentic learning experience on proactive security control. We also found that most learners agreed that the labware was very effective in the learning of data protection knowledge while developing secure mobile applications.

5   Related Work

Jeffery [3] argued that users are not well informed of the permissions their apps use, leading to many insecure applications. They proposed on increasing awareness of the permissions of mobile apps use users feel more secure. They examined effects of the previous and current Android permissions models, as well as a new permissions model enforced at runtime based make users significantly more informed than install-time based models.

Willcox et al. [4] analyzed Mobile Cross-Platform Tools (CPTs) that are being used by mobile app developers to drastically reduce the development time and cost. However, the CPT software layers and translation affect the security of the produced applications. Despite CPTs (e.g., Cordova) have been updated to reduce the attack surface allowing developers to increase the security, the study found that adoption of security best practices and mechanisms remain poor, leading to a significant number of insecure applications.

Peruma et al. [6] studied the trends in app quality and security to create higher quality software, and more effectively battle malware. They collected and reverse engineered 70,785 Android apps from the Google Play store, along with 1,420 malicious apps from other sources. Apps are analyzed using static analysis tools to record information including permissions, size (LOC), defects and permission misuse. Their concludes indicated that app categories not only substantially differ in terms of permissions misuse but also, there is no significant correlation between an app's quality and security.

Geethanjali et al. [7] developed obfuscation technique (AEON), to encrypt code segments and perform runtime decryption during execution. The encrypted code runs in an embeddable Java source interpreter (outside the java virtual machine), therefore, circumventing the scrutiny of Android runtime verifier.

Theisen et al. [10] reported the popularity of software security education offering for both on campus and MOOC students. They found in their student student's performance for on campus software security courses are better than MOOC students, which essentially reinforces that systematic development of resources. Despite the goal of the work was to compare performance of students between traditional degree programs and open source online courses, it is noticeable that using massive amount of videos is of not effective (used in MOOC learning modules). Thus, developing hands-on practices are more effective towards enforcing software security among programmers.

Walden and Frank [10] reported a course on secure software development having 10 module in 2006, with the goal of making it as a capstone course towards secure software engineering. The described modules include security requirements, design principles and patterns, risk management, secure programming (data validation, cryptography), code review and static analysis. The course focused on the web application. Since many curriculums do not have the opportunity to offer secure software development capstone course, our proposed hands on labware may be easily integrate into related computing courses including non-security courses such as databases, operating systems, web development and mobile application.

Peruma et al. [12] focused on labware devoted to android security hands on practice for experiential learning. Their labware is based on real world example apps having vulnerabilities and providing examples of securing them such as intents, xml, JavaScript, broadcast, data storage, protection against the denial of service attacks. The PLASMA lab does not focus on proactive security control based learning where malicious inputs are used to demonstrate vulnerability exploitation and securing applications with coding examples. Our developed labware can also be applied to both web and mobile applications.

The SEED security labs [13] a large collection of labware, which relies on the availability of special virtual machine. Though it got popularity and adopted in many schools who has no option to develop resources on their own, the labware falls short of practical mitigation of security bugs, particularly visibly pointing the source code having vulnerabilities.

Pistoia et al. [11] developed privacy-enforcement system for Android and iOS to detect and repair leakage of private data originating from standards and application-specific sources. ASTRAEA enables visually configuring custom sources of private data, but it relies on application-level instrumentation. The project performs an enhanced form of value-similarity analysis to detect and repair data leakage even when sensitive data has been encoded or hashed and displays the results of the privacy analysis on top of a visual representation of the application's UI.

6 Conclusion

The overall goal of this paper is to address the needs and challenges of building capacity with proactive controls for software security development and the lack of pedagogical materials and real-world learning environment in secure software development through effective, engaging, and investigative approaches through real world oriented hands-on labware. We proposed and developed two innovative learning modules for software security proactive control based on several real world scenarios to broaden and promote proactive control for secure software development in computing education. The initial evaluation find the module effectively helped students learning security proactive control better. Our effort will help students and software developers know what should be considered or best practices during mobile and web software development and raise their overall security level for software development. Students can learn from the misuse of vulnerability cases and insecure mistakes of other organizations. Simultaneously, such cases should be prevented by mitigation actions described in secure protected use cases for building secure software.

References

1. Projects/OWASP Mobile Security Project - Top Ten Mobile Risks (2019). www.owasp.org/index.php/OWASP_Mobile_Security_Project
2. OWASP Top 10 Proactive Controls (2018). https://www.owasp.org/index.php/OWASP_Proactive_Controls
3. Palmerino, J.: Improving android permissions models for increased user awareness and security. In: Proceedings of the 5th International Conference on Mobile Software Engineering and Systems (MOBILESoft), pp. 41–42, May 2018
4. Willocx, M., Vossaert, J., Naessens, V.: Security analysis of cordova applications in Google play. In: Proceedings of the 12th International Conference on Availability, Reliability and Security (ARES), pp. 1–7, August 2017
5. Computer Science Curricula 2013 Curriculum Guidelines for Undergraduate Degree Programs in Computer Science. https://www.acm.org/binaries/content/assets/education/cs2013_web_final.pdf
6. Peruma, A., Krutz, D.E.: Understanding the relationship between quality and security: a large-scale analysis of Android applications. In: Proceedings of the 1st International Workshop on Security Awareness from Design to Deployment, SEAD 2018, pp. 19–25, May 2018

7. Geethanjali, D., Ying, T.L., Melissa, C.W.J., Balachandran, V.: AEON: android encryption based obfuscation. In: Proceedings of the 8th ACM Conference on Data and Application Security and Privacy, pp. 146–148, March 2018
8. Biswas, S., Sajal, M.M.H.K., Afrin, T., Bhuiyan, T., Hassan, M.M.: A study on remote code execution vulnerability in web applications. In: International Conference on Cyber Security and Computer Science (ICONCS 2018) (2018)
9. Amorso, E.: Recent progress in software security. IEEE Softw. **35**, 11–13 (2018)
10. Theisen, C., Williams, L., Oliver, K., Murphy-Hill, E.: Software security education at scale. In: Proceedings of 2016 IEEE/ACM 38th IEEE International Conference on Software Engineering Companion, Austin, TX, USA, pp. 346–355 (2016)
11. Pistoia, M., Tripp, O., Ferrara, P., Centonze, P.: Automatic detection, correction, and visualization of security vulnerabilities in mobile apps. In: Proceedings of the 3rd International Workshop on Mobile Development Lifecycle, pp. 35–36, October 2015
12. Peruma, A., Malachowsky, S.A., Krutz, D.E.: Providing an experiential cybersecurity learning experience through mobile security labs. In: Proceedings of IEEE/ACM 1st International Workshop on Security Awareness from Design to Deployment, Gothenburg, Sweden, 27 May 2018, pp. 51–54. ACM, New York (2018)
13. Hands on labs for security Education. http://www.cis.syr.edu/~wedu/seed/index.html
14. NIST Special publication. https://pages.nist.gov/800–63-3/sp800-63b.html

Capacity Building for a Cybersecurity Workforce Through Hands-on Labs for Internet-of-Things Security

A. Ravishankar Rao[1(✉)] and Daniel Clarke[2]

[1] Fairleigh Dickinson University, Teaneck, NJ 07666, USA
`raviraodr@gmail.com`
[2] Mount Sinai Medical Center, New York, NY, USA
`danieljbclarke@gmail.com`

Abstract. There is growing concern about a cybersecurity skills gap nation-wide, as an insufficient number of students are entering this field. Hence, innovative approaches to capacity building are required. We advocate an approach that taps into public awareness and usage of internet-of-things (IoT) devices, which constitute a major source of security vulnerabilities. There is increasing student curiosity about these devices, and a growing tinkering movement.

We tap into this zeitgeist by proposing educational courseware to give students hands-on laboratory experience in designing secure embedded systems. Our initial experience at offering this courseware on the Raspberry-Pi computing platform resulted in significant student interest and engagement. We propose building on this momentum by creating further laboratory exercises focused on IoT devices in the medical and healthcare domain. This is a rich domain for security applications due to patient privacy concerns, the need to secure medical devices, and the need to share patient medical records.

The newly proposed labs will deploy RFID and barcodes to tag medical devices used in hospitals, sensors to measure patient vitals, biometrics to identify patients, and blockchain to create immutable patient records. These labs can be easily implemented with cheap sensors on the Raspberry-Pi platform, and can be replicated easily. By using this approach, we can build a national talent pool in the area of secure embedded devices.

We are in the process of developing these proposed labs. We plan to test them in a classroom setting in Fall 2019. Upon successful testing, we plan to release our labs to interested institutions.

Keywords: Cybersecurity · Curriculum development · Internet of things

1 Introduction

The Cyber Security National Action Plan [1] states: "The need for cybersecurity professionals to fill vacancies in government and industry across the United States is acute. Programs that successfully identify under-served and under-utilized potential students are critically needed to accelerate growth of the profession". Hence, capacity building in cybersecurity is an important challenge that needs to be addressed jointly by

© Springer Nature Switzerland AG 2020
K.-K. R. Choo et al. (Eds.): NCS 2019, AISC 1055, pp. 14–29, 2020.
https://doi.org/10.1007/978-3-030-31239-8_2

educational institutions, industry, and the government. The pool of college students entering into cybersecurity is a subset of a larger pool of STEM (Science, Technology, Engineering and Mathematics) students. However, fewer than 40% of the students in the US who enter college with the intention of majoring in a STEM field complete a STEM degree [2]. Hence, a holistic approach is required that increases the appeal of STEM fields to students while improving graduation and retention rates. The goal of the current paper is to demonstrate that capacity building in both STEM and cyber-security fields can be enhanced through the development of hands-on laboratory exercises that excite and engage students [3–6].

Current steps being taken by the Department of Defense include a program to identify talented undergraduate students and bestow scholarships on them to pursue careers in cybersecurity. By tying the scholarship award to a mandatory period of service, the government is assured that the selected students will begin their careers in areas which urgently need their skills. There are other government agencies including the National Science Foundation (NSF) that are also engaged in capacity building efforts. The NSF aims to increase the number of investigators who are capable of conducting fundamental education research in STEM fields.

We address the following two requirements of workforce development identified by the US Department of Defense [7]:

(1) How can we create a cybersecurity workforce that is capable of integrating cybersecurity into all aspects of computing and engineering?
(2) How can we create a cyber IT workforce that can design, build, configure, operate, and maintain IT, networks and capabilities?

We propose to achieve requirement (1) by introducing cybersecurity and its associated learning modules into engineering programs especially in the area of embedded system design. We have begun initial efforts in this direction. We address requirement (2) through the creation of new hands-on lab exercises where learners can design IoT devices with a "security mindset".

Cybersecurity is becoming increasingly important, and there is substantial demand for skilled workers in this field [8, 9]. Yet, few courses teach students hands-on skills in this field that combine hardware and software. A possible reason is that understanding cybersecurity requires a knowledge of a wide range of computer related skills, span-ning multiple topics in electrical engineering, computer engineering and computer science. Deep knowledge of hardware components is important, as security flaws and backdoors may occur at the chip level [10]. In addition, it is important to understand computer networking [11], and the interplay between hardware and software. For instance, the precise functioning of the commonly used stack has been the source of many vulnerabilities [12]. The use of specific applications such as SQL can themselves cause security vulnerabilities, which can be exploited by attacks such as SQL injection [13]. Sometimes multiple machines are recruited in a coordinated fashion to create botnets [14], which are then used in attacks such as the Dyn Distributed Denial of Service (DDoS) attack. A complete understanding of the cybersecurity space is daunting even for well-trained professionals and national security organizations. For instance, the currently evolving situation around the security of Huawei chips is causing disagreement amongst the security agencies in the US, UK and Germany.

There is no internationally accepted consensus on whether chips made by Huawei contain deliberate security backdoors [15]. Similarly, when Bloomberg Businessweek reported on a possible backdoor planted in Chinese-made chips [16], Amazon, Apple, and even the US Department of Homeland Security denied that such a backdoor existed [17]. The upshot of this paragraph is that the field of cybersecurity can appear intimidating to students who are aiming to enter it. Hence, care needs to be exercised by educators to present the field in a realistic and enticing way.

Since cybersecurity is applicable to all sectors of the economy, students can also be simultaneously exposed to sectors of strong growth including finance, manufacturing, and healthcare. At the college level, most students have already experienced aspects of the US healthcare system, including visiting doctors, and filling prescriptions. However, they may never have considered how their personal data is acquired, transmitted, stored and processed. Hence, a specific goal of this paper is to highlight the acquisition and use of healthcare data through a cybersecurity lens. Accordingly, we propose a series of laboratory exercises that use simulated patient data to create a visceral feel for cybersecurity applications in the healthcare space. There are several open questions in the information technology space for healthcare, including the acquisition and use of biometric data for patient identification [18], the maintenance of patient privacy within an organization and sharing of patient records securely across multiple organizations [19]. The protection of patient data across multiple IT systems creates several security challenges. Consequently, the intersection of cybersecurity, patient data and medical devices is witnessing significant growth [20, 21]. Blockchain is being proposed as a technology for sharing patient data while maintaining privacy [22].

Hence, the goal of our proposed laboratory exercises is to inculcate a "security mindset" amongst the students. The advantage of this approach is that it will make the field of cybersecurity more attractive and more accessible to students through concrete and relatable applications. Furthermore, when students acquire and process their own data, such as personal health related data, they are able to develop a deeper appreciation of the need to secure and protect such data.

Due to the limited time students have for undergraduate education, and the ever-expanding frontiers of multiple fields, we strongly advocate weaving specific technology application areas through STEM courses. We have successfully used such an approach by highlighting the importance of data science in a core undergraduate course, entitled "Modern Technologies: Principles, Applications and Impacts", taught by the first author at Fairleigh Dickinson University (FDU). We determined that this approach helped the students develop a "data habit of mind" from their first undergraduate year onwards, and our experience is detailed in Rao et al. [23]. We use a similar approach in this paper to create a "security mindset" in the students.

A major contribution of the present paper is to describe our positive classroom experience with engaging students through hands-on laboratory exercises using embedded systems and internet-of-things (IoT) devices. This achieves multiple objectives, as a major source of security vulnerabilities is the pervasive use of IoT devices, ranging from internet-enabled refrigerators to smartwatches. Every device connected to the internet is a potential source of security vulnerabilities. Hence, the creation of hands-on lab exercises in the cybersecurity of embedded devices could excite and engage students. This should also help ameliorate the perceived dullness of coursework in STEM fields [24], while enhancing the students' skill base.

2 Background

The technology landscape is very diverse in terms of the types of devices, ranging from simple embedded devices to supercomputers. Similarly, the education landscape is very diverse, in terms of the types of students who go to college and the caliber of the colleges themselves. Hence we need a nuanced assessment of the capabilities of the current generation of students. A "one size fits all" approach is not suitable, as an approach that works at top-tier colleges may not work at mid or lower tiered colleges. We present our current viewpoint about both technology and education in the realm of cybersecurity. We review existing literature in the area of instructional material and research on teaching emerging technologies. This is followed by a review of the hardware and software components used in creating the laboratory material.

2.1 The Internet-of-Things and Its Growing Impact on Cybersecurity

Figure 1 shows that consumer oriented smart-home devices can inadvertently and easily spread personal information can. Embedded devices are purchased from retailers and immediately connected to the public internet without any precautions. The testing agency Dark Cubed [25] has observed several anomalies and unexplained communications as shown in Fig. 1. The simple act of operating these devices leads to a leakage and distribution of personal data. According to Gartner, the expected number of IoT devices will be 20 billion by 2020. The 5G rollout occurring in 2019 will only accelerate the adoption of IoT devices. This growth is very concerning to the security community. In fact, in preparation for the 2020 Olympics in Tokyo, the government of Japan plans to hack into their own citizens' IoT devices to warn them of vulnerabilities [26].

Fig. 1. IoT devices can inadvertently spread personal information [25]. Many of these IoT devices are insecure right from the time they are installed. This leads to security problems worldwide. (Reproduced with permission from Pepper IoT).

This challenge provides opportunities for companies who can design more secure products and for educational institutions to build capacity by imparting necessary skills to students. Due to the availability of inexpensive computing devices, many younger students have become interested in tinkering with them. Though many retail stores are closing in the USA and UK, Raspberry Pi bucked this trend by opening its first retail store in Cambridge, UK in Feb. 2019 [27]. We can interpret this as a sign of growing public interest in such computing devices. At the author's institution Fairleigh Dickinson University (FDU), we have seen increasing student interest and curiosity in exploring and tinkering with emerging technologies. Two of the author's undergraduate students are technology hobbyists, and taught themselves how to fly and use drones. They applied this technology to create a humorous short film that won the top award at CampusFest-2017, a nationwide film festival held across campuses at US universities. The film can be viewed at https://www.youtube.com/watch?v=Mrt2zNum3Js. One undergraduate student operated a successful business that fixed broken cellphones. Others performed bitcoin mining. Educational institutions can guide these students deeper into the underlying technologies, especially in cybersecurity.

IoT devices seeing pervasive usage include smartphones, fitness trackers, smart watches, and smart home appliances. This greatly increases the total number of security exposures. Phishing attacks are steadily increasing, affecting all users of these devices. Users are typically unaware about how their privacy can be breached. A disturbing incident concerned a fitness tracker app used by US Army personnel which revealed the location of secret army bases [28]. This story demonstrates that the use of IoT devices such as a cellphone can introduce unexpected security issues.

Many such problems can be fixed after they are discovered, typically by installing security patches. However, prevention is better than cure. We can increase awareness by fostering a "security mindset" in the users of these technologies. A natural way of channeling the growing student interest in tinkering with IoT devices is to focus their attention on security vulnerabilities. Users need to understand the mechanisms and ploys used by attackers, so they can stay alert and watchful. We can train students in STEM fields to grasp the technicalities behind cyber-attacks. Hence, it is desirable to:
(1) Provide detailed hands-on exercises to students in the area of cyber-security
(2) Inculcate a "security mindset" in the students.

2.2   Instructional Material

There is a lack of *integrated* course material that can guide students to study emerging technologies such as cloud computing [29], the internet-of-things [30], security [31], cryptography [32], and blockchain [33]. Individual books in these areas [32, 33], may not be accessible for undergraduate students, especially for students at mid-tier universities or community colleges. Many available textbooks are theoretically oriented, and contain limited hands-on exercises. Publishers are being squeezed by rapidly changing technologies and are struggling to create relevant textbooks. In order to combat declining revenues, many publishers restrict access to e-Books by making them expire after a semester [34]. Students are reluctant to keep paying fees to extend their access to e-Books. The implications of this situation are examined in Rao [3].

A possible solution is to have a government agency or non-profit organization created curated online content, as is being done by the National Security Agency at http://clark. center.

In addition, free educational material is available on the internet, including MOOC (massively open online courses) platforms such as coursera.org and edx.org [35]. The Massachusetts Institute of Technology (MIT) has made a large number of courses available freely (https://ocw.mit.edu/index.htm). However, free online courses are quite limited in specialized areas such as cybersecurity. In response to the huge demand for course materials in areas such as blockchain and cryptocurrency *undergraduate* students in University of California Berkeley offered a course entitled "Blockchain Fundamentals", on EdX.org in 2018 [35]. Of course, making the courses available is only the first step. Ensuring that students complete these MOOC courses is a challenge, as very few actually complete them. Another issue that is relevant in the field of cybersecurity is that MOOC courses rarely contain hands-on laboratory material, especially involving hardware components.

Due to these considerations, individual instructors play an outsized role in spreading knowledge of modern technologies by including cutting-edge material in their courses. Educators such as Hamblen [36] have described their experiences in offering hands-on laboratory exercises to students in the internet-of-things. These are customized courses offered at specific institutions. The challenge is to continuously update such courses based on new hardware, software and applications. The experiences and student observations shared by these educators are useful to the education community at large. For instance, research [37] shows that students are motivated to a greater extent by laboratory work, and many prefer to work in teams.

Only recently, in the space of cybersecurity education, Dark et al. [38] created of an online repository called CLARK (Cybersecurity Labs and Resource Knowledge Base), intended to be populated with instructional material from multiple institutions. The website is functional at https://clark.center/home. The NSA has funded this promising development, which could gain momentum if sufficient institutions contribute material, and students start utilizing it.

2.3   Our First Experience with Creating Secure Embedded Systems Courseware

The first author teaches the current Embedded systems course at FDU, EENG7709, which combines instruction in hardware and software on inexpensive computing devices used in control applications. With the explosion of the internet-of-things, this has become an important course, especially due to the potential for cybersecurity related issues. The diversity of students interested in this course poses several challenges. Both graduate and undergraduate students with different majors and skills currently register for the course. Typically, electrical engineering majors possess limited knowledge about software engineering and the Linux operating system. Whereas computer science majors have limited skills in bread-boarding and working with electrical devices such as IoT sensors. Due to these reasons, hands-on laboratory work is a central part of this course. Finally, the course modules need to fit into a 16-week semester while covering hardware, software and laboratory work. The pace and depth should be reasonable to interest and challenge students while not overwhelming them.

Hardware for the Laboratory: Manufacturers such as Texas Instruments (ti.com) and Keil (keil.com) produce development boards which contain embedded processors packaged with software integrated development environments (IDEs). We favor the Raspberry Pi Model-3B, shown in Fig. 2, as it provides a useful platform to experiment with IoT technologies, and has a low cost of $35. The Raspberry Pi board contains a processor and graphics chip and program memory (RAM). It also includes multiple GPIO (general purpose I/O) pins and connectors for external devices.

An attractive feature of the Raspberry Pi is the inclusion of several built-in sensors for physical variables such as temperature, humidity and acceleration. We reviewed the use of the temperature sensor in our previous work [4]. In addition, a camera can be mounted on the Raspberry Pi board to extend its usability, as shown in Fig. 2.

Fig. 2. The camera board V2.1 connected to the Raspberry Pi Model 3-B. The camera uses a high quality Sony IMX219 sensor. We chose the camera as it motivates several useful applications such as security, and also blockchains that utilize image data.

Software for the Laboratory: The Raspberry Pi runs a flavor of the Linux operating system, known as Raspbian [39]. This provides students with a ready platform to run and practice basic Linux commands. Students were introduced to basic Python programming. They manipulated the camera through a command-line driven API (application programming interface), which allowed them to capture images at regular intervals for a camera security application. They automated the acquisition and storage of images through custom Python scripts.

Blockchain as a Target Application: There is a current wave of interest in blockchain technologies and bitcoin. Though students are excited about these developments, many have very little knowledge and understanding about these technologies. Through exposure to simple laboratory exercises, students can develop basic knowledge relatively quickly. We considered the application of blockchain to monitor food supply chains [40], where it is important to ensure that there is no tampering of data entries containing product images. We modeled this application with a device that (a) captures an image, (b) combines it with metadata including the name of the merchant, location, date and time, (c) computes the cryptographic hash value of the product image data concatenated with the metadata, and (d) saves the product image, metadata and the hash value in a single block. This block can be replicated and saved in multiple locations via distributed cloud storage. If there is suspected tampering of the underlying image data or metadata, we can re-compute the hash value, and compare it against the original computed value stored in the distributed copies. We successfully taught this essential concept behind the blockchain to the students in a single lecture.

Classroom Experience in Fall 2018: We report our experience in a 3-credit course on Embedded Systems, EENG7709 taught by the first author at FDU in Fall 2018. The class consisted of 11 students with majors in Electrical Engineering or Computer science. There were three undergraduate seniors and eight graduate students. None of the students had any prior exposure to cryptography. We introduced students to the basics of encoding, hashing and blockchains for 60 min before conducting the lab. We gave students sample code to run and modify. We observed that all students could create the SHA-256 values of images that they captured within 30 min.

For the lab report, we asked students to capture images, and compute the SHA-256 values. They also modified the image and computed the SHA-256 values. Most students took pictures of themselves, which we cannot reproduce due to privacy concerns. Figure 3 shows that any modification of an image changes its SHA-256 hash value. This demonstrates the immutability feature of records in a blockchain, where any changes are detected by comparing hash values. Due to the compelling visual nature of this demonstration, students could learn and assimilate core concepts within a short lesson.

THE UNTOLD STORY OF NOTPETYA, THE MOST DEVASTATING CYBERATTACK IN HISTORY

SHA-256 hash of the original image:
505a0bbd51358d8caf5cde3994b8005c5
4d9935864c6b740cb12a29ad8555e76

THE UNTOLD STORY OF NOTPETYA,

SHA-256 hash of the altered image:
f15f86827771177b4cf3c49c19838df1
916bc25e00ae8d5b5c3d55d81c65e9f9

Fig. 3. Shows that an alteration of an image (e.g. by cropping it) causes its SHA-256 hash value to change. The image on the left shows the cover of the Wired magazine issue in October 2018 that discussed the Notpetya cyberattack. The image on the right is cropped.

Students were very excited to learn about cryptography and blockchains, as captured through a few reactions in Fig. 4. These have been reproduced from the paper in [3]. In future coursework, we will use additional real-world applications of blockchain technology. We recommend that instructors should offer students immediate hands-on practice after a short theoretical lecture.

"I found this lab to be interesting in the fact that I was able to have a better understand in how hashing works and how it is applied to block chain. We can see this today with bit coin (sha256) and how this lab relates to it. This lab was not too difficult you just need to do some of your own research to have a better understanding of the code and how it all works."	"Before this lab, I didn't know about hashing. I haven't taken computer networks yet, so I didn't know what the concept entail. This lab helped me not only to learn about Blockchain, but to also learn about hashing and what it is. Taking multiple pictures and generating different hashes showed me that not one single image is the same, and that this process is very useful when you want to verify the legitimacy of a file because if it is tampered with, the hash of that file would be different than that of the one you actually need."

Fig. 4. Sample student reactions to the laboratory exercises.

3   Methods for Capacity Building in Secure Embedded Systems

In this section, we propose an innovative approach to capacity building in secure embedded systems by devising a series of hands-on laboratory exercises. These exercises combine the Raspberry-Pi platform with several inexpensive sensors that can model the creation and use of healthcare data. This illustrates the application of security concepts to an important sector of the economy, and provides an opportunity for students to collect and use their own data. Prior to our award of the NSA-CNAP grant in 2017 [41], a few graduate course modules were available that could be used to create a secure embedded systems specialization as shown in Fig. 5(A).

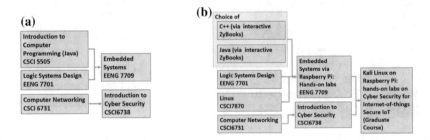

Fig. 5. (A) Original modules at the graduate level. (B) Proposed target modules for secure system design at the graduate level.

Through the NSA-CNAP grant, we are moving to create a graduate specialization in Secure Embedded Systems as shown in Fig. 5(B). We aim to incorporate this specialization into FDU's MS degree programs in Computer Engineering and Cybersecurity and Information Assurance. A similar approach can be used at the undergraduate level.

We upgraded the EENG7709 course as shown in Fig. 6. We used the zyBooks platform (zybooks.com) to teach Python programming. Several hands-on lab exercises were used to teach the students about cybersecurity issues in embedded systems. Students were receptive and appreciated this new approach [42]. The students were very excited to use hands-on labs with Raspberry-Pis. As compared to a theoretical approach, students were better able to relate to concepts in embedded systems and cybersecurity [4, 6]. We conducted lab exercises on random number generation, dictionary attacks, simulated phishing attacks, and SQL injection attacks. We are pleased to acknowledge that the NSA-CNAP grant greatly accelerated the introduction of new course material by providing both the hardware and student research assistant support for writing and testing software.

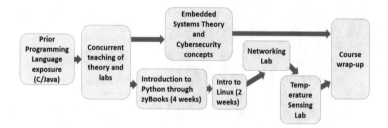

Fig. 6. The overall structure of the newly designed 14-week EENG7709 Embedded Systems course taught in Fall 2017. The new material consisted of introductions to Python and Linux, conducted on the Raspberry Pi platform.

Our capacity-building strategy builds on this momentum, and introduces additional engaging course content. This exposure will help students attain cybersecurity proficiency in secure embedded system design. Ultimately, they will become smarter engineers and designers of the future who appreciate the security challenges surrounding embedded devices.

We have set up an isolated cybersecurity lab using a "private" wi-fi network of 22 Raspberry Pi machines as described in the paper by PI-Rao et al. [3]. This allows students to experiment freely without worrying about bringing down the university network. We describe a sequence of six new labs that will help educate students about cybersecurity issues in embedded devices, helping to build capacity. The first two labs give students an understanding of authentication mechanisms and the vulnerability of IoT devices. The next four labs have applications in the medical and healthcare space.

Lab 1: Simulated Phishing Attack that Circumvents Two-Factor Authentication
This introductory lab will teach the students about mechanisms behind simple cyber-attacks and provide sample code that they can run. Currently, two-factor authentication

is considered an online security "best practice". However, this method is still prone to phishing attacks [43]. We will simulate such an attack by creating a fraudulent website that collects a user's login credentials and also the one-time authentication code generated by a two-factor service. This shows that there are ways to circumvent even a method that is considered to be reasonably secure. This drives home the point that no security tool is perfect, and that users must always exercise caution.

Lab 2: Using Shodan to Scan IoT Devices and Identify Vulnerabilities

Before beginning this lab, students will be given a background in ethical issues in cybersecurity, including punishable criminal offenses involving cybercrimes. Though they will learn about techniques for identifying vulnerabilities, they will be expressly forbidden from exploiting them. We will demonstrate how the Shodan website (www. shodan.io) can be used to search for IoT devices. Shodan is a search engine that scans the Internet and parses metadata sent back by connected devices, as shown in Fig. 7.

Fig. 7. (A) By using Shodan, users can search for IoT devices on the internet. (B) A search result showing open ports on an industrial control system.

This metadata can be used to search the National vulnerability database, https://nvd. nist.gov/vuln/search and the Exploit database, https://www.exploit-db.com/ to identify potential attack surfaces that could be used by hackers. As preventive measures, students will be taught to use firewalls, blacklisting and whitelisting. Students will also learn how such vulnerabilities can be exploited by DDoS attacks.

Lab 3: Using RFID and Barcodes to Tag Medical Devices

The FDA established a unique device identification (UDI) system for medical devices sold in the USA. We can simulate the use of such UDIs through RFID sensors embedded in barcode labels. RFID sensors are increasingly being used by hospitals for asset identification and tracking. Students will learn key components of such a network by hands-on experience with RFID and barcodes as shown in Fig. 8.

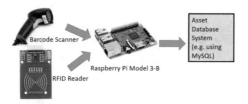

Fig. 8. Students will use a barcode scanner and RFID reader to tag and identify simulated medical devices. We will store information in an asset database system implemented in MySQL.

Lab 4: Using Sensors to Measure Patient Vitals

In many hospitals today, patient vitals are measured and entered by a healthcare provider into the computer. This could introduce potential human errors [18]. Students will create a system that directly enters such information into the computer. As an example, we will use a pulse oximeter, which uploads pulse and oxygen levels directly into a simulated electronic medical record (EMR) as shown in Fig. 9. We will use end-to-end encryption to protect the data which is transmitted to secure cloud storage.

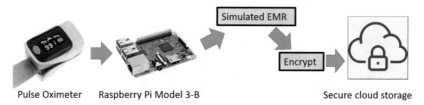

Pulse Oximeter   Raspberry Pi Model 3-B   Secure cloud storage

Fig. 9. We will use a pulse oximeter (e.g. Nonin or Contex CMS-50F) with USB and/or Bluetooth connectivity, allowing the device to be attached to the Raspberry Pi.

Lab 5: Using Biometrics for Patient Identification

Currently, most hospitals identify patients by their name and birthdate. This appears to be causing increasing problems in some cases as multiple patients may have the same name and birthdate. A recent article in the Wall Street Journal [18] about the Texas healthcare system observed that "there are now 2,833 Maria Garcias, with 528 of them having the same date of birth." Interestingly, there is no nationally standardized approach to this problem. Hence, hospitals are creating their own solutions, with some starting to deploy biometrics. We will simulate a patient identification system by using fingerprints as the biometric. As shown in Fig. 10, a simple fingerprint reader is attached to a Raspberry-PI. The scanned fingerprint is converted to a private key to access medical records. A simple medical record will consist of the patient identifier combined with the oxygen and pulse readings obtained in **Lab 4.** This sequential progression of the labs illustrates how the skills learnt in one lab can be applied to subsequent labs. The use of biometric information can also provide a line of defense against cyberattacks targeting electronic health records.

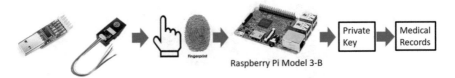

Raspberry Pi Model 3-B

Fig. 10. Biometric information (e.g. a fingerprint) can be used to access patient records.

Lab 6: Using Blockchain to Create Immutable Patient Records

There is a current wave of interest in applying blockchain technology to multiple domains, including healthcare. Though the technology implementations of blockchain are evolving rapidly, we have crafted a simple lab to illustrates the key concepts. As described earlier, we used a camera module attached to a Raspberry-PI to collect images and compute its SHA-256 value. Any alterations to the image would be detectable as the SHA-256 value would change. This was shown in Fig. 3. We will modify this lab to use a simulated electronic health record created in Labs 4 and 5 above. The insertion of this record in a blockchain, as shown in Fig. 11, protects it from tampering. Furthermore, multiple copies of this record will be made available on multiple Raspberry-PI machines, which further protects the data in the event of a cyberattack.

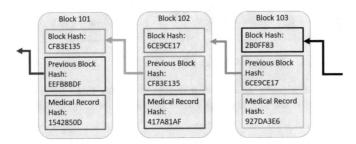

Fig. 11. A simple depiction of a blockchain. For the "medical record", we will use data related to patient vitals and biometrics that the student collected in Labs 4 and 5.

We are currently implementing the labs 1–6 described above. We plan to offer these labs in an Embedded Systems course in Fall 2019 at FDU. Upon successful testing of these labs in a classroom environment, we plan to make them available to interested institutions. This is subject to continued support of this work from FDU and potential funding agencies.

4   Conclusion

There is a widening gap between what the students learn in traditional courses and the cutting edge of industrial technologies. Students are already dropping from STEM majors at a high rate, and this widening gap is likely to exacerbate the problem. This necessitates rapid changes in curricula and the investigation of new pedagogical techniques. Course material delivered in traditional lecture formats may not grab the attention of students and motivate them towards graduation. Many courses still lack significant hands-on activities that encourage experimentation.

Since IoT devices are a major source of security vulnerabilities, we focused on developing courseware in secure embedded systems. A promising application area is the protection and security of medical devices and patient data. Hence, we propose a capacity building approach through hands-on laboratory activity in the medical and healthcare space.

Our research and teaching experience shows that such hands-on activities can be introduced advantageously if sufficient care is placed on accommodating students with varied backgrounds. Self-contained lectures and labs need to be designed to deliver essential information as efficiently as possible, while leaving room for student experimentation. Such a course may be taught at an undergraduate or graduate level, by adjusting the technical depth of the material. The course material should utilize hands-on exercises in every class to make students understand the essential key concepts.

We presented modifications introduced to a traditional graduate level embedded systems course taught at Fairleigh Dickinson University in Fall 2018. We used hands-on laboratory exercises on the Raspberry-Pi platform to teach the students about real-world internet-of-things applications. We introduced new lab exercises involving a camera module, which allowed students to capture and store images on a cloud storage platform such as Google drive. We also introduced students to cryptographic hashing and its use in blockchains. Students were very excited about these hands-on laboratory activities in emerging technologies such as the internet-of-things, cloud computing and blockchain. These activities enabled students to view, run and modify the code to perform the necessary operations. This allows them to internalize the required steps and develop a deeper appreciation of emerging applications.

Finally, we proposed a series of six new labs that build on this existing courseware. These labs include topics such as using RFID and barcodes to tag medical devices used in hospitals, using sensors to measure patient vitals, using biometrics for patient identification and using blockchain to create immutable patient records. We chose these specific labs because they can be easily implemented with cheap sensors on the Raspberry-Pi platform, and can be replicated across multiple educational institutions. By using this approach, we can build a national talent pool in the area of secure embedded devices.

References

1. https://www.gsa.gov/portal/content/129694
2. American Association for the Advancement of Science, "Achieving Systemic Change: A Sourcebook for Advancing and Funding Undergraduate STEM Education," Ed. C. Fry. Association of American Colleges and Universities (2014). http://www.aacu.org/sites/default/files/files/publications/EPKALSourcebook.pdf
3. Rao, A.R., Dave, R.: Developing hands-on laboratory exercises for teaching STEM students the internet-of-things, cloud computing and blockchain applications. In: IEEE Integrated STEM Education Conference, Princeton, NJ (2019)

4. Rao, A.R., Clarke, D., Bhadiyadra, M., Phadke, S.: Development of an embedded system course to teach the internet-of-things. In: IEEE STEM Education Conference, ISEC, Princeton, pp. 154–160 (2018)
5. Rao, A.R.: A novel STEAM approach: using cinematic meditation exercises to motivate students and predict performance in an engineering class. In: 2017 IEEE Integrated STEM Education Conference (ISEC), pp. 64–70. Princeton University (2017)
6. Rao, A.R., Clarke, D., Yeskepalli, D., Mallu, M.-R.: Teaching cybersecurity concepts through Internet-of-things applications based on the Raspberry Pi. In: Colloquium for Information Systems Security Education (CISSE), New Orleans (2018)
7. DoD Cyber Scholarship Program, 28 February 2019. https://dodcio.defense.gov/Portals/0/Documents/Cyber/dodcyspfastfacts.pdf
8. Libicki, M.C., Senty, D., Pollak, J.: Hackers Wanted: an examination of the cybersecurity labor market. Rand Corporation (2014)
9. Kornelis, C.: The Hot, lucrative market in IT security talent. Wall Street J. (2019)
10. Skorobogatov, S., Woods, C.: Breakthrough silicon scanning discovers backdoor in military chip. In: International Workshop on Cryptographic Hardware and Embedded Systems, pp. 23–40 (2012)
11. Singhal, A., Ou, X.: Security risk analysis of enterprise networks using probabilistic attack graphs. In: Network Security Metrics, pp. 53–73. Springer, Heidelberg (2017)
12. Anley, C., Heasman, J., Lindner, F., Richarte, G.: The Shellcoder's Handbook: Discovering and Exploiting Security Holes. Wiley, Hoboken (2011)
13. Halfond, W.G., Viegas, J., Orso, A.: A classification of SQL-injection attacks and countermeasures (2006)
14. Rashidi, B., Fung, C., Bertino, E.: A collaborative DDoS defence framework using network function virtualization. IEEE Trans. Inf. Forensics Secur. **12**, 2483–2497 (2017)
15. Pancevski, B., Germano, S.: In rebuke to U.S., Germany considers letting Huawei In. Wall Street J. (2019)
16. Robertson, J., Riley, M.: The Big Hack: How China Used a Tiny Chip to Infiltrate U.S. Companies. Bloomberg Businessweek, 4 October 2018
17. Naughton, J.: The tech giants, the US and the Chinese spy chips that never were… or were they. The Guardian (2018)
18. Gormley, B.: Hospitals turn to biometrics to identify patients. Wall Street J. (2019)
19. Li, M., Yu, S., Zheng, Y., Ren, K., Lou, W.: Scalable and secure sharing of personal health records in cloud computing using attribute-based encryption. IEEE Trans. Parallel Distrib. Syst. **24**, 131–143 (2013)
20. Fu, K., Blum, J.: Inside risks controlling for cybersecurity risks of medical device software. Commun. ACM **56**, 35–37 (2013)
21. Perakslis, E.D., Stanley, M.: A cybersecurity primer for translational research. Sci. Transl. Med. **8**, 322ps2 (2016)
22. Kshetri, N.: Blockchain's roles in strengthening cybersecurity and protecting privacy. Telecommun. Policy **41**, 1027–1038 (2017)
23. Rao, A.R., Desai, Y., Mishra, K.: Data science education through education data: an end-to-end perspective. In: IEEE STEM Education Conference (ISEC), Princeton (2019)
24. Seymour, E.: Talking About Leaving: Why Undergraduates Leave the Sciences. Westview Press, Boulder (2000)
25. Takahashi, D.: Smart devices aren't so bright when it comes to security, 29 January 2019. https://venturebeat.com/2019/01/29/pepper-iot-smart-devices-arent-so-bright-when-it-comes-to-security/
26. Cimpanu, C.: Japanese government plans to hack into citizens' IoT devices. Zdnet (2019). zdnet.com

27. Raspberry Pi opens first High Street store in Cambridge. BBC (2019). BBC.com
28. Hern, A.: Fitness tracking app Strava gives away location of secret US army bases. The Guardian (2018)
29. Armbrust, M., Fox, A., Griffith, R., Joseph, A.D., Katz, R., Konwinski, A., et al.: A view of cloud computing. Commun. ACM **53**, 50–58 (2010)
30. Atzori, L., Iera, A., Morabito, G.: The internet of things: a survey. Comput. Netw. **54**, 2787–2805 (2010)
31. Singer, P.W., Friedman, A.: Cybersecurity: What Everyone Needs to Know. Oxford University Press, Oxford (2014)
32. Lindell, Y., Katz, J.: Introduction to Modern Cryptography. Chapman and Hall/CRC (2014)
33. Swan, M.: Blockchain: Blueprint for a New Economy. O'Reilly Media, Inc. (2015)
34. Mckenna, L.: Why Students Are Still Spending So Much for College Textbooks. The Atlantic, 26 January 2018. https://www.theatlantic.com/education/archive/2018/01/why-students-are-still-spending-so-much-for-college-textbooks/551639/
35. Mearian, L.: UC Berkeley puts blockchain training online; thousands sign up. Computerworld, 19 June 2018. https://www.computerworld.com/article/3282791/blockchain/uc-berkeley-puts-blockchain-training-online-thousands-sign-up.html
36. Hamblen, J.O., Van Bekkum, G.M.: An embedded systems laboratory to support rapid prototyping of robotics and the internet of things. IEEE Trans. Educ. **56**, 121–128 (2013)
37. Callaghan, V.: Buzz-Boarding; practical support for teaching computing based on the internet-of-things. In: 1st Annual Conference on the Aiming for Excellence in STEM Learning and Teaching, Imperial College, London & The Royal Geographical Society, pp. 12–13 (2012)
38. Dark, M., Kaza, S., Taylor, B.: {CLARK}–the cybersecurity labs and resource knowledge-base–a living digital library. In: 2018 {USENIX} Workshop on Advances in Security Education ({ASE} 2018) (2018)
39. Harrington, W.: Learning Raspbian. Packt Publishing Ltd (2015)
40. Arsyad, A.A., Dhadkah, S., Köppen, M.: Two-factor blockchain for traceability cacao supply chain. In: International Conference on Intelligent Networking and Collaborative Systems, pp. 332–339 (2018)
41. Cybersecurity Workforce Education - CNAP Initiatives' Number H98230- I 7- I -032. "Developing Hands-on Exercises for Secure Embedded System Design & Security Data Analytics for Computing and Engineering Students. CNAP-CAE CNAP-CAE2017 Grant# H98230-17-1-0321. National Security Agency (2017)
42. Rao, A.R., Clarke, D., Mohammed, N.: Creating an anchor hands-on cybersecurity course using the Raspberry Pi. In: Colloquium for Information Systems Security Education (CISSE), New Orleans (2018)
43. Wolff, J.: Two-Factor Authentication Might Not Keep You Safe. New York Times (2019)

An Analysis of Cybersecurity Legislation and Policy Creation on the State Level

Adam Alexander, Paul Graham, Eric Jackson, Bryant Johnson,
Tania Williams[(✉)], and Jaehong Park

University of Alabama in Huntsville, Huntsville, AL 35488, USA
{aha0007,pag0006,eaj0010,bej0003,tania.williams,
jae.park}@uah.edu

Abstract. To best create an effective cybersecurity strategy, it is imperative to understand the policy discussions and trends on a federal and state level. Effective cybersecurity legislation is vital to maintaining our country's infrastructure and protecting our citizenry. Since cybersecurity is often decided on the state level, states need to be aware of the trends in cybersecurity legislation. The purpose of this research was to conduct an analysis of cybersecurity policy from across the United States in an effort to assist the State-level understanding on their cybersecurity risk profile. This analysis included an examination of common trends in cybersecurity legislation. It involved researching cybersecurity policies from all 50 states and the federal government. After creating this baseline, the next phase of the research was to find and record relevant metadata for each policy. This data contained additional data, such as did it pass, who were the supporters, was it revised and other information that is useful to cybersecurity policy creators. The final goal of the research was to provide a searchable tool that could be utilized to fashion a successful cybersecurity bill and a summary of cybersecurity trends from 2011 to Spring 2018.

Keywords: Cybersecurity · Policy · Legislation · United States · States · Federal government

1 Introduction

It is critical that individual states enact policy dealing with cybersecurity. The National Governors Association, in hopes of addressing the cybersecurity deficit found in states across the nation, drafted A Compact to Improve Cybersecurity [1]. This compact includes a commitment to build cybersecurity governance, to prepare and defend the state from cybersecurity events, and to grow the nation's cybersecurity workforce. However, meeting such a commitment is difficult without an understanding of existing attempts of cybersecurity legislation from across the country.

As technology advances and cyber threats continue to grow, updating our country's cybersecurity policy is an important and daunting task. Our collective security infrastructure is woefully out-of-date and security policies differ from state to state. Therefore, the governor of Indiana signed executive order 17-11 in January of 2017, creating a council to "develop, maintain and execute an implementation plan for accomplishing strategic cybersecurity objectives that are specific, measurable, achievable, and relevant to the strategic vision" of the state [2].

The role of this research was to provide the state with an analysis of existing cybersecurity policy from across the United States proposed from 2011 to present (as of Spring 2018). The research identified trends in policy (whether a policy was adopted or not after proposal). This research will serve as a baseline for the State of Indiana when crafting their policy and will provide valuable insight to other states who might choose to use the research.

In order to assist the States of Indiana in fulfilling the compact by developing their cybersecurity policy, we conducted a policy analysis using the following research questions:

- What policy has been passed successfully/unsuccessfully in other states from 2011 to present (Spring 2018)?
- Who were the supporters of the policy?
- What type of support did the proposed policy receive, and if it did not pass, why?
- How can such information be presented to Indiana stakeholders in a clear and concise manner?
- What trends are evident among the states regarding cybersecurity policy?

In order to ensure that all policy was evaluated systematically, we developed a data collection form. Additionally, we organized the research by the 20 existing Indiana committees, streamlining the examination and evaluation of the data. We examined similar trends analysis research and found, while research exists, the scope of the research was narrower. For example, Lowry examined the regulation of mobile payments but only dealt with federal law, making the reporting of such trends much easier [3]. Additionally, we were able to locate studies of trends resulting from one piece of legislation but did not find any previous work dealing with trends regarding state legislation. We provided a baseline for other large scale legislative trends analysis. Additionally, our database of national cyber-related policies provides a valuable resource for other states as they seek to improve their cybersecurity posture.

2 Related Works

In 2007, the government of Estonia was hit by a cyber-attack that paralyzed the country, shutting down its largest bank, rendering credit cards useless, knocking media outlets offline, and crippling the country's telephone communications [4]. Could such an attack happen in the United States? Former cybersecurity czar Richard Clarke maintains that "few national governments have less control over what goes on in its cyberspace than Washington" and that "America's ability to defend its vital systems from cyber-attack ranks among the world's worst" [5]. This threat of cyber-attack is not limited the federal government. Individual states also must consider the threat of weak cybersecurity.

States, which hold databases full of health records, driving records, criminal records, professional licenses, tax information, and birth certificates, must have procedures in place to protect this personally identifiable information. The states also often have jurisdiction of cyber-related crimes and are entrusted with cybersecurity education [6]. As Glennon notes, "Every state has enacted laws directed at protecting state governments and businesses specifically from cyberintrusions" [6]. On top of this, states also bear much of the burden of regulation; however, as Sales states, law and policy of cybersecurity are undertheorized and most governments concern themselves with criminal law but are reluctant to see cybersecurity management in regulatory terms [5].

Bosch also notes issues with regulation, stating a reliability standard, such as those created through the Federal Power Act, "does not fully address Smart Grid cybersecurity from an interoperability perspective" [7]. Alternatively, he notes the difficulty of crafting the standards to begin with, citing the failed GRID Act of 2010, which the federal legislative branch could not agree on how the grid's cybersecurity concerns should be addressed [7].

As every state is unique, so must each state take a different approach to cybersecurity. Schneider, in his call for government support of cybersecurity, noted as social values differ, governments should not expect uniform sets of cybersecurity goals; instead government interventions designed to achieve goals in some geographic region . . . must also accommodate the diversity in goals and enforcement mechanisms found in other regions [8]. When states craft their cybersecurity legislation, it is necessary to build on the experience of other states and to understand national policy trends.

As Godara notes, crime has seen a "revolutionary shift from the main actor, the criminal, to certain non-actors in the cyber world called 'intermediaries'. To what extent an intermediary can be held liable for the crimes committed in cyber space is a question which is mooted all over the world" [9]. Godara's research compares legislative and judicial trends in different countries. Her work was limited to rulings regarding intermediary liability in the United Kingdom, United States, and India. When examining legislation in the United States, her approach was to limit her study to federal court cases and sought to analyze fewer than ten rulings.

Bulger, Burton, O'Neill, and Staksrud also examine legislative trends in their examination of how different countries seek to protect children online [10]. In their research, they examined the United States, South Africa, and the European Union. The research targeted key crimes and then reported each country's laws regarding these crimes. Again, the authors chose to research only federal laws and did not examine legislation from individual states.

Neither Godara nor Bulger et al. considered failed legislation when examining these trends [9, 10]. While both research examples relate to trends in cybersecurity, they do not provide an approach to handling the large volume of legislation relating to cybersecurity produced by individual states.

3 Data Collection and Analysis Process

In this section, we first discuss how we collected bills, what kind of metadata we used to identify and classify those bills. Then, after brief discussion on the database we used, we discuss how we examined and analyzed those collected bills.

3.1 Finding and Classifying a Bill

First, we examined digital archives to look for proposed legislation relating to cyber-security. As stated before, each state usually had a digital archive of bills the researcher could examine using a keyword search. Once that location had been exhausted, secondary locations were searched. For each policy found, we recorded the following information:

- Location (1 of 50 states, Washington D.C., or the U.S. Congress)
- Type of policy (see classifications below)
- Bill name and/or number
- Source (where the bill can be found)

For the policy type, the following classifications were used:

- Government Service
- Finance
- Defense
- Energy
- Water/Wastewater
- Communications
- Healthcare
- Elections
- Economic Development
- Workforce Development

- Personal Identifiable Information
- Public Awareness and Training
- Education
- Emergency Services and Exercise
- Cyber Sharing
- Cyber Organizations (Center)
- Cyber Pre-Thru Post Incident
- Legal/Insurance
- Local Government
- Other critical infrastructure

These classifications were originally the 20 groups that make up the Indiana Executive Council on Cybersecurity and provided an clear way for the end user to reference trends and policies when using the final document as reference.

Data from primary online sources comprised the bulk of the information collected for the trends analysis. Most states provided some type of searchable archive. However, in cases where such databases were not available, we utilized second party databases to collect policy information. These second party databases included sites such as Find Law and Legiscan.

3.2 Creating a Collaborative Database

While many tools were available for storing and managing the research, we sought one that would allow us to collaborate seamlessly and would allow us to share our data with end users without requiring specialized software or paid licensing. We also sought a product that was versatile enough to allow for linking fields together and even sharing data from one table to another using foreign keys. The tool also needed to have several sorting and filtering options. We used an online product called Airtable to meet our needs [11]. This tool, which creates sortable tables of the metadata can be found at the following address: https://airtable.com/shrCcYzKJGH1jyvrx.

After deciding on a tool, we fine-tuned the database design, listing necessary fields and then organized them to streamline the data entry process, identifying primary and foreign keys.

We formatted our information to prepare it for analysis. While reading the bills, the following information was collected in the database: Bill number, State, Type of policy, Type of legislation, Originator (senate, house, joint, or governor's office), Year introduced, Status, Link to online source, Related legislation, Description, Political party affiliation, Bill sponsor, and Link to vote count.

3.3   Trends Analysis

The next step was to begin the preliminary analysis of the data. Each state had its own cybersecurity policies. The number of each classification for every state was analyzed to discover what was most important to that state. We also made an effort to determine states that were currently active in developing cybersecurity programs.

Additionally, vetoed bills and failed legislation were examined. Some states, while successful in passing legislation in the house and senate, failed to garner the support of the state's governor. Since the reasons for such occurrences could be valuable, we wanted to analyze these instances. Failed legislation also merited special consideration. If a certain classification had a high number of bills written but the bills did not pass to become policies, then it can be inferred, while enough people thought the bill would be a good idea, an even greater number of people had negative thoughts about the bill to keep it from passing. This trend was explored to find out why.

We considered the influence of federal legislation. While states are responsible for crafting their own legislation, we wished to determine if the federal government's actions played a role in determining when and what cybersecurity topics were addressed on the state level.

States who proved to be cybersecurity pioneers were identified. Cybersecurity is more of a priority for some states than others. By examining the progression of cybersecurity legislation by state per year, patterns showing states who exhibited steady policy creation were evidenced. The states showing consistent policy creation over time were determined to be cybersecurity pioneers.

Bipartisan policy creation was also considered. One of the primary goals in the trends analysis was to determine factors that played a role in the successful passage of legislation. This included the success of a political party in getting a bill adopted. As data collection progressed, it became evident that bipartisan efforts garnered different results than partisan efforts.

Following the trends analysis, the final step was the analysis of results. The following questions were addressed:

- Are there states that could be considered pioneers to cybersecurity legislation?
- To what degree does the federal government's actions influence state legislation?
- Are there paths that a bill takes that influences its success?

4   Cybersecurity Legislation Analysis

We identified 500 pieces of legislation relevant to cybersecurity within our eight year sample size. We surveyed 454 policies from all fifty states and Washington, D.C., as well as an additional 46 policies from the federal government.

4.1   States Currently Active in Passing Cybersecurity Legislation

In order to determine which states are actively developing their cybersecurity program, all 50 states were examined and the number of policies by year were recorded by state, as shown in Fig. 1.

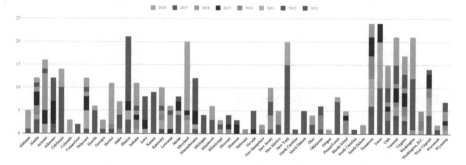

Fig. 1. The quantity of policies developed by each state per year between 2011 and 2018

Looking at the state policy by year, it was apparent that most states had between 1–10 cybersecurity policies. There were seven out of fifty states that had 20 or more policies.

The dates of the policies were also important. If most policies were proposed before 2016, then the state would not be considered as developing their cybersecurity program. Of the seven states with a large range of policies, only four states created most of their policies from 2016 until now. The four states are Illinois, Maryland, New York, and Vermont.

While a single policy can have multiple policy types, it is still worthwhile to look at the number for each type. Illinois, New York, and Vermont had a high number of legal/insurance policies which would support the argument that most of the new policies being created by developing states were of the type legal/insurance. Vermont also had a high number of government service policies, especially in 2018. Figure 1 shows these two states have a high number of policies spread out over the whole sampling period (2011–2018).

4.2   Vetoed Bills

In five instances, proposed legislation made it through both the senate and the house; however, the legislation failed to be finalized by a state's governor.

Two of the bills were vetoed by California governor Edmund G. Brown, Jr. Both were introduced in 2017 and were unanimously passed by the state's assembly and senate. Bill AB1306 detailed the scope of the California Cybersecurity Integration Center, which was established by Governor Brown's executive order in 2015 [12]. Brown, in his Governor's Veto Message, expressed concern "that placing the Center in statute as this bill proposes to do, will unduly limit the Center's flexibility as it pursues its mission to protect the state against cyberattacks" [13]. As for vetoed bill AB531, which required the department of technology's office of information security to evaluate existing security policies and develop plans to address deficiencies, Brown stated that the bill's objectives were already required by AB 670 [14].

A bill was vetoed by Governor Susana Martinez from New Mexico. It received 36 to 3 majority votes of support in the state's senate and 37 to 5 majority votes of support in the state's house. HB 364, while dealing primarily with limiting the prescription of contact lenses and glasses, did deal with cybersecurity by restricting a resident's access to online services. Martinez stated in her House Executive Message No. 57 that the bill limited the use of emerging technologies related to the issuance of contact lenses and glasses [15]. She cited this as the reason she chose to veto the bill.

The other two bills were vetoed by Governor Douglas Ducey of Arizona. Bill SB1434 was vetoed in 2016 after receiving unanimous votes from both the senate and the house. The governor indicated that he vetoed the bill, which dealt with consolidated purchasing and shared services of technology, stating he felt the bill added an extra layer of bureaucracy [16]. HB2566, dealing with password policy, encryption standards, and data security, was vetoed in 2015. It had passed the senate with a vote count of 17 to 11 and passed the house with a vote count of 56 to 1. Ducey stated that his administration had already addressed the concerns outlined in the bill [17].

4.3   Failed Legislation

Figure 2 shows the twenty classifications used to identify bills and the status count of the policies classification. Although a policy can have multiple classifications, this explores the number of times a classification has a relation to a legislation record.

The label "In Progress" is for classifications that are identified to be introduced and still up for discussion, and "Failed" are bills that are inactive, died in chamber, died in committee, or vetoed. Of the twenty classification types used to identify the bills, most classification types tended to have more failed policies than passed bills. We identified that legislation related to Cyber Sharing, Economic Development, and Education have much higher failure rates than the other classifications. The seven classifications that were an exception include: policies dealing with cyber organizations, elections, emergency services and exercise, finance, government service, local government, and water/waste-water. Furthermore, policies that were related to Elections and Water/Wastewater have greater rates of success than the other classifications. Notably, out of the six state legislations dealing with Water/Wastewater, five were passed successfully, one remains in progress, and zero failed.

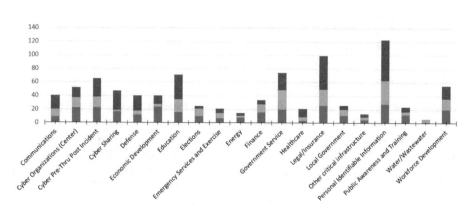

Fig. 2. The quantity of each policy type surveyed that is either still in process, was passed into law, or was failed for any reason

4.4 Influence of Federal Legislation

Figure 3 separates the federal legislation from the state legislation and shows the percentage each topic was covered in bills introduced at those levels within a time frame. In this figure, our eight year sample size was divided into two separate four year periods to show some slight changes in policy creation.

Much of the federal legislation from the U.S. Congress is focused on Defense, Cyber Pre-through-Post Incident, and in Cyber Sharing between organizations. Federal legislation in those categories are consistently higher than all other categories surveyed since 2011. For example, from 2011 to 2014, 61.1% of the federal legislation survey dealt at least some with Cyber Sharing. While those topics were addressed by some at the state level, our data does not show them being addressed by a large amount of states until 2017. Federal legislation appears to be driving state legislation to fill in the gaps where there are security concerns not addressed by the U.S. Congress at all.

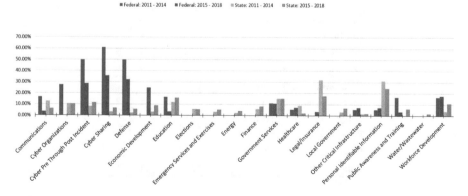

Fig. 3. The percentage of state and federal policies introduced in 4 year periods (2011–2014, and 2015–18) that deal each surveyed category

In contrast to the federal legislation, state legislation heavily focused on topics such as Education, Personally Identifiable Information, Government Services, Legal/Insurance concerns such as defining cybersecurity crimes. These were topics that the U.S. Congress did not have many pieces of legislation on at all.

4.5 Cybersecurity Pioneers

Table 1 shows the number of policies when grouped by state and year. When analyzing the states and the number of policies they have proposed, it is easy to see that most states are not creating new policies. Of the 50 states, only 16 of them have at least 10 new policies since 2011. We used 10 policies as a cut off point since 10 policies provides enough sampling to determine the regularity of policy creation. Pioneering states were Alaska (12), Arizona (16), California (14), Delaware (12), Hawaii (11), Illinois (21), Indiana (11), Maryland (20), Massachusetts (12), New York (20), Tennessee (24), Texas (24), Vermont (21), Virginia (21) Washington (21), and West Virginia (14) These states appear to be in 3 different classifications.

Table 1. The quantity policies and their types that were passed between 2016 and 2018 in the states with the highest surveyed volume

States with high number of policies 2016–2018				
Policy type	IL	MD	NY	VT
Communications			5	3
Cyber organizations	2	1	5	
Cyber pre through post incident	1	1		5
Cyber sharing	1	1	3	
Defense		2		
Economic development		5	5	1
Education	2	3	4	
Elections	1	2	1	
Emergency services and exercises			5	
Energy		1	3	3
Finance	1	2		
Government services	3	2	3	4
Healthcare			1	
Legal/insurance	3	3	7	5
Local government	2		2	
Other critical infrastructure	1		1	
Personal identifiable information			3	4
Public awareness and training	1	1	5	
Water/wastewater			2	
Workforce development	2	5		
Total	20	29	55	25

Early Policy Creation; However the State Has Not Produced Much Legislation of Late: In this category, the state created several policies earlier than 2014 and then less after 2014. These states have dropped in their proactive approach to cybersecurity and are not considered as pioneers. For example, Texas created the first bills for various types of policy. While creating several of bills early on, they have not been active in bill creation since 2015. The states of Tennessee, Texas, and West Virginia meet this criteria. Even though their number of policies are high, their concern for cybersecurity seems to have lessened.

Large Policy Creation; However, Most of the Policies Have Been Created over the Last 3 Years: This grouping shows states that have created most of their cybersecurity policies over the past 3 years (2016–2018). These states, while recently producing more legislation, did not have the early policy adoption to be considered pioneers. Arizona, California, Delaware, Hawaii, Illinois, Indiana, Maryland, Massachusetts, New York, and Washington match this criteria. The higher policy producers worth nothing are Maryland (15 policies in 2018 alone), New York (20 policies in the past two years), and Washington (20 policies in the past two years also).

Steady Policy Creation: These high-producing policy creators consistently created bills over the sample years (2011–2018). As they consistently produced more cybersecurity policies than other states over the same sample time, it would suggest the states were pioneers in cybersecurity policy creation and not as reactive to other states through the years. As Fig. 1 Number of Policies by State per Year shows, Alaska, Vermont and Virginia are the only states that match this criteria. Vermont has the most policies at 21 followed by Virginia at 17. Alaska did not have near as many with 12.

4.6   Bipartisan Success

Of the 454 examples of state level cybersecurity legislation found, 109 records were bipartisan attempts. Of those attempts, 29 pieces of joint legislation were listed as actively being considered, meaning the outcome of the legislation was yet to be determined, and 45 of the bills that were introduced passed. When excluding legislation in progress, the resulting bipartisan success rate was 56%. In addition to bipartisan efforts, there were 5 records introduced by council, with all 5 passing. This success rate is significantly higher than partisan sponsored cybersecurity legislation on the state level, where, of the bills that were no longer actively being considered, only 88 passed, indicating a success rate of 40% (see Fig. 4).

Cybersecurity topics that garnered the most state level bipartisan sponsorship included those relating to personal identifiable information (22 records), government services (19 records), legal (17 records), and cyber pre through post incident (16 records). There were no examples of bipartisan sponsorship relating to general policies.

Idaho and Kansas were the two states with the most bipartisan sponsored legislation, both having 7 records with bipartisan support. Iowa, Texas, Washington, and Wyoming also were close in this category, having 6 instances each of utilizing bipartisan

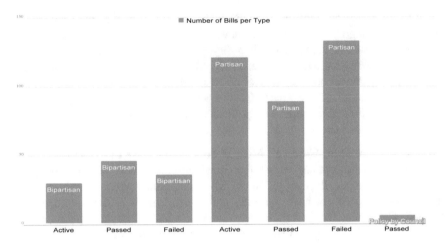

Fig. 4. Success of state level bipartisan legislation attempts as opposed to partisan legislation attempts

sponsorship for cybersecurity legislation. States with no bipartisan support of cybersecurity legislation included Arkansas, California, Georgia, Louisiana, Missouri, Montana, New Mexico, New York, North Carolina, Oklahoma, and Wisconsin. Washington, D.C., also had no records in this area.

This data is being stored at the following link using Airtable. Please follow the link below to view the tool [11], https://airtable.com/shrCcYzKJGH1jyvrx.

5   Challenges

5.1   Varying Terminology

One problem with our research was how verbiage varied from state to state. For example, one state might choose to use the term *cyber security*, while other states might use terms such as *computer crime* or *online security*. To ensure that each state was researched thoroughly and consistently, the researchers agreed on a list of keywords to use in their search.

5.2   Determining Relevance

Also, the relevance of the proposed legislation to the targeted analysis data was also a challenge. Desired topics were often buried deep within unrelated information, resulting in researchers having to read and index bills that were, at first glance, not relevant to the desired data set.

5.3   Tracing a Bill's Origin

Another problem dealt with how bills are created. At times a bill originates in the house, and at other times it can be created in the senate. Bill numbers vary depending on the origin, and they can actually compete with each other. Also, a bill will stall in a committee, or the current legislature may elect not to take up a discussion on the bill. A new bill can be created the following year in order to try to create the policy. These bills must be linked in the research to provide a good picture on policy creation.

Oftentimes a generic bill will pass and become policy. After passing the first bill, a second bill will revise the original policy to provide clarification or additional direction. The original bill and the following bills must be linked in the research also.

6   Conclusion

Excluding federal legislation and active legislation, we found 305 examples of state level legislation relating to cybersecurity. Of those, 138 records passed and 167 failed or were determined to be inactive, demonstrating a success rate of 45%.

Policies concerning elections and water/wastewater had higher success rates than other classifications. Policy topics that exhibited higher than average failure rates were related to cyber sharing, economic development, and education.

During the time period sampled, there seemed to be little correlation between federal cybersecurity policy efforts and those of the states. If fact, the two entities tended to complement each other, with federal policy having a much different focus than the states. For example, federal policies dealt more with defense, while state policies dealt more with education.

States showing consistent push in cybersecurity legislation were Vermont and Virginia. These states created policy steadily over the time period and met the criteria to be considered pioneers in cybersecurity legislation.

We determined that one factor that seemed to increase a piece of legislation's chance of success was the willingness of legislators to cross party lines in initiating new legislation. Bipartisan bills had a success rate of 56%, while bills introduced along party lines only had a success rate of 40%. Popular bipartisan topics included personal identifiable information, government services, legal, and cyber pre through post incident. When compared to the overall success rate of 45%. It is evident that bipartisan support is a favorable predictor of a bill's chance of passage.

7   Future Work

In order for the research to continue to be useful, it is critical that the database be maintained. As new cybersecurity related legislation is proposed and considered, it should be catalogued in the base. By keeping the database current, the picture of national cybersecurity trends will become more granular, and the increased data will allow for better trend analysis.

Additionally, it would be beneficial for future researchers to expand the research by correlating the passage of legislation to related major cyber events. For example, researchers could determine if the Equifax breach resulted in an increase of proposed legislation related to personally identifiable information. If a correlation is evident, this could serve as a predictor of future proposed legislation.

Researchers could also attempt to measure the impact of key successful legislation. An example of this future work could be in the area of workforce development. Researchers could ascertain if states that adopted workforce development legislation have seen an increase in available professionals.

Furthermore, a thorough examination of failed legislation would aid legislators when crafting legislation. By surveying bill sponsors, researchers could identify key barriers to cybersecurity legislation, allowing policy makers the ability to better craft and propose bills. Also, researchers could compare failed legislation from one state to similar successful legislation in another state to determine why similar legislation failed in one state but found success in another.

Acknowledgement. We would like to thank the State of Indiana and INSuRE Program for their support and valuable inputs.

References

1. National Governors Association: Meet the threat: a compact to improve State Cybersecurity (2017). https://www.in.gov/cybersecurity/files/NGA%20Cyber%20Compact.pdf
2. Holcomb, E.J.: Exec. Order No. 17-11. Continuing the Indiana Executive Council on cybersecurity. State of Indiana Executive Department, 9 January 2017. https://www.in.gov/cybersecurity/3812.htm
3. Lowry, C.: What's in your mobile wallet? An analysis of trends in mobile payments and regulation. Fed. Commun. Law J. **68**(2), 353–384 (2016). http://www.fclj.org/wp-content/uploads/2016/09/68.2.4-Carolyn-Lowry.pdf
4. Sales, N.A.: Regulating cyber-security. Northwest. Univ. Law Rev. **107**(4), 1503–1568 (2013)
5. Clarke, R.: War from cyberspace. Natl. Interest **104**, 31–36 (2009). https://nationalinterest.org/article/war-from-cyberspace-3278
6. Glennon, M.J.: State-level cybersecurity. Policy Rev. **171**, 85–102 (2012)
7. Bosch, C.: Securing the smart grid: protecting national security and privacy through mandatory, enforceable interoperability standards. Fordham Urban Law J. **41**(4), 1349–1406 (2014)
8. Schneider, F.: Impediments with policy interventions to foster cybersecurity. Commun. ACM **61**(3), 36–38 (2018)
9. Godara, S.: Role of 'intermediaries' in the cyber world: a comparative study of the legislative policies & recent judicial trends. VIDHIGYA J. Legal Aware. **8**(1), 69–80 (2013)
10. Bulger, M., Burton, P., O'Neill, B., Staksrud, E.: Where policy and practice collide: comparing United States, South African and European Union approaches to protecting children online. New Media Soc. **19**(5), 750–764 (2017)
11. State of Cybersecurity, Airtable: https://airtable.com/shrCcYzKJGH1jyvrx

12. Brown Jr., E.G.: Exec. Order No. B-34-15. Establishing the California Cybersecurity Integration Center, CA.Gov (2015). https://www.calhospitalprepare.org/sites/main/files/file-attachments/b-34-15_cal-csic.pdf
13. Brown Jr., E.G.: Governor's Veto Message, California Legislative Information, 11 October 2017. https://leginfo.legislature.ca.gov/faces/billStatusClient.xhtml?bill_id=201720180AB1306
14. Brown Jr., E.G.: Governor's Veto Message, California Legislative Information, 14 October 2017. http://leginfo.legislature.ca.gov/faces/billStatusClient.xhtml?billid=201720180AB531
15. Martinez, S.: House Executive Message No. 57, New Mexico Secretary of State, 7 April 2017. https://static.votesmart.org/static/vetotext/61925.pdf
16. Ducey, D.A.: Re:Senate Bill 1434, Office of the Governor, May 18, 2016. https://azgovernor.gov/sites/default/files/sb_1434_veto_letter.pdf
17. Ducey, D.A.: RE: House Bill 2566, Arizona State Legislature, 9 April 2015. https://www.azleg.gov/govlettr/52leg/1R/HB2566.pdf

A Survey of Cyber Security Practices in Small Businesses

Eric Imsand[1(✉)], Brian Tucker[2], Joe Paxton[2], and Sara Graves[1]

[1] Information Technology Systems Center, University of Alabama in Huntsville,
Huntsville, AL, USA
eric.imsand@uah.edu
[2] Office of Operational Excellence, University of Alabama in Huntsville,
Huntsville, AL, USA

Abstract. Small businesses are a unique class of organization with challenging cyber security problems that are frequently overlooked. These firms are being increasingly targeted for cyber-attack. These firms are particularly vulnerable to cyber-attack due to the often-valuable information they handle coupled with overworked and undertrained IT support. The University of Alabama in Huntsville is conducting an on-going survey of the cyber security practices of small businesses seeking to implement the NIST SP 800-171 cyber security standard. The data gathered indicates that small businesses in this field are likely to engage in poor security practices arising from common cyber security misconceptions.

Keywords: Information assurance · Compliance · NIST SP800-171 ·
Small business

1 Introduction

Recent studies have shown that businesses, particularly small businesses, are increasingly at risk for a cyber-attack. Perhaps most vulnerable are small, professional "white collar" firms. Such companies have for years operated without a unifying cyber framework. While retail outlets have been forced to adopt cyber frameworks like the Payment Card Industry (PCI) Data Security Standard [6], white collar firms have been applying best-effort cyber security, often based on the knowledge of the most technically savvy employee in the office. This has disastrous consequences for small businesses. A recent survey found that small businesses typically suffer ~60% greater monetary damages than large businesses [7].

Criminals are not the only group that has taken notice of the current situation. The US government contracts with thousands of small businesses for a variety of services ranging from basic food service to complex scientific and engineering design. Motivated by a need to protect the data entrusted to small businesses, the US Government has drafted a new cyber security standard to ensure the protection of government data on private contractor networks. This standard, NIST SP800-171 [1], was first drafted in 2015. The first agency of the US government to adopt this standard was the US Department of Defense. In 2015, the US DoD modified their acquisition

K.-K. R. Choo et al. (Eds.): NCS 2019, AISC 1055, pp. 44–50, 2020.
https://doi.org/10.1007/978-3-030-31239-8_4

rules, commonly known as DFARS, to include a provision requiring private companies that handle sensitive but unclassified data to protect that data by implementing SP800-171. The DFARS does not make any allowances for the implementation of NIST SP800-171. All companies must adopt the standard regardless of size.

The adoption of NIST SP800-171 has proven challenging for the majority of businesses, but perhaps most challenging for small businesses. Though some firms may choose to engage a service provider for IT support, it is very common for such companies to utilize a "do it yourself" (DIY) approach to IT. Clearly such an approach leaves these companies exposed to the dangers posed by a constantly-evolving cyber threat landscape. Understanding the beliefs and practices of this group of companies is critical for the cyber security research community. To address this knowledge gap, the authors have conducted an extensive survey of small businesses that are implementing the NIST SP800-171 standard. To the best of the authors' knowledge, the survey is the first effort to gauge the cyber security practices and beliefs of small "white collar" businesses.

NOTE: This study has been conducted as part of a larger grant from the US Department of Defense Office for Economic Adjustment (OEA). The DoD OEA has provided funding to help small businesses comply with the new NIST standard. The companies surveyed here have paid a small co-pay ($350) to participate in the program. At the end of the program the companies receive a detailed report describing their deficiencies against the SP800-171 standard. As described in Sect. 2, a gap analysis is a critical part of implementing SP800-171.

The remainder of this paper is organized as follows. Section 2 presents a brief overview of NIST SP800-171. Section 3 provides a description of the investigation performed by the authors. Section 4 provides preliminary results of this survey as well as key findings and observations. Section 5 features a set of recommendations for the cyber security community based on the observations presented in Sect. 4.

2 Overview of Federal Cybersecurity Publications

The changes to the DFARS require companies to achieve "adequate" cyber security through the implementation of NIST SP800-171 [4]. To aid in this process, the government has published a variety of supplemental documents, most notably NIST Handbook 162.

2.1 NIST SP800-171

NIST SP800-171 was created by the US government to provide baseline cybersecurity among contractors performing work for the US government. Beginning in 2017, all private contractors performing work for the US Department of Defense were required to implement NIST 800-171 on all of their corporate networks [3] which contained sensitive data that belonged to the US government. NIST 800-171 has been designed to be broader than a cyber security baseline for only DoD contractors, however, as it has been announced that all federal contractors will be required to implement NIST

800-171 in the near future [5]. The importance and prominence of this standard will only increase in the future.

NIST 800-171 organizes 110 controls into 14 broad families covering all of the traditional information assurance activities. The standard requires companies to engage in methodical, ongoing cyber security assessment and remediation. NIST 800-171 is not overly prescriptive, giving companies broad leeway to satisfy the controls in ways that are consistent with their organizational practices.

Organizations needing to comply with NIST 800-171 generally follow a three-step process:

1. Assess their corporate network against all 110 controls
2. Document the results of the assessment in a document called the *System Security Plan*
3. For all controls the company is not currently satisfying, create a plan to satisfy the control and document the plan in a second document named the *Plan of Action*

Additionally, NIST 800-171 has requirements for periodic re-assessment, meaning that companies must re-evaluate their networks on a regular interval.

2.2   NIST Handbook 162

The NIST 800-171 controls are written in a broad way to provide companies maximum flexibility in implementation. While providing flexibility, this approach has also created some confusion among companies who are unsure whether their implementations satisfy the requirements. To aid companies that are implementing the standard, NIST has published a second document, *NIST Handbook 162: NIST MEP Cybersecurity Self-Assessment Handbook for Assessing NIST SP800-171 Security Requirements in Response to DFARS Cybersecurity Requirements* [2] ("MEP Handbook"). This document is designed to function as a supplement to NIST 800-171. For every control listed in 800-171, the MEP Handbook provides an explanation of the control, as well as 1 to 10 questions that companies can use to assess their implementation.

To illustrate the relationship between NIST SP800-171 and the MEP Handbook, consider the following example from NIST 800-171 control 3.1.1 (Table 1):

Table 1. Comparison between SP800-171 and MEP handbook

NIST SP800-171 control text	MEP handbook assessment questions
• Limit system access to authorized users, processes acting on behalf of authorized users, or devices (including other systems).	• Does the company use passwords? • Does the company have an authentication mechanism? • Does the company require users to logon to gain access? • Are account requests authorized before system access is granted? • Does the company maintain a list of authorized users, defining their identity and role and sync with system, application, and data layers?

3 Methodology

As described in Sect. 1, the companies surveyed for this study were all small businesses seeking to implement NIST SP800-171 in order to satisfy the requirements found in the DFARS. In exchange for participating in this study, the companies received a detailed "gap report" that outlined the company's deficiencies against SP800-171. As described in Sect. 2, performing this type of assessment is a critical requirement towards implementing the standard.

This study was conducted in two phases. During the first phase, the authors created a very basic questionnaire for each company. The questionnaire asked questions about the structure of the company (e.g. "How many dedicated IT employees does your company have?"). The questionnaire also featured high-level questions based on each family of NIST SP800-171 controls. The first phase survey was sent to each company and the company completed it off-line and returned the results by e-mail. The results of the initial (phase one) survey were not used for actual data analysis. Instead, the authors reviewed the responses in order to gain a basic understanding of the company and its environment in advance of the second phase of the study.

During phase two, two members of the research team visited each company and conducted an in-depth interview with relevant company representatives. Each interview lasted between 5–6 h. Companies were instructed to provide representatives from relevant teams from within the company, e.g. technical and managerial.

The authors used the MEP Handbook as the basis for this study. The basic approach is as follows:

- Two members of the research staff visited each company in-person
- During the site visit, relevant company personnel were interviewed
- The interview consisted of the 357 questions featured in the MEP Handbook. Each question was discussed in-depth to ensure that company representatives understood what was being asked.
- Based on the answers provided by the company personnel, both researchers independently graded each response with one of the following scores:
  - Implemented
  - Not Implemented
  - Partially Implemented
  - Does Not Apply
  - Alternative Approach

Ratings on each control were validated using multiple sources of knowledge. These include:

- Subject matter expertise
- Q&A sessions with NIST representatives
- Q&A sessions with representatives from DCMA (Defense Contract Management Agency)
- Collaboration with other industry SMEs engaged in similar work.

After completion of the interview, each researcher reviewed the scores assigned by the other researcher. In cases where different scores were assigned to the same question, the research team members discussed the company's response and settled on a final score for that question. Then the results of the MEP Handbook questions were "rolled up" to assign a final score for each control. For the purpose of this study, a company must satisfy all of the questions in the MEP Handbook in order to satisfy the NIST 800-171 control.

4 Results and Analysis

To date, a total of 11 companies have been fully surveyed and received their returned gap report. Though data collection is ongoing, some interesting trends are emerging.

Overall, the businesses surveyed were already complying with, on average, $\sim 19\%$ of the NIST SP800-171 controls. To some degree, this result was expected. The NIST controls are designed to push companies towards a methodical, deliberate management of cyber security. The fact that small businesses do not have complex management of cyber risk is not surprising.

To gain a high-level understanding of the level of compliance demonstrated by the surveyed companies, each of the 110 controls was tabulated to see how many companies were complying. Then, these scores were averaged against the other controls in that family. Table 2 contains an aggregated average of the compliance for each of the 14 families identified in NIST 800-171.

Table 2. Rates of compliance among surveyed companies for each SP800-171 family

NIST control family	Average compliance among surveyed companies
3.1 Access control	22.73%
3.2 Awareness and training	15.15%
3.3 Audit and accountability	21.21%
3.4 Configuration management	9.09%
3.5 Identification and authentication	19.01%
3.6 Incident response	15.15%
3.7 Maintenance	19.7%
3.8 Media protection	4.04%
3.9 Personnel security	36.36%
3.10 Physical protection	42.42%
3.11 Risk assessment	3%
3.12 Security assessment	4.55%
3.13 Systems and communications protection	18.18%
3.14 System and information integrity	25.97%

The results of the survey thus far paint a grim picture. Out of 110 controls, only 10 were being practiced by more than half of the surveyed companies. The 10 controls that were most often practiced by the surveyed companies are listed in Table 3.

Table 3. Most frequently observed SP800-171 controls among surveyed companies

SP800-171 control	Observed frequency
3.1.2: Limit system access to the types of transactions and functions that authorized users are permitted to execute	63.64%
3.1.5: Employ the principle of least privilege, including for specific security functions and privileged accounts	63.64%
3.3.8: Protect audit information and audit tools from unauthorized access, modification, and deletion	63.64%
3.5.4: Employ replay-resistant authentication mechanisms for network access to privileged and non-privileged accounts	72.73%
3.5.11: Obscure feedback of authentication information	81.82%
3.7.2: Provide effective controls on the tools, techniques, mechanisms, and personnel used to conduct information system maintenance	54.55%
3.10.5: Control and managed physical access devices	63.64%
3.14.2: Provide protection from malicious code at appropriate locations within organization information systems	63.64%
3.14.3: Monitor information system security alerts and advisories and take appropriate actions in response	54.55%
3.1.2: Limit system access to the types of transactions and functions that authorized users are permitted to execute	63.64%

For most of the controls listed in Table 3, the companies that were observing these controls were doing so accidentally, i.e. the controls were in practice to address non-cyber related concerns such as meeting NISPOM requirements. Based on the responses given during the interviews, control 3.14.2 (requiring companies to run anti-virus software) was the only control found to be widely adopted for purely cyber reasons.

5 Conclusion

While the collection of data is ongoing, the initial results gathered thus far indicate that small, professional firms have been adopting a very limited approach to cyber security due at least in part to common cyber security misconceptions. The data indicates that the overwhelming majority of the companies surveyed are far outside compliance with the NIST SP800-171 standard. Though there are a few technically challenging controls found in the specification, the majority of SP800-171 prescribes activities that can reasonably be described as "industry best practices".

This finding is of particular note to the cyber research and education community. Small businesses frequently choose to perform their own information assurance activities. Early results indicate that these activities are driven by common misunderstandings like the infallibility of anti-virus. Combating these misconceptions should be a priority for the cyber education and research community.

References

1. National Institute of Standards and Technology: NIST Special Publication 800-171 Revision 1: Protecting Controlled Unclassified Information in Nonfederal Systems and Organizations. https://nvlpubs.nist.gov/nistpubs/SpecialPublications/NIST.SP.800-171r1.pdf. Accessed 27 Feb 2019
2. National Institute of Standards and Technology: NIST Handbook 162: NIST MEP Cybersecurity Self-assessment Handbook For Assessing NIST SP 800-171 Security Requirements in Response to DFARS Cybersecurity Requirements. https://nvlpubs.nist.gov/nistpubs/hb/2017/nist.hb.162.pdf. Accessed 27 Feb 2019
3. National Institute of Standards and Technology: DFARS Cybersecurity Requirements. https://www.nist.gov/mep/cybersecurity-resources-manufacturers/dfars800-171-compliance. Accessed 27 Feb 2019
4. US Department of Defense: Defense Federal Acquisition Supplement, 252.204-7012. https://www.acq.osd.mil/dpap/dars/dfars/html/current/252204.htm#252.204-7012. Accessed 3 Mar 2019
5. Educause Review: CUI Requirements in Federal Contracts Aren't FAR Away. https://er.educause.edu/blogs/2018/5/cui-requirements-in-federal-contracts-arent-far-away. Accessed 3 Mar 2019
6. PCI Security Standards Council: Payment Card Industry (PCI) Data Security Standard Version 3.2.1. https://www.pcisecuritystandards.org/documents/PCI_DSS_v3-2-1.pdf?agreement=true&time=1553228375592. Accessed 3 Mar 2019
7. Rosenburg, J.: Vulnerable to attack: businesses should boost cyber defenses. https://www.fifthdomain.com/industry/2019/03/13/vulnerable-to-attack-businesses-should-boost-cyber-defenses. Accessed 21 Mar 2019

Improving the Efficiency of a Cyber Competition Through Data Analysis

A Guide to CyberForce Competition™ Red Team Recruitment and Structure

Amanda Joyce$^{(\boxtimes)}$ and Steven Day

Argonne National Laboratory, Lemont, IL 60439, USA
{atheel,days}@anl.gov

Abstract. Identifying success metrics for large-scale competitions can be complex. There are a vast number of moving parts that are not always intricately connected to the overall experience of the competition. Characterizing and individualizing the success metrics for each component of the competition can lead to a larger overall impact. This paper will outline the CyberForce Competition™ Red Team history, recruitment strategies, and scoring metrics utilized throughout the competition as well as identifying metrics that can provide insight into what makes a "successful" Red Team.

Keywords: Cyber defense competition · CyberForce Competition™ · Department of Energy · DOE · Argonne National Laboratory · Red Team · Competition · Workforce development

1 CyberForce Competition™ Background

The U.S. Department of Energy (DOE), hosts interactive, scenario-based competitions to give student teams a hands-on cybersecurity experience and raise awareness of the nexus between critical infrastructure and cybersecurity. The DOE looks to identify how to fill the cybersecurity gap within the energy sector through the CyberForce Competition™. The CyberForce Competition is not just a red-blue team exercise. It takes into account the needs and satisfaction system users, competing demands upon the attention of network defenders, the need to develop innovate defense strategies to mitigate risk with limited resources, and realistic models of operational technology systems.

Prior to competition day, the competition staff provide the Blue Teams (collegiate students) an energy-based scenario, a vulnerable cloud-based environment, and remote access to their industrial control system (ICS) device. The Blue Teams have two to four weeks prior to the competition day to harden their cloud-based infrastructure and detail their mitigation strategies in a written report. They are provided physical access to their ICS device the day prior to the competition once onsite at their hosting national laboratory.

© Springer Nature Switzerland AG 2020
K.-K. R. Choo et al. (Eds.): NCS 2019, AISC 1055, pp. 51–60, 2020.
https://doi.org/10.1007/978-3-030-31239-8_5

The core of the competition is the 8-hour attack phase. During this time, Blue Teams must defend their infrastructure from the Red Team (attackers) as well as maintain acceptable service delivery to their Green Team (users and operators). Additionally during the competition, anomalies (real world diversions) are opened which allow teams to submit for additional points if the task is completed. Student are also asked to provide a five-minute lightning talk which is given to a panel of security experts during the event to explain why their innovative defensive strategies are worthy of additional points. At the end of the competition, one team is announced the national winner and local winners are announced for each laboratory.

The CyberForce Competition Red Team has rapidly grown in an effort to stay proportionate to the Blue Team growth. With the inception of the competition in 2016, the Red Team had 13 volunteers at a single site. In the December 2018 competition, the Red Team had 170 volunteers at seven sites. With large growth and dispersion around the United States, recruitment of Red Team volunteers has been a crucial key for success in achieving the overall goal of evaluating the degree to which Blue Teams are exhibiting defensive behaviors.

2   History of Red Team

The competition began in 2016 at Argonne National Laboratory. The competition hosted eight teams from the Illinois and Iowa region. The recruitment of Red Team volunteers was insignificant and the volunteers were provided with minimal strategic or technical direction by the competition staff. Fast forward to 2018, the number of volunteers is much greater and the strategic objectives for the competition have been narrowed and refined. Table 1 outlines the growth over the four competitions to-date.

Table 1. Growth in Red Team volunteers

Event date	Number of participating labs	Number of Blue Teams	Number of Red Team volunteers
April 2016	1	8	13
April 2017	1	15	15
April 2018	3	25	85
December 2018	7	64	170

The recruitment goal of Red Team volunteers has since been a ratio of two Red volunteers per Blue Team. This has allowed room for potential last-minute volunteer drops. This goal was developed based on the first two competitions' Red Team recruitment numbers as well as observations made throughout recruitment of missing critical skill sets not being emphasized (i.e., ICS subject matter expert).

3 Red Team Recruitment

For the first two competitions, the Red Team was centralized at Argonne as that was the only competition site. The goal of the team was to disrupt, distract, and impede system services of as many teams as possible. Red Team received no help from the competition organizing committee, other than provided competition network address ranges, which were provided on the Friday before the attack phase commenced. There was very little strategic direction and no scoring rubrics or objectives that were asked to be met.

The competition in April 2018 increased in laboratory participation with Oak Ridge and Pacific Northwest National Laboratories joining Argonne as well as introduced a cloud-based environment rather than a virtualized one. This meant that the Red Team was no longer in one centralized location, but spread throughout the three laboratories. Due to the increase in volunteers and locations, a rework on recruitment to identify critical skill gaps was completed. Each laboratory was responsible for recruiting Red Team members for their respective site. The goal was to recruit two Red Team members for every one Blue Team competing. The competition committee requested potential Red Team volunteers to disclose the following information pertinent to the competition.

- Full time occupation
- Years of experience
- Cybersecurity-related Certifications
- Previous participation in cyber defense competitions
- Comfort level with specific cyber-attack and defense tools and techniques
- Comfort level assessing various operating systems (e.g., Windows, Linux, Mac).

The most recent competition, held in December 2018, saw the largest growth yet for the competition. Four additional laboratories and over 40 more teams were added. Argonne, Brookhaven, Idaho, Lawrence Berkeley, Oak Ridge, Pacific Northwest, and Sandia National Laboratories were hosts to the December 2018 competition. Similar to April 2018, the national laboratories had a distributed Red Team. The recruitment goal stayed the same, at roughly two Red Team volunteers for every Blue Team. This meant recruitment goals were minimally 140 volunteers for the Red Team (not factoring in needs for the green team). Recruitment surpassed that goal with 172 volunteers of which, 55 participants had either previously participated in a CyberForce Competition or another competition similar and the majority of the participants found the opportunity through their laboratory's internal promotions or staff.

The largest struggle that the competition sees with this method of recruitment is self-evaluation. Until the Red Team members have actually become aware of the components, operating systems, and devices they are working with during the competition, they are making generic evaluations on what their comfort level is. Additionally at times, we have found that some volunteers are interested in learning about the Red Teaming process and while familiar with specific operating systems or devices, they are not proficient in attacks and tools. During each recruitment phase, we encourage participation but ensure that volunteers are aware that this exercise is largely an individual effort with minor mentoring or coaching from the Red Team leads. Some laboratories have implemented mentoring Red Teams that are paired up with much more experienced Red Teamers and have the groups interact and ask questions. Since each competition is a completely new setup, the Red Team preparation will never fully decrease.

4   Reconnaissance and Attack Phases

The Red Team reconnaissance and attack phases are very important to the competition. The reconnaissance or scouting phase ultimately determines the flow of the competition based on how much information Red Team gathers prior to the attack phase. More detailed information collected about teams' systems means a more defined and quicker attack path based on known vulnerabilities and overall system assessment.

In the first three competitions, a criticism from Red Team volunteers was the inability to do pre-competition reconnaissance. In 2016, Red Team was not allotted any prior time for reconnaissance of team environments. In other words, they went into the attack phase blind. This led to a large delay in Blue Teams seeing any form of "action" by the Red Team. Additionally, there was no interaction prior to the competitions among Red Team volunteers.

In 2017, the structure of the competition provided Red Team volunteers a four-hour window to do reconnaissance on the Friday prior to the competition. A communication channel was setup for sharing data and code throughout the competition via Mattermost, an open source, private cloud based communication platform. Red Team volunteers were able to quickly share code with everyone and communicate with competition staff.

In April 2018, the structure changed again. The Red Team was allotted an eight-hour recon window the day before the competition. Competition organizers provided the competition network address space and an opportunity to the Red Team to perform passive scans overnight. The volunteers were provided their own Slack channel (public cloud collaborative communication platform) to begin conversations prior to the competition, Red Team leaders utilized Slack as a means to ensure all laboratories were in sync, and some laboratories had pre-competition conference calls to talk strategy.

In December 2018, the Red Team received the same eight-hour window but they were provided documentation and blueprints of the Blue Team's original (pre-secured) network topology and systems. This change was done to ensure that the competition's focus was on the Blue Team competitors and less on the skill of the Red Team members. Red Team also was provided their own Slack channel for communication across all laboratories. Red Team leads and lab-specific Red Teams had pre-competition calls about timing, exploits, and strategy.

Additionally, over the years of the competition, Red Team's interaction with the Blue Teams has changed. Initially, it was a one-to-one ratio between Red Team volunteer to Blue Team. Scans were performed and attacks were established by a Red Team member for a specific Blue Team, which didn't allow for a big picture outlook or effective Red Team coordination.

In April 2018, there was an average of three Red Team volunteers per Blue Team. In December 2018, the competition had an average of 2.75 Red Team volunteers per Blue Team. With the increase in Red Team volunteers, it allowed volunteers to seek for skillsets complimentary to theirs. This increase in bandwidth allowed for a more advanced thought process for potential compromising of Blue Team system and the ability to actually attack.

4.1 Rules and Attack Sophistication

The types of attacks performed by the Red Team have become more sophisticated as the size of the Red Team has grown. Having stronger leadership, a larger volunteer team base, and a more advanced level of skill allows for a more strategic and better-defined approach to the teams goals. One example of this is the growth from manual analysis reconnaissance analysis in the first two years, to a trusted party with the organizing staff.

In 2016 and 2017, the Red Team has very minimal rules of engagement. The main scope of the rules included:

- No pre-competition reconnaissance, including social engineering,
- No physical tampering with Blue Team hardware, and
- No presence in the competition space except for opening and closing remarks.

Since 2018 and the movement of the competition to the cloud, additional rules were implemented to ensure participant safety, inherent cloud provider rules, and a level playing field. While there has been additional automation developed into the structure of the competition environment to assist with quicker re-imaging, Red Team members still have to be cautious to only attack the competition network and not the cloud platform. This concern was slightly alleviated by utilizing a virtual private network (VPN) that allowed competitors to access the competition instance, the concern remained the same if the participant was working off the VPN.

Red Team members also had to be aware that specific types of attacks such as denial of service (DoS) attacks or excessive network intensive fuzzing are not permitted within the cloud environment. While these types of attacks are no longer permitted, moving the Red Team to a trusted party and providing the pre-built vulnerable systems to them, capturing data sooner and exploiting systems was more efficient in accessing Blue Teams. With a trusted Red Team, the emphasis is placed on Blue Teams to harden and defend their networks rather than the emphasis placed on Red Team learning a new system and developing exploits in an 8-hour window.

5   Scoring

The Red Team score was entirely subjective for both 2016 and 2017. This meant the score for Red Team was based on personal thoughts of how the Blue Teams did and may have included some bias towards or against teams. Two Red Team members may have scored the same team very differently based solely on their skills and knowledge in attacking.

The Red Team portion of the total team points was large and so it was important to ensure objectivity and defensibility in it. In 2016, the score was out of 300 points: 100 points for security actions taken prior to competition, 100 points for response to attacks by Red Team, and 100 points for a catchall category that included sportsmanship. In 2017, the score was out of 500 points and had no categories associated with it. Red Team was to take into account the types of attacks, the number of attacks, and complexity of attacks.

5.1   April 2018

Upon review of the 2016 and 2017 Red Team scores, the scoring was overhauled for April 2018 to decrease subjectivity where possible. Table 2 provides a high-level outline of the rubric that was developed for this purpose. In addition to the information below, each area had a description of what constituted success, and documentation the Red Teamer was expected to collect.

Each item was scored on scale of 0–2. As Red Teamers largely consisted of security professionals with real-world experience, these point definitions were built around that experience. Thus, a zero score was defined as being significantly worse than what would be found in a "typical" real-world environment, a score of one as being typical, and a score of two being better than typical and/or flawless. Additionally, a description was given for each area/score pairing. After totaling these points, the score was scaled up to 500 points for the morning and 1000 points for the afternoon, for total possible points of 1500.

Table 2. April 2018 Red Team rubric outline

Category	Scoring area	Morning score	Afternoon score
Identify	Recognize types of vulnerabilities and associated attacks	X	
	Apply system, network, and OS hardening techniques	X	
	Recognize and respond to vulnerabilities in information systems	X	X
Protect	Restrict logical access to the network and network activity	X	X
	Protect individual components from exploitation	X	X
	Prevent unauthorized disclosure or modification of data	X	X
Respond	Detect security events and incidents		X
	Maintain functionality during adverse conditions		X
	Restore the system after an incident		X
Sportsmanship	Secure the team network in a manner practicable in a real-world environment, both technically and politically		X
	Demonstrate professional and ethical conduct throughout the competition		X

The items included on this rubric were drawn from a combination of a subset of the Knowledge, Skills, and Abilities in the National Institute of Science and Technology (NIST) National Initiative for Cybersecurity Education (NICE) Workforce Framework, and NIST Special Publication 800-82r2: Guide to Industrial Control Systems Security. The overall categories were drawn from the NIST Cybersecurity Framework (CSF). The NIST NICE Framework 800-82 were analyzed to determine the work roles and controls relevant to CFC learning objectives, and assessable from a Red Team perspective.

To allow the Red Team members to refine their understanding of each Blue Team over the course of the day, the rubric was split into two, with a morning and afternoon section. The afternoon section was worth twice as many points; with the assumption, the Red Team members would be best-informed to assess teams at the end of the day. However, a past challenge in Red Team scoring was Red Teamers remembering events from the full attack phase period, and Blue Teams being surprised by a large "swing score" at the end of the day.

For this reason, the morning score was added to ensure a mid-day check-in by Red Team, to give Blue Teams some indication of their standing, and capture early impressions. For similar reasons, the afternoon rubric focused more heavily on the latter half of the NIST CSF process (i.e., Protect and Respond) while the morning section focused on Identify and Protect. A sportsmanship score was maintained at the end of the day to allow Red Teams to reward teams that were particularly realistic and professional in their conduct, from the Red Team perspective.

To attempt to promote consistency between Blue Teams over the course of the day, an attack sophistication scale was developed and Red Teamers were instructed to move sequentially from the low end to the high end with each team they were assessing. In reality, getting the Red Teamers to comply with such a scale is difficult, especially as the number of volunteers grows. However, it provided structure upon which Red Team leaders could build a cohesive strategy.

While this method reduced some of the subjectivity, it did not assist in ensuring that all teams had a consistent caliber of Red Team volunteers. Additionally, Red Team volunteers were then partnered with another volunteer that complemented their skill sets before being assigned a Blue Team. This method also did not take into account volunteers failing to show up, or showing up last minute without registering; while some labs may have done initial shuffling, this cannot be said of all laboratories. This created major challenges for the Red Team leaders to make effective Red Teamer to Blue Team assignments and balance available skillsets.

Additionally, instructions to Red Teamers to reference the rubric throughout the course of the day, take notes on key points, and collect evidentiary screenshots or other supporting material were difficult to communicate and enforce. Thus, many rubrics were completed in a very hasty manner, in some cases literally as the Red Team member was walking out the door. While a professional vulnerability assessment or penetration test tends to consists of a large amount of time allocated to both require-ments gather and documentation, these phases are very difficult to incorporate into a competition environment. Many volunteers are traveling from out of town, have other time commitments, and even things such as foul weather can impede additional coordination. In practice, it is not realistic to assume any time from the volunteers beyond the Saturday of the event, and possibly a few hours on the day period. However, periodic calls between Red Team leaders, or a more hierarchical structure with corresponding levels of advance preparation, may provide a way to mitigate some of this chaos.

5.2   December 2018

After the April 2018 competition, the Red Team volunteer comments were reviewed and a large change occurred for December 2018: The Red Team leads now became a trusted entity with the competition staff. This meant that the inherent vulnerabilities provided to the Blue Teams to find and harden were equally shared with Red Team leads from each laboratory. Between the Red Team leads and the technical committee, exploits were scripted and tested on test environments to ensure that proper measures were taken. This meant that for both the morning portion and afternoon portion, the Red Team volunteers had scripted and assigned exploits that would be run on every team.

This accounted for 2/3 of the score or 50 points each. Each exploit was scored based on success (worked or did not work) with 20 total exploits for the day.

In addition to scripted attacks, Red Team volunteers had a form based on additional intrusions, which made up a subjective portion of the score. In review of submissions and comments from Red Team, there seemed to be confusion on the threshold for submitting items into the intrusion report.

6   Data and Scoring Correlations

Since the competitions has changed scoring rubrics and point valuations over the last four competitions, this section will larger look at the raw red team data for the December 2018 competition. December 2018 had 62 participating teams that were scored throughout the day on different parts of the competition. Table 3 below outlines the overall competition low, high, and mean scores as well as each laboratories' scores related to their specific teams.

Table 3. Raw Red Team scores for December 2018 competition

Laboratory	No. of teams scored	Low red score	High red score	Mean red score	Delta
All labs	62	0	1500	1025.4	
Argonne	16	0	1500	1071.88	+46.48
Brookhaven	5	450	1075	730	−295.4
Idaho	6	525	1425	845.83	−179.57
Lawrence Berkeley	4	1100	1450	1318.75	+293.35
Oak Ridge	10	375	1425	1092.5	+67.1
Pacific Northwest	13	300	1475	1067.31	+41.91
Sandia	8	500	1225	953.13	−72.27

In review, Argonne and Pacific Northwest scored closest to the mean of the entire competition but they also held over 46% of the total teams at the competition. The largest differences came from Lawrence Berkeley and Brookhaven who held just over 14% of the total teams. In Table 4, we add additional data on the percentage of local teams that scored within the top 10, 25, and 50% of highest scoring teams.

Table 4. Percentage of local teams in top scoring teams

Laboratory	Total teams	Percentage of local teams in top 10%	Percentage of local teams in top 25%	Percentage of local teams in top 50%
Argonne	16	19%	31%	56%
Brookhaven	5	0%	20%	20%
Idaho	6	0%	17%	17%
Lawrence Berkeley	4	25%	50%	50%
Oak Ridge	10	0%	20%	60%
Pacific Northwest	13	15%	31%	69%
Sandia	8	0%	13%	38%

Interesting facts that can be identified is that for Brookhaven, Idaho, and Sandia all had less than 50% of their local teams in the top 50% of high scores. This can be review that Brookhaven had a mean Red Team score of 730 which was 295.4 points less than the competition average, Idaho had a mean Red Team score of 845.83 which was 179.57 points less than the competition average, and Sandia had a mean Red Team score of 953.13 which was 72.27 points less than the competition average. While there is not enough data, it can be gleamed that success in Red Team scoring can largely emphasize where a team may fall in the overall competition.

7 Future Competitions and Conclusion

The goal of the competition is to continue to grow and improve recruitment efforts and scoring every competition. The next competition is scheduled for November of 2019. The organizing committee is currently operating with the understanding that the number of Blue Teams and sites will expand again for this next iteration. This drives the need for both quantity of Red Team volunteers, but even more so scalable and defensible processes and methodologies. Future competitions will also have a larger workforce development emphasis.

As the competition continues to evolve and grow year after year, so does the Red Team. A team that was not considered a high priority in the first iterations of the competition has evolved into an end-to-end process of recruiting, selection, organization, coordination, data gathering, and continuous improvement.

One potential pool of Red Team volunteer recruits is past competitors. Since Blue Teams are collegiate, the hope is that once Blue Team members graduate from their institution, they still have an interest in participating in the competition as a Red Team volunteer. They would be familiar with how the competition is laid out and would potentially bring in added knowledge from being on the other side.

Red Team scoring continues to be improved and will continue to evolve year after year. Utilizing feedback from Red Team members, leads, and Blue Teams, the scoring committee will look to provide more clear guidance and outlines of scoring for all participants. For the 2019 competition, the goal is to ensure that Red Team leads have all the information necessary to assist their Red Teams.

Additionally, we continue to look into what the appropriate pre-competition allotment time for Red Team members to have that does not negatively affect Blue Teams. There is always a review of the rules of engagement to ensure we are not overly restricting Red Team, provide an educational experience for all participants, and that rules evolve based on post-competition survey results. Over the next years, Red Team data will continue to be reviewed for potential gaps and areas of improvement.

Acknowledgement. Argonne National Laboratory's work was supported by the U.S. Department of Energy, Office of Science, under contract DE-AC02-06CH11357.

Interdisciplinary Cybersecurity: Rethinking the Approach and the Process

Johanna Jacob, Michelle Peters[✉], and T. Andrew Yang

University of Houston Clear Lake, Houston, TX 77058, USA
petersm@uhcl.edu

Abstract. The need for cybersecurity professionals continues to grow and education systems are responding in a variety of ways. This study focusses on the "interdisciplinarity" of cybersecurity that contributes to the emerging dialogue on the direction, content and techniques involved in the growth and development of cybersecurity education and training. The study also recognizes the contributions of other disciplines to the field of cybersecurity by the discussion of relevant theories that contribute to understanding security in the context of legal, economics and criminology perspectives. Finally, quantitative analysis (security metrics) is done to understand the existing knowledge of security behaviors and beliefs among students from technical and non-technical majors, helps measure the interest fostered towards an academic pathway in cybersecurity and substantiates on the need for providing a level of cyber education for all individuals appropriate to their role in the society.

Keywords: Interdisciplinary cybersecurity · Cybersecurity Education · Cybersecurity · Collaborative cybersecurity

1 Introduction

1.1 Background and Problem

The term "cybersecurity" has been the highlight of academic literature for many years. With a significant rise in the proliferation of technology and the innovation that comes along with it, cybercrime has equally penetrated all aspects of human endeavor. The rising number of breaches and threats to personal, organizational and national safety have led to an increased focus on the defensive measures. It has in fact become the highest priority items on the global policy and national security agendas [1]. According to Cyber Security Business report, Cyber Crime damage costs will hit $6 trillion annually by 2021 [2]. In response, the Cybersecurity Policy Review [3] demands for a national strategy to develop awareness and incorporate a cyber-secure workforce that is adequate in expertise and skills to be cyber-ready against the potential threats faced by the nation. There is a serious need for cybersecurity talent [1] to secure the infrastructure of federal and private entities against the growing cyber risks.

A 2017 survey by Statista reports [4] that the greatest cybersecurity problem of the United States was hacking by foreign governments. The challenges posed by technology misuse and abuse are manifold and requires an equal contribution from

© Springer Nature Switzerland AG 2020
K.-K. R. Choo et al. (Eds.): NCS 2019, AISC 1055, pp. 61–74, 2020.
https://doi.org/10.1007/978-3-030-31239-8_6

computer science and social science researchers to better understand the dynamics of the attack and perpetrator, and to propose a feasible solution to combat it. To exemplify this, consider phishing emails. Phishing emails can be blocked by email server software based on rules and classification strategies that are configured on the server end. However, it may still penetrate through to the end user. Potential recipients must be able to identify and understand these phishing messages as a threat to reduce the chances of being victimized. One needs to understand the behavioral and attitudinal differences that led some to respond to fraudulent messages while some others do not. On a much larger scale, it is important for organizations to understand the attack, the attacker and the dynamics around them.

Holt [5] points out that it is critical to situate a cybercrime threat or vulnerability in a multidisciplinary context. A holistic approach to cybersecurity is one that considers the many disciplines that produce cybersecurity professionals – technical and non-technical alike, in a coherent fashion. Such an approach respects the relative contributions of the different subfields and recognizes that, prospective cybersecurity professionals must develop an expertise within their individual subfield while simultaneously understanding how their work fits into rest of the field.

However, such an approach to cybersecurity has been stove piped for decades in the education system of the nation. For instance, the disciplines of computer science and engineering are focused on developing algorithms and secure devices that support sensitive systems, and data/information processing while information technology and information assurance focus on better techniques, tools and process to protect information from being misused. While there is a higher emphasis on understanding the technical nature of the cyber environment, the networked systems, operating systems and the security threats around them, there is little to no emphasis on the human actors and their decision-making process that plays vital role in a cyber-attack being successful [6]. Knowing this will allow institutions or organizations to tailor educational programs accordingly.

1.2 Significance

This study will significantly recognize the contributions of other disciplines to the field of cybersecurity by the discussion of theories that contribute to the understanding security in the context of legal, economics and managerial perspectives. A quantitative analysis is done to understand the existing knowledge of security behaviors and beliefs and measure the interest fostered towards an academic pathway in cybersecurity. The results of the analysis will help to understand the demand and need for a collaborative cybersecurity program in the Department of Computer Science.

2 Literature Review

The breathtaking pace of change in computing and technology and its widespread adoption in virtually every human endeavor has led to the dawn of a never seen era of Interdisciplinarity. Nearly all field of human activity require an understanding and application of that field within the context of one or more other fields. As Way [7] quotes it, "Interdisciplinarity is the combining of two or more disciplines into a single, cross-discipline learning experience". This section will highlight the importance of an interdisciplinary education in cybersecurity followed by contributing theories from disciplines as criminology, legal studies and economics and detail on the theoretical framework which baselines the quantitative study.

2.1 Cybersecurity and Criminology

Thousands of Cyber-attacks are being launched against internet users across the world. In fact, cyber-attacks have become arduously frequent and highly expensive to individual users, businesses, organizations, economies and other infrastructural entities. In 2016, Symantec [8] discovered more than 430 million unique pieces of new malware, 91% of these were originated by employing phishing techniques.

It is globally realized that humans are the weakest link in cybersecurity. Most of the system security organizations work on the premise that human factor is the weakest link in cybersecurity. In fact, humans have moved ahead of machines as the top target for cybercriminals. There were 3.8 billion internet users in 2017, up from 2 billion in 2015 [9]. According to Cybersecurity Ventures [10], there will be 6 billion internet users by 2022 and more than 7.5 billion internet users by 2030. This vast increase in the number of internet users raises concern in terms of vulnerabilities and emerging threats by ideologically motivated offenders to cause harm and further their political and social agendas.

However, a lack of empirical research on cyber-attackers limits our knowledge of the factors that affect their behavior. As Sandeep [11] denotes, the "interaction between computers and humans is not a simple mechanism but is instead a complex interplay of social, psychological, technical and environmental factors operating in a continuum of organizational internality and externality". Within the field of criminology, numerous theories exist to elucidate why crime occurs, why certain people engage in deviant behavior while others refrain from it and ways to help predict future crime behaviors and practices [12]. This below section presents some of the theories in the light of cybercrime as follows:

Aker's Social Learning Theory. Precisely used to explain a diverse body of criminal behaviors, this theory encompasses four fundamental avenues namely, differential association, definitions, differential reinforcement, and imitation. The theory reinforces the idea that individuals develop motivation and skills to commit crime by associating themselves with those who are involved in crime (deviant peers). With respect to cybercrime, research indicates that this theory can help elaborate the issues of software piracy. Burruss et al. [13], found that individuals who associate with software piracy peers learn and consequently follow the deviant conduct. Not only does the social learning theory explain for software piracy but also posits to other cybercrimes because of its ability to explain the rationalizations, skills and behavior that the criminals are reinforced with through their association with, and observation of others [13]. Thus, the main idea behind this theory is understanding the motives of delinquent peers and their functions in the context of various cyber-crimes.

Routine Activity Theory. Developed by Cohen and Felson, this theory posits that the behavior of most victims is repetitive and predictable, and that the likelihood of victimization is dependent on three important elements - motivated offenders, suitable targets and the absence of capable guardians [14]. While the motivated offender is someone willing to commit a crime if an opportunity presents itself, the target is the one that the motivated offender values (e.g., credit card information) and the capable guardian is a person or an entity that obstructs the offender's ability to acquire the target.

Situational Crime Prevention Theory. The situational crime prevention theory is a strategy that addresses specific crimes by manipulating the environment in a way that increases the risk to the offender, while reducing the potential reward for committing the crime [14]. Unlike other criminology theories, this theory does not postulate on why the offender did the crime. Rather, it tends to focus more on the reducing the crime opportunities. Hardening the targets of crime by encrypting sensitive information, implementing access control mechanisms, securing off-site data and performing background checks on employees and restricting unauthorized installations on computers are some of the examples of this theory. Situational Crime Prevention Theory is used to reduce cyber stalking and other online victimization crimes. Criminal behavior cannot be explained by one theory but requires a conjunction of various theories to recompense for what each individual theory failed to explain. However, while criminological theory in the physical realm enjoys a rich history with diverse contributions and clear paradigm development and shifts, explanatory research and studies with respect to digital and electronic crime and information security success remains relatively undeveloped.

2.2 Cybersecurity and Economics

The economics of cybersecurity or "cyber economics" as the newly evolved name, is one of the thriving interdisciplinary facets of growing cyber security issues in the United States. Conservatively, a total of 15 billion US dollars are spent every year by organizations in the United States to secure their communication and information systems [7]. Though the investments are higher, the economic impacts of cyber-attacks and breaches have set to surpass the cost of investment by large. In 2009, the cost of cyber-attacks was estimated by the then President of United States, Barack Obama, to be 1 trillion dollars per year or translated as 6% of the Gross Domestic Product of the United States [15]. However, the estimates have appeared to vary widely. In 2010, internet crime cost totaled to 560 million USD, out of which Phishing, one of the top social engineering attacks, accounted to 120 million dollar per quarter [7].

In order to effectively learn and understand the economically complex cyber-attacks, it is important to understand the interconnections and complexities in our economy that cyber attackers use to cause greater destruction. In lieu of this, the following economical concepts are discussed as below,

Economic Redundancies. The first feature of our economy that is crucial to cyberattack consequences is the way systems can substitute for other systems. These redundancies are usually the main factor limiting the consequences of a cyberattack. To deal with redundancies, cyber attackers employ combinations of cyberattacks designed to produce Intensifier Effects. These are simultaneous attacks on different systems or businesses that could otherwise serve as substitutes for each other. When several systems could serve as substitutes, a successful cyberattack on the first of these systems will generally have extremely limited consequences. Further successful attacks on those systems that can substitute will produce only very small increases in destructiveness.

This continues until the capacity of the remaining systems is no longer enough to allow them to take over for the systems that have been attacked. The consequences of the cyberattacks will then go abruptly from being small to being huge. This has important implications for the planning of almost any cyberattacks. In this regard, Economic redundancies, and the potential for intensifier Effects to overcome them, will be a major consideration in choosing targets [15].

Economic Interdependencies. The second economic feature that's crucial to cyber-attack consequences is the way production is organized into value chains. For instance, one company might turn ore into metal. Another company will turn the metal into mechanical parts. Another company will incorporate the mechanical parts into airplanes. This interdependency is the basis for any kind of economic cooperation. But on the other hand, these interdependencies provide enormous opportunities for cyber attackers to find ways to exploit. The reason is that mechanisms that companies employ to coordinate their value chains can also be used to make compensating adjustments if part of the value chain is disrupted [15]. The below flowchart diagrams the economic activities. The systems that make up the value chain are represented as channels that flow into each other. To exploit such value-chain attacks, cyber attackers need to employ a combination of cyber-attack to produce a Cascade Effect. By this mechanism, a successful attack on one set of businesses will affect numerous other businesses up and down the value chain [15].

Economic Near Monopolies. Businesses and enterprises that are monopolies in their area of service are prone to a higher range of cyber-attacks. Because near monopolies produce large inputs through limited means, they give attackers opportunities to produce limited effects with limited means. To take advantage of such monopolies, cyber attackers employ combinations of attacks specifically designed to produce Multiplier effects. The sort of companies that could be attacked to produce Multiplier Effects would make especially tempting targets, because they are small sized. And their budgets for cybersecurity are small.

From the discussion of the above economic concepts, the structural analysis of an economy is a powerful tool for cyber attackers and it eventually becomes a more essential tool for cyber defenders [15]. An effective cyber defense program or training cannot be satisfied with identifying a few individual cyber-attack scenarios. Taking proper accountability of economics in security thinking requires an adjustment in outlook. Economics is therefore a powerful analytical tool to defend against cyber activities.

2.3 Cybersecurity and Legal Studies

The need for a comprehensive approach to cyber security deriving from the architecture of the internet and emerging cyber threats and incidents requires a systematic development, interpretation and application of legal areas and instruments. With politically motivated cyber incidents on the rise, cyber security has grown into an immediate area of concern for national governments and international organizations. In this regard, an approach combining considerations of threat, deterrence and response from different areas of authority and responsibility are significant to cater to the defensive actions against the attacks. This has led to the discussion of a coordinated legal approach. From a legal perspective, this means that the national legal approaches to data and consumer protection and due diligence will determine law enforcement and national defense capabilities [16]. Understanding these legal policies in the light of cybersecurity adds a holistic perspective to defending and responding to such attacks. Some of the categories of legal studies in the light of cybersecurity have been briefed in the following section.

Computer Crime Laws. These laws deal with a broad range of criminal offenses committed using a computer or similar electronic device as identify thefts, online stalking, bullying, sex crimes etc. This law typically includes procedural and legal ramifications for prohibition, investigation and prosecution of criminal activity [16]. Its application extends to a wide range of fields as computer hacking, viruses, internet gambling, encryption, online undercover operations, internet surveillance etc.

Information Privacy Laws. Information privacy laws includes the development of constitutional, tort, contract, property, and statutory law to address emerging threats to privacy. Laws under the information privacy law deal with privacy in the media, law enforcement, and online transactions, medical and genetic privacy and for personal privacy [16].

Homeland Security Law and Policy. These policies concern the Department of Homeland Security and the adoption of the Homeland Security Act of 2002 [16]. The laws under the Homeland Security define legal responses and actions for protection of critical infrastructure, information sharing, liability for terrorist attacks, risk insurance, threats to electronic infrastructure and combating the finance of terrorism.

Counterterrorism Laws. These set of laws provide an analysis of legal mechanisms in the fields of criminal, civil, military, immigration, and administrative law used by the U.S. government to combat domestic and international terrorism. The laws also in detail charts out the effectiveness of government actions and alternatives for achieving public safety goals and the effect of such actions on U.S. citizens and citizens of other countries.

Intelligence Laws. These set of laws identify and analyze current legal questions that face intelligent practitioners. They also include constitutional, statutory and executive authorities that govern the intelligence community. A comprehensive defense to cyber-attacks includes a strong contribution from a legal perspective. Instead of addressing a specific threat, cyber threats should be regarded as a spectrum where different stages and effects of cyber incidents are aligned. Depending on the motivation, effects and actors, a cyber-incident will be categorized as a breach of law short of cyber-crime, crime, national security relevant incident or cyber warfare [16]. An interdisciplinary, holistic education in Cybersecurity is borne out of understanding and applying these laws in context to the security issues learnt.

2.4   A Multi-modular Approach to Interdisciplinary Education in Cybersecurity

Based on the discussion of cyber related interdisciplinary theories in the above section, the need for a comprehensive approach to cybersecurity is essential as it covers the information society and the challenges tackled leading to a palliative understanding rather than a stove piped approach [17].

In this regard, the newly developed model provides an opportunity to explore technical and non-technical content in a four-year program by integrating disciplinary and interdisciplinary electives at different levels. The model called as "Multi-discipline, Multi-level, Multi-thread model" allows potential candidates to specialize in subjects of Cybersecurity along with relevant interdisciplinary subject matter. Figure 1 shows a diagrammatic representation of the proposed model. The model would accommodate electives from other disciplines that are relevant to the Cyber domain.

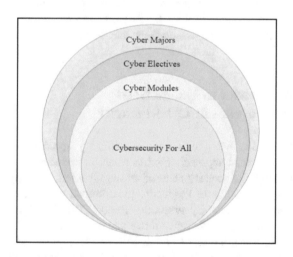

The model works on a top-down approach, allowing different pedagogical methods to be employed in each level of advancement.

Elaborating the model, the following are taken into consideration,

Fig. 1. Multi-discipline, multi-level and multi-thread model

Cybersecurity for All. The model is designed as a prototype to foster an inclusive, interdisciplinary approach in Cybersecurity Education. Although many four-year institutions have stringent requirements for general education, the idea of putting cyber into general education courses applies to any college or university. This approach that is named as "Cybersecurity for all" includes an introductory cybersecurity course that envisions a taxonomy for cyber education across the entire spectrum of curriculum, including non-technical, non-computing fields of study.

Cyber Modules. Cyber Modules will be the foundational element towards the "Cybersecurity for all" framework. The modules can be incorporated into courses to infuse knowledge about security measures and protocols. The modules are reusable, interdisciplinary and can be aggregated as a unit (or thread) to be pluggable into different disciplines, threads and electives. This will enable significant addition of cybersecurity into the core courses as well as in general education classes such as

International Relations, Legal Studies, Business Administration, Management, Psychology etc.

Cyber Electives (Interdisciplinary). Cyber electives include a myriad of courses that could be adopted into the curriculum to infuse a holistic education in Cybersecurity. In addition to the electives offered in Computer Science or Information Technology, the cyber electives contain interdisciplinary electives from across the spectrum of courses. These courses will include electives from Legal Studies, Economics, Criminology, Business and Psychology.

Cyber Majors. Cyber Majors includes majors such as Computer Science, Information Technology, Software Engineering and Computer Information Systems that allow students to select chosen sub-topics within their desired major. The majors must present core cyber subjects across their curriculum. The major must be culminated by a Capstone Project in the Cyber Domain that gives students, an exposure to implement the knowledge gained through the coursework [18]. Other than the above elements, the model also greatly motivates enrichment opportunities by fostering student research groups, clubs and chapters of renowned cyber associations that is inclusive of all majors.

3 Understanding Security Metrics to Gauge Need for an Interdisciplinary Program in Cybersecurity

In the above section, we discussed the need for an interdisciplinary approach in Cybersecurity Education. However, such an approach requires a careful understanding and an abstraction of the dynamics of the security state of an organization. Security attacks are emerging as commonplace events in academic, government, public and private sectors. According to Cisco's Midyear Cybersecurity Report [18], "Business Email Compromise (BEC) has become a highly lucrative threat vector for attackers. U. S. $5.3 billion was stolen due to BEC fraud between October 2013 and December 2016 while ransomware exploits cost US$1 billion in 2016." With emerging challenges and threats, it becomes imperative for preparing the talent in the pipeline with the required exposure in terms of training and skills.

Incorporating an interdisciplinary instructional design for students from technical and non-technical majors depends greatly on understanding the perceptions of cybersecurity risks, vulnerabilities, and practices that students bring to the classroom. Students are not categorized as "clear slates" when it comes to cybersecurity. Rather, students carry an initial understanding of security practices and risks that have been shaped through various means (e.g., social media, course offerings) and personal experiences. The purpose of this study was to abstract the existing knowledge of security, awareness of threats and vulnerabilities, and the interest fostered towards a career path in cybersecurity education, and workforce across students from technical and non-technical majors. This abstraction will help develop a deeper understanding and guide the development of curriculum, tools, labs and aid in the decision-making process of incorporating cyber awareness within the organization.

Table 1. Student population at University of Houston - Clear Lake

	Students (n)	Percentage (%)
Degree		
Undergraduate	6,064	71
Graduate	2,478	29
Gender		
Male	3,176	37.2
Female	5,366	62.8
Enrollment by College		
College of Education	1,486	16.6
College of Business	2,564	30
College of Human Sciences and Humanities	2,228	26.1
College of Science and Engineering	2,219	26
Race/Ethnicity		
White	3,228	37.8
Hispanic/Latino	2,776	32.5
Black	689	8.1
International	894	10.5
Other	955	11.1

3.1 Case Site

A case study of the University of Houston – Clear Lake (UHCL) was used for this paper. The population consisted of undergraduate students from the College of Business, College of Human Sciences and Humanities, and College of Science and Engineering at UHCL; a Hispanic-serving institution (HIS) with a current enrollment of 8,677 students. Table 1 displays the student population of UHCL along with race/ethnicity and classification of students according to degrees for the previous academic school year (2017–2018).

3.2 Participant Demographics

From the above population, a purposeful sample of students across different majors were selected to participate in the survey. Altogether, 228 students participated in the survey. Table 2 displays the participant demographics regarding gender, age, and race/ethnicity that took the selected classes. "n" represents the frequency, i.e., the number of students that fall in that particular category and "%" represents the percentage value for the same.

Most students were female comprising of 56.4% (n = 128). Male participants comprised of 43.9% (n = 100) of the sample population. About age classification, participants in the 18–24 age group constituted the majority of all the respondents, comprising of 66.7% (n = 152), followed by students in the 25–34 category, comprising of 28.1% (n = 64) of the total sample. Regarding Ethnicity, most of the survey respondents identified themselves as White or Caucasian, comprising of 36.9%

(n = 84). The Hispanic/Latino numbers were also close to that of White/Caucasian, comprising of 36.9% (n = 86).

3.3 Materials, Methods and Procedure

For purposes of this study, a survey design was employed. A purposeful sample of undergraduate students majoring in Economics, Computer Science, Information Technology, Legal Studies, Management, and Criminology at UHCL were administered the researcher-constructed Integrated Approach to Cybersecurity Education Survey to assess student perceptions on security behavior and beliefs, and measure the interest gathered towards an interdisciplinary approach. The data were analyzed using descriptive statistics (frequencies, percentages), and a two-tailed paired samples t-test.

Table 2. Overall participant demographics

	Crim.		CS		Econ.		IT		Legal Studies		Mgmt.	
	n	%	n	%	n	%	n	%	n	%	n	%
Gender												
Male	23	39.7	10	20.8	22	41.5	26	86.7	1	12.5	18	56.3
Female	35	60.3	37	77.1	31	58.5	4	13.3	7	87.5	14	43.8
Race												
Asian	4	6.9	8	16.7	6	11.3	4	13.3	0	0	1	3.1
Black	3	5.2	5	10.4	6	11.3	0	0	0	0	3	9.4
Hispanic	24	41.4	18	37.5	15	28.3	8	26.7	7	87.5	12	37.5
Native Amer.	0	0	0	0	1	1.9	0	0	0	0	1	3.1
Other	0	0	2	4.2	0	0	0	0	0	0	0	0
2 or more	7	12.1	0	0	3	5.7	2	6.7	0	0	3	9.4
White	20	34.5	15	31.3	22	41.5	16	53.3	1	12.5	12	37.5
Age												
18–24	41	70.7	28	58.3	40	75.5	13	43.3	5	62.5	25	78.1
25–34	13	22.4	17	35.4	9	17	15	50	3	37.5	7	21.9
35–44	3	5.2	3	6.3	4	7.5	1	3.3	0	0	0	0
45–54	1	1.7	0	0	0	0	1	3.3	0	0	0	0
55–64	0	0	0	0	0	0	0	0	0	0	0	0

Note. Crim. = Criminology, CS = Computer Science, Econ. = Economics, IT = Information Technology, Mgmt. = Management.

3.4 Findings and Discussion

Security Concern. In general, the perception of internet users' security and trust have strong impact on carrying out their day to day activities on the internet. The results of the analysis demonstrate that the users' perception generally meet the expectation of their security concerns and lean towards the secure side.

While only 9% of students from technical majors expressed their unconcern over their security on the internet, more than 25% of students from the non-technical majors expressed the same. From the data obtained, it is understood that students from technical majors were less unconcerned about their security on the internet than the students from the non-technical majors.

It also helps understand that a relatively average number of students from the non-technical majors could be susceptible to the attack of internet usage due to their expressed unconcern (Fig. 2). This is also posited by Shropshire et al. [19], that there is a strong connection between the intent to comply with security rules and the traits of agreeableness and conscientiousness which means that accurate knowledge of security concerns is influenced by past experiences of making security decisions and executing the same.

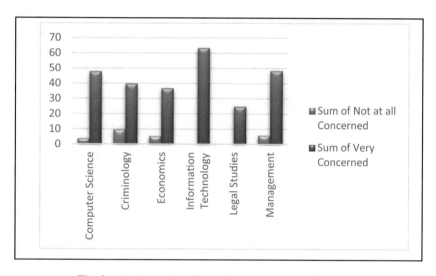

Fig. 2. An abstraction of security concerns from the survey

Protection from Viruses and Defensive Actions. The set of questions catered to understand the perception of the respondents in terms of their security behaviors are of concern in this section. This helped understand the types of security behaviors that participants exhibited. The actions were clustered into two categories – Behaviors that place trust-in-software and behaviors that trust-in-self.

While respondents in the first cluster agreed to have their anti-virus, firewall and security products up to date, most of the students from the technical and the non-technical majors claimed to do this to protect their devices against hackers. The second cluster of behaviors place trust in themselves, about their restraint to accessing websites, and carefulness to open email attachment or click on malicious downloads. 69.1% of the students fell in this category. It is also of importance to note that the results of the two-tailed paired t-test indicated a statistically significant mean difference from the pre- and post-survey ($p < .001$); implying that the respondent's perceptions about defensive actions against viruses were changed following an intervention that was administered as part of the survey. The intervention included a short lecture on Security Practices.

Password Practices. Passwords are a key part of many security technologies and they are the most commonly used authentication method. For a password system to be secure, users must make conscious decisions about what passwords to use and where to re-use passwords.

From the results of the analysis, 33.3% of students from Criminology agree that they would use the same passwords for all the websites for consistency and ease while 66.6% of students in the Management class agreed to write down the passwords in some form, so they can look it up. This exhibits a sheer contradiction to the best practice in the field that passwords must be long, random and unique to each account. In this regard, Das et al. [20] estimated that 43–51% of user's re-use passwords across accounts and Ur et al. [21] denotes that people re-use passwords because they have never personally experienced negative consequences stemming from re-use. This sheds some serious concern on incorporating and educating student of novel password practices.

3.5 Security Metrics Versus Awareness

Based on the findings, there is a serious need to develop and implement a security awareness program spanning technical and non-technical majors in the case site. There has been an exponential increase in the usage of internet, particularly among millennials and older generation. A fact sheet from the Pew Research Center [7] quotes that "Millennials have often led older Americans in their adoption and use of technology and this largely holds true today. But there has also been a significant growth in tech adoption in recent years among older generations". This denotes how reliance on internet usage in fulfilling personal and academic tasks demonstrates a paradigm shift. However, this increasing global population is one of the main contributing factors to changes in cyber threats.

In coping with the cyber threat landscape that has transitioned from the use of savvy hacking skills to sophisticated and well-planned strategies, cybersecurity awareness is deemed essential for internet users like youngsters as a counter-measure strategy to combat silent privacy invasion.

Cybersecurity awareness is defined as a methodology to educate internet users to be sensitive to the various cyber threats and the vulnerabilities of computers and data to these threats. Shaw et al. [22] defines cybersecurity as, "the degree of users' understanding about the importance of information security, and their responsibilities to exercise sufficient levels of information control to protect the organization's data and networks". These definitions help imply two significant things, alerting internet users of cybersecurity issues and threats, and enhancing internet users' understanding of cyber threats so they can be fully committed to embracing securing during internet use.

4   Conclusion

From the analysis of the data gathered, a significant understanding of the security culture among students from technical and non-technical majors are understood. This understanding establishes the baseline state of security in an academic institution with security policies but no security awareness programs in place. Also, the study greatly emphasizes that the importance of awareness cannot be ignored if security is a goal. The understanding of the human element assists in defining the security metrics and awareness strategies of an organization. This in turn deepens the understanding of the foundations required for the design and development of a cybersecurity awareness program that enhances a security culture, reduces the lackadaisical attitude of end users that cause them to be the weakest link in the security chain. Deployment of such a program across diverse disciplines and majors helps educate students on how to address specific threats and increases their resilience in defensive actions against them.

Acknowledgement. This research has been supported by the National Science Foundation (under grant #1723596) and by the National Security Agency (under grant # H98230-17-1-0355).

References

1. Newsweek Educational Insight: the Cybersecurity Threat - Fighting Back, 9th May 2017/2018. http://www.newsweek.com/insights/leading-cybersecurity-programs-2017
2. Morgan, S.: Top 5 cybersecurity facts, figures and statistics for 2018, 5 May 2018. https://www.csoonline.com/article/3153707/security/top-5-cybersecurity-facts-figures-and-statistics.html
3. Donovan, B.C.S., Daniel, M., Scott, T.: Strengthening the federal cybersecurity workforce. In: Strengthening the Federal Cybersecurity Workforce, ed: Obama White House (2016)
4. Statista. U.S. government and cyber crime - Statistics & Facts. https://www.statista.com/topics/3387/us-government-and-cyber-crime/
5. Holt, T.J.: Cybercrime Through an Interdisciplinary Lens. Taylor & Francis, Milton Park (2016)
6. Ramirez, R.B.: Making cyber security interdisciplinary: recommendations for a novel curriculum and terminology harmonization. Massachusetts Institute of Technology (2017)
7. Way, T., Whidden, S.: A loosely-coupled approach to interdisciplinary computer science education. In: Proceedings of the International Conference on Frontiers in Education: Computer Science and Computer Engineering (FECS), p. 1. The Steering Committee of the World Congress in Computer Science, Computer Engineering and Applied Computing (WorldComp) (2014)
8. Symantec Enterprises, 2018 Internet Security Threat Report (2018). https://www.symantec.com/security-center/threat-report
9. Tirumala, S.S., Sarrafzadeh, A., Pang, P.: A survey on Internet usage and cybersecurity awareness in students. In: 2016 14th Annual Conference on Privacy, Security and Trust (PST), pp. 223–228. IEEE (2016)
10. Reid, G.: How Many Internet Users Will The World Have In 2022, And In 2030? (2018). https://cybersecurityventures.com/how-many-internet-users-will-the-world-have-in-2022-and-in-2030/

11. Mittal, S.: Understanding the human dimension of cyber security. Indian J. Criminol. Criminalistics **34**, 141–152 (2016)
12. Jaishankar, K.: Establishing a theory of cyber crimes. Int. J. Cyber Criminol. **1**(2), 7–9 (2007)
13. Holt, T.J., Burruss, G.W., Bossler, A.M.: Social learning and cyber-deviance: examining the importance of a full social learning model in the virtual world. J. Crime Justice **33**(2), 31–61 (2010)
14. Pratt, T.C., Holtfreter, K., Reisig, M.D.: Routine online activity and internet fraud targeting: extending the generality of routine activity theory. J. Res. Crime Delinquency **47**(3), 267–296 (2010)
15. Borg, S.: Economically complex cyberattacks. IEEE Secur. Privacy **3**(6), 64–67 (2005)
16. Wilk, A.: Cyber security education and law. In: 2016 IEEE International Conference on Software Science, Technology and Engineering (SWSTE), pp. 94–103. IEEE (2016)
17. Jacobson, D., Rursch, J., Idziorek, J.: Security across the curriculum and beyond. In: Proceedings of the 2012 IEEE Frontiers in Education Conference (FIE), pp. 1–6. IEEE Computer Society (2012)
18. CISCO. Cisco 2017 Midyear Cybersecurity Report (2017). www.cisco.com/c/dam/global/es_mx/solutions/security/pdf/cisco-2017-midyear-cybersecurity-report.pdf
19. Das, A., Bonneau, J., Caesar, M., Borisov, N., Wang, X.: The tangled web of password reuse. In: 12 Proceedings of the 2014 Network and Distributed System Security Symposium (NDSS) (2014)
20. Melicher, W., Ur, B., Segreti, S.M., Komanduri, S., Bauer, L., Christin, N., Cranor, L.F.: Fast, lean and accurate: modeling password guessability using neural networks. In: Proceedings of USENIX Security (2016)
21. Chen, C.C., Shaw, R., Yang, S.C.: Mitigating information security risks by increasing user security awareness: a case study of an information security awareness system. Inf. Technol. Learn. Perform. J. **24**(1), 1–15 (2006)
22. Caulkins, B.D., Badillo-Urquiola, K., Bockelman, P., Leis, R.: Cyber workforce development using a behavioral cybersecurity paradigm. In: International Conference on Cyber Conflict (CyCon U.S.), pp. 1–6 (2016)
23. National Initiative for Cybersecurity Careers and Studies. https://niccs.us-cert.gov/
24. Jacob, J., Wei, W., Sha, K., Davari, S., Yang, T.A.: Is the nice cybersecurity workforce framework (NCWF) effective for a workforce comprised of interdisciplinary majors? In: International Conference Scientific Computing, CSC 2018 (2018)
25. Ramirez, R., Choucri, N.: Improving interdisciplinary communication with standardized cyber security terminology: a literature review. IEEE Access **4**, 2216–2243 (2016)
26. Rege, A.: Multidisciplinary experiential learning for holistic cybersecurity education, research and evaluation. In: USENIX Summit on Gaming, Games, and Gamification in Security Education (3GSE 15) (2015)
27. Fink, A., Litwin, M.S.: How to Assess and Interpret Survey Psychometrics. Sage, Thousand Oaks (2003)
28. Rahim, N.H.A., Hamid, S., Mat Kiah, M.L., Shamshirband, S., Furnell, S.: A systematic review of approaches to assessing cybersecurity awareness. Kybernetes **44**(4), 606–622 (2015)

Developing and Implementing a Cybersecurity Course for Middle School

Yeşem Kurt Peker$^{(\boxtimes)}$ and Hillary Fleenor

Columbus State University, Columbus, GA 31907, USA
peker_yesem@columbusstate.edu

Abstract. Investing in raising a generation with more security-aware minds is vital in preventing many of the security incidents that threaten individuals, organizations, businesses and nations today. In addition, there is a high need for well-trained cybersecurity professionals to protect our networks, computer systems, and infrastructure. Early exposure to basic cybersecurity concerns and concepts is key to developing awareness, piquing student interest, and laying foundation for more complex skill development. In this work, authors present their work developing a middle school cybersecurity curriculum for U.S. grade 8 students. This includes standards, objectives, and lessons for implementation within a year-long business and computer science course. The authors also share their experience and results from piloting the curriculum with nearly sixty grade 8 students in two classes at a Title I middle school in Columbus, GA during the 2017–2018 school year. A comparison of pre and post test results on the content show a significant growth in mastery of key cybersecurity concepts. In addition, artifacts from the course indicate an increased student interest in cybersecurity.

Keywords: Cybersecurity curriculum · K-12 education · Middle school

1 Introduction

A good understanding of cybersecurity threats and vulnerabilities as well as measures to prevent these is essential for everyone in the technology driven, highly connected era we live in today. Also, there is a growing need in the nation and world-wide for well-trained computing professionals including cybersecurity specialists. This need for trained computing professionals is reaching critical proportions in the field of cyber-security. A global study by Intel found that 82% of IT professionals worldwide report a shortage of skilled cybersecurity professionals with 71% "of respondents citing this shortage as responsible for direct and measurable damage to organizations whose lack of talent makes them more desirable hacking targets" [1]. The Bureau of Labor Statistics (BLS) lists the projected growth of Information Security Analysts for 2014–2024 as 18% (11% higher than average growth in all occupations) [2].

As humans remain to be the weakest link in cybersecurity, it is imperative to focus on younger generations and invest in raising more security-aware minds. Early exposure to basic cybersecurity concerns and concepts is key to developing awareness, piquing student interest, and laying the foundation for more complex skill development. Primary and secondary schools are essential for exposing students to cybersecurity topics at

© Springer Nature Switzerland AG 2020
K.-K. R. Choo et al. (Eds.): NCS 2019, AISC 1055, pp. 75–84, 2020.
https://doi.org/10.1007/978-3-030-31239-8_7

earlier ages in order to achieve levels of expertise needed by industry, governments, and other organizations as well as to shape their minds to be more aware of the threats and vulnerabilities in the cyber space. Students of today are the parents, educators, professionals, and leaders of tomorrow and the impact of their actions and behavior is significant for the future. In addition, as we focus on increasing the awareness and skilled workforce in cybersecurity, we need to make sure to diversify our technical workforce. Currently minorities and females are not fairly represented in the cybersecurity field.

In this paper, we share our experience and results developing and implementing a cybersecurity curriculum in a middle school in southwest Georgia. This project was possible through a collaboration between Columbus State University and the Rothschild Leadership Academy, a middle school. It included development and delivery of a middle school cybersecurity curriculum. Through our project, over 50 students had exposure to cybersecurity. The curriculum went beyond the common topics of online hygiene and included technical concepts as well. All students involved in the project were minority students. Our results indicate a significant growth in their knowledge of cybersecurity. The paper is structured as follows: In the Background section, we provide the status quo of cybersecurity curriculum in Georgia middle schools motivating our work and collaboration. In Sect. 3, we describe the development of the curriculum process and list the standards we set for cybersecurity in middle schools. Section 4 describes the implementation specifics and Sect. 5 presents the results from our work. Finally, in Sect. 6, we share lessons learned from our experience for those who may be interested in a similar kind of endeavor.

2 Background

At the time when we began this project there were no courses at the middle school level (grades 6, 7, and 8) in the state of Georgia, USA devoted to computer science topics. Each grade level had a "Business and Computer Science" course that coupled business standards with business computer applications such as Microsoft Office. Programming, cybersecurity, and other computer science essentials were not part of the course curriculum [3]. These courses were taught by teachers with a business education certification. The state of Georgia is currently undergoing a shift that includes the creation of computer science curriculum and requirements of teacher credentials for teaching these courses. The team on the project is actively participating in this transition.

Our project started with collaboration of the Teacher Education Department and the School of Computer Science at Columbus State University, and the Muscogee County School District on a grant proposal to NSA's MEPP program. The proposed project had three main goals:

1. Provide a good foundation for students in cybersecurity knowledge and skills,
2. Enhance students' ability to behave in a secure manner online as well as to secure their environment, and
3. Increase their interest in cybersecurity education

The project included the development of a cybersecurity curriculum for middle schools that covers fundamentals of cybersecurity laying a foundation for more advanced topics in high school, training of the teacher who will help implement the curriculum in cybersecurity, and delivery and assessment of the curriculum. The proposal included funding for purchasing 30 laptop computers for the middle school, tuition and fees for the teacher to take the introductory information security course offered in the CS Department in the summer and stipends for the faculty and the teacher who were involved in developing and delivering the curriculum. The team that developed and delivered the curriculum consisted of a tenure track assistant professor in computer science with a background in cybersecurity and an expertise in cryptography, a lecturer/outreach coordinator with a master's in computer science, a master's in education, and previous experience teaching middle school; and a current middle school teacher certified in business education. The curriculum was piloted with 60 students who were all minority students (African American) with slightly over 50% females.

2.1 Literature Review

At the time we developed the curriculum we could not find a cybersecurity curriculum for a middle school course in the literature. We did find cybersecurity themed competitions, modules, as well as after school and summer camp activities aimed at middle school students such as [3–5]. Without any previous work to use as a model, we relied on the principles developed by NSA's GenCyber [5] program to help provide a framework for the curriculum.

3 Curriculum Development

The curriculum was developed to be implemented as part of the "Business and Computer Science" course taught over the course of a school year. This was necessary for piloting the program, as, at the time, there was no standalone course on the topics of cybersecurity or computer science among the middle school pathways in Georgia. With that in mind, the team ensured the standards they created would mesh well with the current standards for the Business and Computer Science course and would flow in a way that would be conducive to student learning. The standards for Georgia for the grade 8 "Business and Computer Science" course can be found online [6]. The team was committed to active learning and other pedagogical best practices such as collaborative learning both inside and outside the classroom, engaging assignments, and engaging students through play and productions. These were forefront in lesson plan development. Our project also included the development of activities geared towards underrepresented groups.

3.1 Standards and Objectives

The team developed nine standards to cover a range of cybersecurity topics as well as a number of level appropriate objectives for each standard. The standards and objectives were developed considering level appropriateness, flow, integration with current

"Business and Computer Science" standards, and for student interest and potential for active learning. Standards and objectives are as follows.

- Students will examine the basics of cybersecurity needs for business, government, and organizations
  - List and define the elements of the CIA triad (confidentiality, integrity, and availability)
  - Explain components of access control: Identification, Authentication, Authorization, Non-repudiation
  - Be able to create a strong password and identify weak passwords
  - List and describe the basics steps in security risk management
  - Discuss the importance of physical security
  - Explore career paths in cybersecurity
- Students will examine the principles of cybersecurity and basic mechanisms used for protecting data and resources
  - Define the cybersecurity first principles: Least Privilege, Minimization, Abstraction, Domain Separation, Process Isolation, Information Hiding, Layering, Simplicity, Modularity, Resource Encapsulation [5]
  - Understand the principles behind encryption
  - Understand basic principles of public key cryptography
  - Understand how hashing is used to check the integrity of data
  - Understand how hashing is used to store passwords
  - Understand how steganography can be used to hide data
  - Demonstrate a basic understanding of antimalware, firewalls; IDS/IPS; VPN, back-up; redundancy systems, updates
- Students will understand the basics of computer organization
  - List and explain the role of the major components of a computer (input devices, output devices, storage devices, CPU, RAM)
  - Understand the binary number system and be able to convert between decimal and binary numbers up to 4 binary digits
  - Understand Windows basics
  - Understand Linux basics
- Students will demonstrate the ability to solve problems related to security
  - Understand hexadecimal number system and be able to convert between 4 binary digits and 2 hexadecimal digits
  - Encrypt and decrypt using Caesar (shift) cipher
  - Demonstrate the use of software tools for password cracking
  - Utilize steganography to hide a message
  - Understand the basics of Forensics
- Students will examine the basics of networking and the Internet
  - Explain IP addresses and the function of a Domain Name Server (DNS)
  - Explain how packets travel through a network
  - Understand the basic elements of HTML and how it is the basic building block of the World Wide Web
  - Utilize tools to understand network traffic

- Students will examine common cyber attacks
  - Describe tools/methods for gathering information from the Internet
  - Explain Denial-of-Service (DOS) and Distributed-Denial-of-Service (DDOS) attacks
  - Explain Man-in-the-Middle attacks
  - Differentiate between the different kinds of malware (virus, worm, trojan, spyware, ransomware, rootkit, adware, backdoor, browser hijacker)
  - Understand basics of mobile security
- Students will demonstrate an understanding of social engineering
  - Define social engineering
  - Demonstrate understanding of phishing attempts in email, on websites, phone calls, in person, and through social media
  - Examine different ways of social engineering (piggybacking, tailgating, spear phishing, whaling, etc.)
  - Understand the dangers of oversharing private information on social media
- Students will demonstrate an understanding of cybersecurity ethics, digital citizenship, and laws governing privacy
  - Explain the difference between a white hat (ethical) hacker (penetration tester) and a black hat (unethical) hacker
  - Define plagiarism and copyright infringement and understand how to avoid them
  - Explore security and privacy laws
  - Explain cyberbullying and its consequences
- Students will examine the impacts of cybersecurity and cybercrime
  - Discuss the economics and social impact of data breaches
  - Discuss the economics and social impact of identity theft
  - Explore the social impact of cyber warfare
    - Between nation states
    - Between a nation state and opposing factions within the same state
  - Explore the role of digital forensics in catching cybercriminals

3.2   Resource Collection

There are a number of pre-existing resources for teaching cybersecurity at the primary/secondary grade levels. The team spent time researching and collecting these resources for inclusion in lesson plans. This list of resources is organized by standard and shared via our school's site for this grant award [7]. The team prepared slides for various topics that can be used for teaching the material in the classroom. Hands-on and other active participation activities were developed for topics when such activities were not available among the resources.

3.3   Additional Activities

In addition to the curriculum and the material to explain the concepts, various activities where students applied their understanding of the material were included in the curriculum. These included preparation of posters and/or other media (fliers, recordings) on cybersecurity, participation in CyberPatriot [4], completion of Columbus State University's Cybersecurity Awareness Module, Completion of Nova Cybersecurity Labs [8], trips to the TSYS Cybersecurity Center at Columbus State University for activities by faculty and students at the University.

4 Course Implementation

4.1 Course Instruction

The Business and Computer Science course containing the piloted cybersecurity content was taught by the middle school teacher member of the team to two classes of 8th grade students (30 students per class) during the 2017/2018 school year. Parents and students were asked to sign a commitment form acknowledging that students would not use any information learned in the course for unethical purposes during a meeting held prior to the start of the school year.

The classes met every day for 50 min. Two days per week were devoted to teaching business standards and two days per week were devoted to teaching cybersecurity standards. Fridays were reserved for assessment and re-teaching. Standards involving business computer applications were integrated into teaching both the business and cybersecurity standards. For example, students created a presentation about a cyber-security topic of their choice using Microsoft PowerPoint. The curriculum included field trips to the University as well as guest instructors throughout the year.

4.2 Assessment

As was stated in Sect. 2, the goals of our project were three-fold. At the completion of the curriculum, we expected students to (1) have increased knowledge in cybersecurity topics; (2) be able to enhance security for their online behavior and secure their environment, and (3) have an increased interest in cybersecurity in general as well as a possible career path.

The assessment instruments consisted of a pre/post test to measure the gain in content knowledge, a pre/post survey to gauge change in their interest in cybersecurity and online safety and security awareness, as well as various portfolio items that students have completed as part of the curriculum.

The pre and post- tests were identical and included 32 multiple-choice questions on the standards of the curriculum. The reason the pre and post-tests were designed to be identical is that the questions were not mere calculation questions where we could change the values to assess the same skillset; most questions involved meaning or application of the cybersecurity concepts in the standard. Also, whether the student got the question right or wrong was not revealed to the student and the time between pre and post-tests were long enough (more than 6 months) that it was highly unlikely that a student would remember the questions on the pre-test when they took the post-test.

The portfolio items for assessment included writing assignments, participation in CyberPatriot [4], completion of modules such as Nova Labs [8], and Cybersecurity Awareness Modules developed by Columbus State University.

5   Results

The results from our project are very rewarding and encouraging despite the fact we could not cover the entire curriculum due to aspects of today's public school environment out of our control such as testing, curricular events, school closures due to inclement weather, etc. The number of instructional days we had with students was about half of what we had originally planned. However, despite the abbreviated instruction time, test results show significant gains and interviews show increased student interest in cybersecurity.

5.1   Cybersecurity Content

We developed a test to assess student knowledge on the cybersecurity topics included in the curriculum. The questions in the test can be obtained by contacting the authors. The test was first administered in the beginning of the school year before students were exposed to our cybersecurity curriculum and again at the end after covering the cybersecurity curriculum. The teacher of the courses was not involved in creating the questions for the test and was not given access to the questions until after the test was administered at the end of instruction. We will refer to the two administrations as pre-test and post-test. The number of students who took the pre-test was 55. Due to various reasons, some students did not continue in the course and the number of students who took the post-test was 48. Figure 1 shows the results (in percentage of students who answered correctly) of the pre- and post-tests.

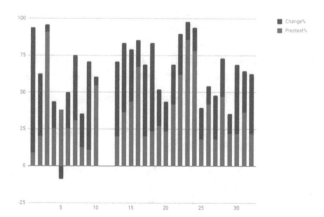

Fig. 1. Changes in the percentages of correct answers for each question from pre to post-test

In our analysis we had to eliminate two questions: Question 11 on file permissions in Linux as the topic was not covered and Question 12 where a binary number needed to be converted to hexadecimal and the correct answer was not included among the options. In the remaining 30 questions, in the pre-test, as expected, students' answers appeared random except for questions on topics that are already part of regular school curriculum such as strong passwords, cyberbullying, online safety, and ethics (questions 3, 22, 23, 24).

On the questions that were on the technical content, the percentage of students who gave the correct answer increased significantly except for one question, question 5, which is about hash functions. We think that this drop in the percentage of correct answers is due to the wording of the question. The question asked "which is FALSE about hash functions" which may not be the best type of question to assess student understanding at the middle school level [9]. The percentage increase in the correct answers for questions 1, 2, 3, 5–10, 13–32 ranged from 5% to 84%.

The increase in questions on topics that are already part of regular school curriculum were significantly lower than those concepts that are new as expected. For example, the lowest increase (4.92%) was on the question about choosing the strongest password among options. The increase in cybersecurity concepts such as the principles of cybersecurity, CIA, what steganography is, symptoms of a malware, how documents travel across the Internet, Denial of Service Attacks, Caesar cipher, security risk assessment, basic Linux commands were more than 40%. The average increase on all questions assessed was 30.15%.

5.2   Cybersecurity Awareness and Interest

As was stated in Sect. 4.2 a pre/post survey was prepared to assess increase in cyber awareness and interest in cybersecurity. Due to circumstances out of our control, the post survey was not administered before the schools closed. The teacher was able to have 12 students complete it after the schools were closed. The number of students who took the pre-survey was 55. Because comparing 55 responses from the pretest with 12 responses from the posttest would not provide a reliable analysis, we omit the comparative analysis of common questions. We rely on the additional questions asking students about the impact the cybersecurity curriculum had on their lives to assess the impact of the course on their online cyber behavior. The questions are:

1. What impacts, if any, did the cybersecurity lessons have on your behavior in the cyber world?
2. Did you share any of your learning from the cybersecurity lessons with people outside the cybersecurity class? Explain.

For question 1, except for one among the 12 who took the post survey, students stated that the course impacted their online behavior. They mentioned the importance of strong passwords and secure websites. One student said, "It impacted me so much. Cyber security is bery [sic] important, you have to make sure that you dont [sic] put no personal information on the internet & always be aware of what you post.".

For question 2, except for one among the 12 who took the exam, students stated that they shared their learning from the course with people outside the class. One student said, "yes i have discussed the things i have learned to others outside of my classroom and they was so surprised on my knowledge of cyber security and that for my age most people don't even know about it so they told to keep doing what i am doing because i am good at it and i want to purse [sic] a career in the cyber world." Another noted "I have been outside enough to get to share my knowledge about cyber security with others, but if I did I would have lots of good things to say about cyber security to the point where my actually want to think about learning more about cyber security them self so that they can also tell others."

Other assessments that indicate student awareness are the portfolio items collected for completion of NSA Day of Cyber Activity, an online cybersecurity awareness module, Nova Labs Cybersecurity Lab [8].

Moreover, the project increased interest in the school community as well. Parents of grade 6 and grade 7 students were asking how their student could take the class. Students from the grade 8 courses as well as several grade 6 and grade 7 students began meeting outside of class with a community volunteer and cybersecurity students from the university to study and prepare to compete in the CyberPatriot [4] contest online.

5.3 Conclusion

Our results indicate that students' knowledge in cybersecurity increased significantly; they are more aware of threats and measures against them in the cyber world; and some of them are considering taking more courses in cybersecurity in the future. We conclude from our study that a small exposure to cybersecurity goes a long way in terms of increasing students' knowledge, awareness, and interest in cybersecurity.

Another conclusion we have is that, for the curriculum to be successful, we need teachers who are properly prepared to teach cybersecurity. Support and additional resources for teachers and collaboration with higher education is essential in effectively teaching cybersecurity, a field that is newly shaping and continuously changing.

This important effort paved the way for inclusion of cybersecurity at a greater scale in middle schools in the state of Georgia. Currently, in the 2018–2019 school year, a larger scale pilot program that includes cybersecurity in addition to two other computing related pathways are being implemented in some of the GA middle schools including the middle school on this project. The team members on this project were actively involved in the creation of the curriculum for the computing pathways for middle schools in GA.

6 Lesson Learned and Next Steps

The project demonstrated that cybersecurity content can be successfully integrated into a business course at the middle school level. It also demonstrated that cybersecurity topics can be taught at a level appropriate and interesting to grade 8 students. However, the middle school teacher of the courses indicated that the workload on her to learn a new content area and prepare to teach the lessons was overwhelming. She suggested that a teacher with computer science background teaching a standalone cybersecurity course would find the task much more manageable. However, she plans to continue to integrate cybersecurity into her grade 8 business courses and expand to 7th and 6th grade business courses as well.

Another lesson learned was the need to work closely with the IT department well ahead of the start of school to ensure equipment with the proper software and settings is in place for students. Due to policies that vary between school districts, this can be a complicated and lengthy process.

The next steps in the project are a continuation of the cybersecurity topics at the middle school involved. However, the topics will be spread among grade 6, 7, and 8 to ensure a more complete coverage. At the end of last school year, the Georgia Department of Education began developing three computer science courses for middle school, including a cybersecurity course. The materials from this project have been shared with schools interested in piloting the new course.

Acknowledgements. This work was partially supported by an NSA MEPP grant, FY17-0002.

References

1. Intel. Global Study Reveals Businesses and Countries Vulnerable due to Shortage of Cybersecurity talent. Intel Newsroom (2016). https://newsroom.intel.com/news-releases/global-study-reveals-businesses-countries-vulnerable-due-shortage-cybersecurity-talent/. Accessed 17 Aug 2017
2. Bureau of Labor Statistics (BLS). Information Security Analyst. Occupational Outlook Handbook. https://www.bls.gov/ooh/computer-and-information-technology/information-security-analysts.htm. Accessed 17 Aug 2017
3. Pusey, P., Gondree, M., Peterson, Z.: The outcomes of cybersecurity competitions and implications for underrepresented populations. IEEE Secur. Privacy **14**(6), 90–95 (2016)
4. US CyberPatriot. Air Force Association's Cyberpatriot National Youth Cyber Education Program. https://www.uscyberpatriot.org/. Accessed 17 Aug 2017
5. GenCyber. Inspiring the Next Generation of Cyber Stars. https://www.gen-cyber.com/. Accessed 17 Aug 2017
6. GeorgiaStandards.org. Middle School Business and Computer Science Grade 8. Middle School CTAE Courses (2015). Middle School Business and Computer Science Grade 8. https://www.georgiastandards.org/standards/Pages/BrowseStandards/ctae-middle.aspx. Accessed 20 Aug 2017
7. TSYS School of Computer Science 2017 NSA MEPP Grant. https://cs.columbusstate.edu/csu_nsa_mepp.php. Accessed 17 Apr 2019
8. PBS. NovaLabs. https://www.pbs.org/wgbh/nova/labs/lab/cyber/. Accessed 17 Apr 2019
9. Chiavaroli, N.: Negatively-worded multiple choice questions: an avoidable threat to validity. Pract. Assess. Res. Eval. **22**(3), 2 (2017)

A Comparison Study of Cybersecurity Workforce Frameworks and Future Directions

Dan J. Kim[✉], Bradford Love, and Sanghun Kim

Information Technology and Decision Sciences,
G. Brint Ryan College of Business, University of North Texas,
1155 Union Circle #305249, Denton, TX 76203-5249, USA
dan.kim@unt.edu

Abstract. The cybersecurity industry is continuing to grow in workforce participation due to the growing popularity of its services in both commercial and government organizations. The government has recognized and created several frameworks that outline the work role standards for the multiple sectors. A comparison of these frameworks would not only provide a perspective of the purpose for maintaining different frameworks, but also provide a model of their relationships and the sectors that they cover. This paper will compare frameworks, outline the differences and similarities among them, and suggest ways to promote the workforce participation even more.

Keywords: Cybersecurity frameworks · Workforce · Knowledge Units

1 Introduction

With the rise of digital media, ecommerce, and cloud services, the cybersecurity industry is becoming globally significant and the workforce behind it is crucial. Both government and commercial organizational leadership have become aware of the importance of protecting their information from adversaries and criminal opportunists because if not addressed accordingly, not only will they face the consequences of their information being leaked to the world, but the credibility of these organizations will falter as well. Some prime examples of these information breaches in the 21st century would be the Equifax hack which affected over 148 million Americans [6], the Yahoo account leak which totaled up to 3 billion accounts compromised [4], and the Facebook security breach that exposed over 50 million user accounts [5].

Unfortunately, many of these breaches may have been avoided if there was a robust cybersecurity workforce helping to protect these organizations, but the emerging industry has yet to have obtained it. Due to these growing concerns, the U.S. government has developed several workforce frameworks that identify the work role specifications and requirements for the cybersecurity industry to help promote growth.

These frameworks are created to provide an understanding of what cybersecurity is, give detailed lists of cybersecurity related tasks, and give guidance on what experience is needed in order to complete such tasks. The workforce frameworks that are going to be discussed below are the National Centers of Academic Excellence in Cyber Defense or Cyber Operations Education Programs (CAE-CD/CO), the NICE Cybersecurity

© Springer Nature Switzerland AG 2020
K.-K. R. Choo et al. (Eds.): NCS 2019, AISC 1055, pp. 85–96, 2020.
https://doi.org/10.1007/978-3-030-31239-8_8

Workforce Framework (NCWF), Department of Defense Directive 8570/8140, and the Cybersecurity Industry Model. The frameworks help map specific segments of the industry and what is to be expected of them when it comes to the cyber workforce. This study identifies different cybersecurity workforce frameworks and compares these frameworks in terms of key elements, strengths and weaknesses.

This comparison will hopefully outline the relationships among the national cybersecurity frameworks to help explain their contributions in the context of the cybersecurity industry workforce.

2 Existing Cybersecurity Workforce Frameworks

The U.S. government has created and recognized several separate frameworks that map the roles, responsibilities, and requirements of the cybersecurity industry. While their primary functions and organization are similar, the audiences and purpose are distinct. The following sections provide an overview of these frameworks.

2.1 CAE-CD/CO Programs

The National Centers of Academic Excellence (CAE) have two education programs called Cyber Defense (CD) and Cyber Operations (CO) which are cosponsored by the National Security Agency (NSA) and the Department of Homeland Security (DHS) [7]. The goal of these education programs is to reduce vulnerability in the national information infrastructure and produce more qualified and trained cyber security experts coming out of higher education [7].

The purpose of the CAE-CD program is to help strengthen the nation's cyber defenses by "promoting higher education and research in cyber defense and producing professionals with cyber defense expertise" [7]. The CAE-CD program has the following designations: CAE in CD Education (CAE CDE) for Associate, Bachelor, Masters, and Doctoral Programs, and a CAE in CD Research (CAE-R) [7]. The way that the CAE-CD program works is that all two-year, four-year, and graduate level institutions in the United States can apply to become a Center of Academic Excellence in Cyber Defense, but they must meet stringent criteria which includes demonstrations such as program outreach, practice of cyber defense at the institution level, and successfully mapping the institution's curriculum to the two-year Core Knowledge Units (KU) [7]. Some of these Knowledge Units can be seen in the Table 1 and requirements specifically for the CAE-CDE designation can be seen in Fig. 1. Criteria does change based on whether a two-year or a four-year+ school is applying as well in which the four-year+ schools must successfully map the institution's curriculum to all the two-year Core KU's, four-year Core KUs, and five Optional KUs of their choice [7]. Once met, schools receive formal recognition from the U.S. Government and may elect to specialize in several possible focus areas [7]. The Centers of Academic Excellence in Cyber Operations program complements the CAE-CD program but provides emphasis on tools and techniques used in the cyber world [7]. The CAE-CO is deeply technical and rooted in computer science, computer engineering, and electrical engineering disciplines, with extensive opportunities for hands-on applications in opportunities such as labs and exercises [7]. Schools applying to become a CAE-CO accredited institution also have stringent criteria just like the CAE-CD involving mapping their institutions curriculum to KU's and many more requirements. The tools and techniques used and taught throughout the CAE-CO program are crucial to intelligence, military, and law enforcement organizations authorized to perform cyber operations [7].

Table 1. Basic Knowledge Units of CAE-CD Education Programs

A. Cybersecurity foundations	H. Cyber threats
B. Cybersecurity principles	I. Advanced algorithms
C. IT systems components	J. Algorithms
D. Basic cryptography	K. Cyber crime
E. Basic networking	L. Data structures
F. Network defense	M. Databases
G. Operating systems concepts	N. IA compliance

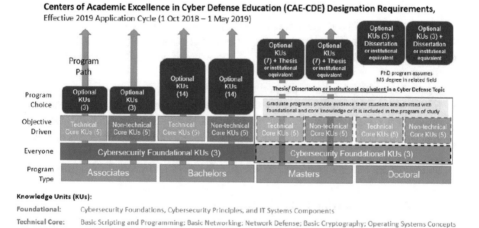

Centers of Academic Excellence in Cyber Defense Education (CAE-CDE) Designation Requirements, Effective 2019 Application Cycle (1 Oct 2018 – 1 May 2019)

Knowledge Units (KUs):

Foundational: Cybersecurity Foundations, Cybersecurity Principles, and IT Systems Components

Technical Core: Basic Scripting and Programming; Basic Networking; Network Defense; Basic Cryptography; Operating Systems Concepts

Nontechnical Core: Cyber Threats; Policy, Legal, Ethics, and Compliance; Security Program Management; Security Risk Analysis; Cybersecurity Planning and Management

Fig. 1. Updated Knowledge Units of CAE-CDE designation requirements [7]

2.2 NICE Cybersecurity Workforce Framework (NCWF)

The National Initiative for Cybersecurity Education (NICE) Cybersecurity Workforce Framework is a NIST-led national partnership between public, private, and academic sectors focused on establishing a codification that outlines cybersecurity work and workers irrespective of where or for whom the work is performed for [8]. The goals of this initiative are to maintain a globally competitive cybersecurity workforce and broaden the pool of skilled workers able to support a cyber-secure nation.

Being developed as a national resource that categorizes, organizes, and describes cybersecurity work, the NCWF uses a consistent, comparable, and repeatable approach to help select and specify cybersecurity roles for positions within organizations [8]. People involved in cybersecurity such as educators, students, employers, employees, training providers, and policymakers use the NCWF to help organize the way they think and talk about the cybersecurity work and workforce requirements.

The NCWF itself is composed of three separate components, Categories, Specialty Areas, and Work Roles. Categories are a high-level grouping of common cybersecurity functions, Specialty Areas are distinct areas of cybersecurity work, and Work Roles are the most detailed groupings of cybersecurity work composed of specific knowledge, skills, and abilities (KSAs) needed to complete tasks [8].

As stated in [4], "The Framework describes cybersecurity work regardless of organizational structures, jobs, titles, or other idiosyncratic conventions." The Specialty Areas can be used as a measurement against an organization's existing occupational structure by comparing the tasks and competencies based on their functions.

Aligning cybersecurity-related degrees, jobs, training, and certifications to the NCWF brings the following benefits:

- Colleges and training vendors can create programs aligned to jobs.
- Students will graduate with knowledge and skills that employers need
- Employers will have a larger pool of more qualified candidates from which to recruit.
- Employees will have more defined career paths and opportunities.
- Employers can set salary ranges for comparable work, because the Framework provides a classification system and common vocabulary for organizing cybersecurity work and workers that share common major functions, regardless of actual job titles.
- Policymakers can set standards to advance the field and promote job growth.

2.3 DOD Directive (DoDD) 8570 and 8140

The Department of Defense (DoD) released the DoD Directive 8570 in 2004, which later was signed into effect on December 19, 2005. The directives objective is to create policies and assigned responsibilities for DoD's efforts in Information Assurance (IA) training, certification, and workforce management [1]. What the directive outlines is that anyone who is involved with intelligence, missions, and security in cyberspace working for or within the United Sates DoD are required to be trained and qualified per the standards given from the directive [1]. The standards and requirements range from certification types needed in order to work in a certain position to the degree of experience required to fill such a role.

Ten years after the signing of 8570, the Department of Defense (DoD) released the Cyberspace Workforce Management directive, DoD Directive (DoDD) 8140.01 on August 11 2015 [9]. This directive is defined for managing workforce development for personnel who support DoD intelligence, security and law enforcement missions in cyberspace. As stated in [9], 8140.01 "reissues and renumbers DoDD 8570.01 to update and expand established policies and assigned responsibilities for managing the DoD cyberspace workforce." The goal of this directive is to unify the overall cyberspace workforce and establish specific workforce elements (cyberspace effects, cybersecurity and cyberspace information technology) to align, manage and standardize cyberspace work roles, baseline qualifications and training requirements [9].

At the time of writing, no specific training requirements for the DoDD 8140.01 have been released like there was for the 8570. Often seen regarding the 8570 and the requirements laid out by it, the "Approved Baseline Certifications" chart in Fig. 2 shows differing roles with multiple levels that require different certifications [3]. This type of manual for the 8140 is still awaiting to be released, but Robert F. Lentz who is a current senior advisor to the Cyber Security Consulting Group (CSCG) and also a former Deputy Assistant Secretary of Defense for Cyber, Identity, and Information Assurance says he expects the 8140 manual to be "more flexible and inclusive than the 8570," and it is said to "emphasize hands-on experience and training" [1].

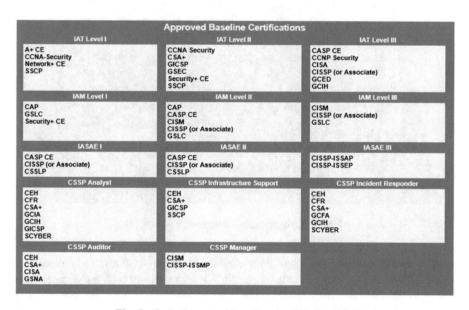

Fig. 2. DoD Approved Baseline Certifications [3]

2.4 Cybersecurity Industry Model

In June 2014, the U.S. Department of Labor published their own model of cybersecurity competencies as seen in Fig. 3 [2]. This model defines the latest skill and knowledge requirements needed by individuals whose activities impact the security of their organization's cyberspace. The purpose of it is to provide a knowledge roadmap to interested citizens who are new to the industry or those with limited experience thus promoting cybersecurity employment and a robust labor force [2].

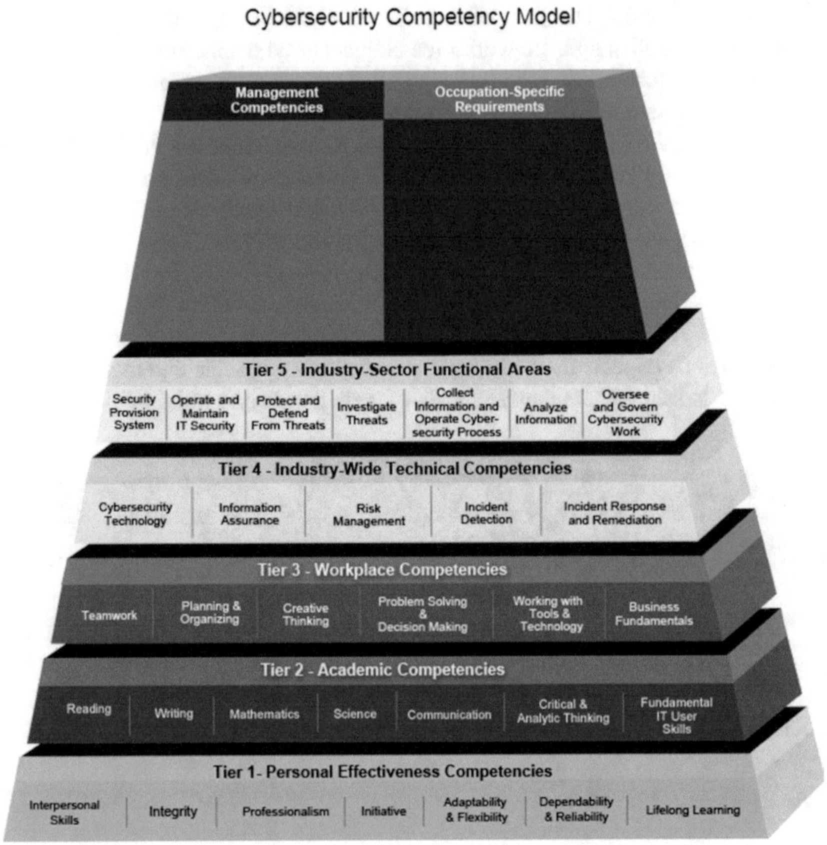

Cybersecurity Competency Model

Fig. 3. Cybersecurity Competency Model [2]

The model incorporates many competencies identified in the already mentioned above NCWF and intends to complement NCWF by including both the competencies and requirements needed by not only the average worker who uses the internet or an organization's computer network, but as well as a cybersecurity professional already in the workforce [2].

The model is part of a general competency clearinghouse maintained by the Employment and Training Administration that is intended to aggregate and clarify industry standards. The stated purpose of this clearinghouse is to provide employers and employment candidates with industry expectations [2]. The model is intentionally broad so that proprietors may add or subtract competencies to fit their own experiences and expectations. The competencies are primarily categorized by tier, with each tier progressively specific to the cybersecurity industry.

3   A Comparison Between the Frameworks

3.1   Primary Characteristics of the Frameworks

An initial analyzation of these frameworks, as seen in Table 2, provides an easily viewable layout of their distinct characteristics, such as initiative organizations, goals, and primary target audience. This diversification of unique characteristics is meant to cover the entire spectrum of academic standards and workforce competencies to optimize the cybersecurity workforce participation. Therefore, it is critical that each framework have an identified and unique purpose and audience that can be used to measure the frameworks effectiveness in promoting industry standards [8].

Table 2. Framework comparison

	CAE-CD/CO	NCWF	DoDD 8570/8140	Cybersecurity industry model
Initiative organization	NSA and DHS	NIST (National Institute of Standard and Technology)	DoD (Department of Defense)	Department of Labor
Goal/objective	Promotes higher education and research; and produces professionals with Cyber Operations (CO) and Cyber Defense (CD) expertise in various disciplines	Describes and categorizes cybersecurity work and identifies sample job titles, tasks, and KSAs (Knowledge, skills, and Abilities)	Provides the foundation for identifying education, training, and certification requirements to support cybersecurity personnel qualification	Provide a comprehensive overview of cybersecurity workforce competencies, including roles, training, and career paths for job seekers.
Audience	• 2-year, 4-year and graduate-level institutions in the U. S. seeking CAE designations	• Employers • Current/Future cybersecurity workers • Training/Certification programs • Educators • Technology providers	• Military, civilian, and contractor cybersecurity personnel • Training vendors, certification bodies, colleges, and continuing education providers	• Current and future workforce (generally unemployed or under employed) participants, as well as employers, who need guidance on work role-KSA congruity
Mapping component	• 3 Foundational KUs • 5 Core Technical KUs • 5 Core Non-Technical KUs • 57 Optional KUs	• 7 Categories (Securely, Provision, Protect and Defend, Investigate, College and Operate, analyze, Operate and Maintain, Oversee and Govern) • 33 Specialty Areas • 52 Work Roles (KSA)	• DoD work roles, tasks, functions and baseline KSAs • Under development awaiting DoD 8140.01 Manual	• Pyramid: tapering from general skills and responsibilities to specialized competency areas at the top • Utilizes NICE Framework (NCWF), KSAs
Highlights	• 100% Academia focused • Technical/Non-technical KUs • 2nd Biggest mapping component	• Backed by NIST • Large target audience • Biggest mapping component	• Small target audience • 8140.01 Manual not released yet • Strict guidelines and requirements	• Similar structure as NICE • Large target audience • Catered toward newer workforce participants

3.2   Goal Comparison

As seen in Table 2, each of the Frameworks have distinct characteristics regarding what they are trying to achieve. The CAE-CD/CO has a clear goal to promote higher education and research while producing professionals with cybersecurity expertise. The CAE-CD/CO works to achieve this with its audience by giving CAE designations to schools that apply and meet the requirements needed. Once designated, students attending these schools can obtain degrees catered to cybersecurity which can be focused on certain specialties that are available depending on what their school has to offer. The transition from academia to a professional workforce in cybersecurity is where the CAE-CD/CO looks to fit.

For the NCWF, the objective is to describe and categorize cybersecurity work and simplify job titles, tasks, and KSAs. What this objective works to do is standardize requirements, experience, and knowledge needed in order to obtain certain work roles in the cybersecurity field. This goal is achieved by the components of the NICE Framework consisting of 7 categories, 33 specialty areas, and 52 work roles all detailed to help current and future cybersecurity workforce participants understand what is necessary for the roles they want to move into.

For the DoD Directive 8570 and 8140, the goal is to provide a foundation for identifying education, training, and certification requirements needed to help support the qualification of cybersecurity professionals who are or want to become involved in the DoD cyberspace. This goal is achieved by the levels and requirements that are articulated in the Approved Baseline Certifications table and in the 8570 manual. An 8140 manual is still awaiting to be released which is said to have even more details and requirements in store.

For the Cybersecurity Industry Model, the goal is to provide a comprehensive overview of the cybersecurity workforce competencies, including roles, training, and other cybersecurity career paths for job seekers. This goal is achieved by the simplicity of the Cybersecurity Competency Model Pyramid which has its base built from general skills and responsibilities and builds up in tiers to specialized competency near the top. This model, along with the KSAs that it borrows from the NCWF, help show basic individuals ranging from just using the Internet, to an entry-level cybersecurity professional, what is needed to not only get started in cybersecurity, but what requirements are needed for set work roles as well.

3.3   Target Audience Comparison

When viewing Table 2, it is clearly visible to see the difference in frameworks when it comes to the targeted audiences. The CAE-CD/CO is focused on promoting higher education and getting more research done on the side of cybersecurity through academia. This clearly shows the NSA and DHS feel that starting in academia and transitioning to the workforce is an effective step to not only increasing participation, but also increasing the quality of future professionals. According to a report by Burning Glass Technologies, up to 84% of cybersecurity job postings require at least a bachelor's degree, while nearly a quarter require a master's degree [8]. With this in mind, the CAE-CD/CO's targeted audience clearly can help increase the workforce.

When viewing the targeted audience of the NCWF, it is more pointed away from higher education and more toward current and future employees and employers. With the NCWF's goal of categorizing and identifying sample job titles, tasks, and much more, it is attempting to set a standardization for all workforce participants so that if a person who is not actively involved in cybersecurity or has ever been involved, they now have the ability to look at the framework and get a better understanding of what they need in order to push into the cybersecurity workforce. This can also be effective for currently involved cybersecurity workforce members due to helping articulate what is needed to help view what is needed to move into other roles in the industry. The training and certification industry as well can use the NCWF to help structure companies are beginning to invest more into protecting their assets and the NCWF is effective at targeting the audience needed to keep those assets secure.

When viewing the targeted audience of the DoD Directive 8570 and 8140, they are more aimed toward individuals who are either in DoD or are wanting to be involved in the DoD cyberspace in the future. The reason for this is because the DoD feels only individuals who have obtained the skills, experience, and certifications required are fit to be working in any capacity with the DoD cyberspace or else they can put not only themselves, but the nation at risk as well. This type of targeted audience does shrink the size of workforce participants specifically for the DoD but is still an exceptional structure to recruit top quality cybersecurity professionals current and future.

When viewing the targeted audience of the Cybersecurity Industry Model, it is very similar to the NCWF's audience except it is a bit more encompassing of all individuals in the workforce and not as pushed toward trainings, educators, and certifications. The Cybersecurity Industry Model works to be to a more simplified audience with very little to no experience in cybersecurity and be a stepping stone into a future career in the cyberspace. This is effective towards newer workforce participants but not more the more veteran ones.

3.4   Highlights Comparison

Each of the frameworks has its strengths and its weaknesses, a comparison of these highlights will help show the distinct advantages and disadvantages each one has.

First, the highlights of the CAE-CD/CO show that it has a unique fit versus the other frameworks due to it being 100% academia focused. No other framework recognized by the government has this much of a focus on that sector. Also, CAE-CD/CO has both technical and non-technical knowledge units which means students will be getting both hands-on and management related experience which is a huge necessity in the cybersecurity workforce. The last point is it has the 2nd biggest mapping component behind the NCWF, which means it is full of all types of knowledge and information for newcomers into the workforce. This large of a component gives newcomers a wider selection of interests to pursue rather than forcing them down a few paths.

Next is going to be the highlights of the NCWF. The first big highlight point is the fact that the NCWF is backed by NIST. NIST is one of the biggest governmental agencies that sets standards for recommended security controls for information systems at federal agencies. This helps legitimize the NCWF and make it appealable to a wider audience. This leads into the second big highlight which is the NCWF's target audience is very large. The Framework doesn't just focus on future workforce participants, it also puts focus on technology, academia, trainings, and many more. This type of targeting is one of the main differentiating factors between the other frameworks. The last highlight is that the NCWF has the biggest mapping component out of all the frameworks. With seven categories containing 33 specialty areas and 52 work roles, the NCWF is a plethora of knowledge and information for all types of workforce participants ranging from brand new to the profession to seasoned veterans.

The DoD Directive 8570 and 8140 also have some distinct qualities that differentiates them from the others. First is going to be the fact that it goes for a smaller target audience. The Directives look for only people who are already in the DoD cyberspace or plan on joining it in the future. This is unlike the other approaches who usually have a larger audience. Also, the newest document from the DoD Directive 8140 does not yet have a manual released like there was for the 8570. This can be seen as a negative for it means that many of the new requirements and information newcomers are looking for is still not available to them even after multiple years. The last highlight is that the Directives have very strict guidelines and requirements compared to the others. These Directives are built more like a level-flow with requirements to get to those levels while the other frameworks are more of knowledge learning and experience gaining.

The Cybersecurity Industry Model is unique in the way it caters very much to a newer audience. Backed by the Department of Labor, it looks to get the very new members of the cybersecurity world and give them a clear understanding of what they need to know which is not done in the other frameworks as well. Another unique highlight is that the model has a very large target audience just like the NCWF. This feature creates the ability to reach out to all edges of the workforce and help recruit quality talent unlike some of the other frameworks that are narrow on their specifications. The last highlight is that this model is built heavily around the NCWF. It can almost be described as a simpler version of the NCWF but built into a pyramid-type breakdown.

4 Discussion – Issues and Suggestions

4.1 Starting Point Discrepancy

Based off the comparison of the frameworks, there are some problems that can be seen. Some of the frameworks are more appealing for what are considered novice members and other frameworks are more appealing for experienced and professionals. This split between them can cause much deter to newer workforce participants because they can get distracted and confused on which framework to follow and which is the most effective for their future career choices. When it comes to recruitment of newer members, a solidified structure that contains all the crucial aspects and requirements needed to advance a career in cybersecurity is a must. A framework consisting of certifications, skills, and experience usually equipped with the first two is the best and easiest road map for current and future cybersecurity workforce participants.

4.2 Academic and Work Experience Discrepancy

When it comes to differentiating between the academia side and the actual work experience side of the frameworks, there is a very blurred line. In the CAE-CD/CO, everything is detailed greatly to the academia side and can help push students in a cybersecurity direction, but at the end of the day a degree from a CAE designated school does not indicate what type of tools or actual experience a person will have coming out.

4.3 Mapping Components

The mapping components of each framework are so very different that a simplified structure is not able to be seen. Workforce participants will have to decipher many of the components to find out exactly what they are looking for. The NCWF and the Cybersecurity Industry Model are both very similar as well which can cause confusion to which one workforce participants should focus on first or more. As stated in other areas of this paper, confusion in the beginning stages for new workforce participants will begin to drive them away due to the exhaustion and frustration of a non-single solidified framework.

4.4 Suggestions

For the entire workforce to begin to prosper, a suggestion would be to use a more solidified framework for all aspects of cybersecurity and incorporate commercial cybersecurity certification programs along with it. Certifications are a fantastic way for managers to know what type of skills and experience newcomers are having when they are joining into the sectors. Once a solidified framework that encompasses all the best aspects of the frameworks already in place is created along with having lots of certifications to help show progression and benchmarks to reach, the workforce participation will most likely grow and the bubble of open cybersecurity jobs with no seat fillers will begin to dwindle.

5 Conclusion

The cybersecurity workforce is being threatened by a shortened number of qualified individuals for a vast number of positions needing them. In order to help battle this problem, some U.S. government agencies have recognized and created several cybersecurity workforce frameworks to recruit new talent. With the creation of these frameworks came a set of standards and expectations to specific members of different industry sectors according to common work role requirements. Although having a strong structure, there are gaps founded between the frameworks that can deter current and future cybersecurity professionals from following and using them. This comparison should ultimately help identify how the frameworks differ, help reveal some of the issues between the frameworks, and suggest solutions to help close those gaps.

Acknowledgement. This research has been supported by the National Security Agency under the grant # H98230-17-1-0355. Materials developed under the terms of this grant will be used by NSA to advance cybersecurity education for the nation.

References

1. "Beginner's Guide to Decoding DoDD 8570, 8570-M, and 8140." CyberVista, 29 November 2018. www.cybervista.net/decoding-dodd-8570-8140/. Accessed 03 Mar 2019
2. Cybersecurity Industry Model. Employments and Training Administration; U.S. Department of Labor (2014). https://www.careeronestop.org/CompetencyModel/competency-models/cybersecurity.aspx. Accessed 03 Mar 2019
3. DoD, DoD Approved Baseline Certifications, extension of Appendix 3 of the DoD 8570.01-Manual. http://iase.disa.mil/iawip/Pages/iabaseline.aspx. Accessed 03 Mar 2019
4. Framework for Improving Critical Infrastructure Cybersecurity. National Institute of Standards and Technology (NIST) (2014). https://www.nist.gov/cyberframework. Accessed 03 Mar 2019
5. Isaac, M., Frenkel, S.: Facebook Security Breach Exposes Accounts of 50 Million Users. The New York Times, 28 September 2018. www.nytimes.com/2018/09/28/technology/facebook-hack-data-breach.html. Accessed 03 Mar 2019
6. Kennedy, M.: Equifax Says 2.4 Million More People Were Impacted By Huge 2017 Breach. NPR, 1 March 2018. www.npr.org/sections/thetwo-way/2018/03/01/589854759/equifax-says-2-4-million-more-people-were-impacted-by-huge-2017-breach
7. National Centers of Academic Excellence. National Centers of Academic Excellence in Cyber Operations. https://www.nsa.gov/resources/students-educators/centers-academic-excellence/#defense. Accessed 03 Mar 2019
8. National Institute of Standards and Technology, Cybersecurity Framework. http://www.nist.gov/cyberframework. Accessed 03 Mar 2019
9. Work, R. Deputy Secretary of Defense. Department of Defense Directive 8140.01 (2015). http://www.esd.whs.mil/Portals/54/Documents/DD/issuances/dodd/814001_2015_dodd.pdf. Accessed 03 Mar 2019

Cyber Security Technology

Unmanned Aerial Vehicles (UAVs) Threat Analysis and a Routine Activity Theory Based Mitigation Approach

Alan Roder[1](✉) and Kim-Kwang Raymond Choo[2]

[1] Digital Forensics Unit, West Midlands Police, West Midlands, UK
alan.roder@ucdconnect.ie
[2] Department of Information Systems and Cyber Security,
University of Texas at San Antonio, San Antonio, TX 78249, USA
raymond.choo@fulbrightmail.org

Abstract. With the evolution of technology in today's modern society comes the global success of unmanned aerial vehicles (UAVs; also known as drones). There are, however, opportunities for UAVs to be abused in criminal or nefarious activities. The aim of this paper is to outline the opportunities for UAVs to be used in criminal activities, and explore the potential countermeasures and their effectiveness. This paper will seek to apply Routine Activity Theory to reduce the opportunities for UAV related crime and theorise strategies to increase accountability. Finally, this paper will attempt to forecast the threat of a criminal act within 2019.

Keywords: Drone · UAV · Threat analysis · UAV crime ·
Routine Activity Theory · Crime mitigation · UAV countermeasures

1 Introduction

Unmanned aerial vehicles (UAVs; also referred to as drones) are increasingly commonplace in our society, including in large events/public gatherings [1]. In addition to UAVs, there are other categories of unmanned vehicles, such as unmanned ground vehicle (UGVs; any remote controlled land-based vehicle), unmanned sailing vehicles (USVs; any remote controlled vessel that sails on the water), unmanned underwater vehicles (UUVs; any remote controlled vessel that sails below the water), and unmanned outer space vehicles (UOSVs; any remote controlled vehicle in outer space).

Whilst laws have been created surrounding the flight of UAVs in public spaces, these are often either not known, misunderstood or ignored by the operator. This disregard of Civil Aviation Authority (CAA) sanctioned laws is not due to any specific intention to break them, rather the misconception of the law, lack of knowledge and/or the feeling of discontent on the part of the operator. Individuals might watch a UAV fly overhead, but they would not consider the possibility that it was a threat, rather that is was of interest and at worst consider it a nuisance. For example, UAV footage obtained from YouTube regularly show the public which is being flown over, waving at the UAV, possibly believing that it is a reality show that they may be (involuntarily)

K.-K. R. Choo et al. (Eds.): NCS 2019, AISC 1055, pp. 99–115, 2020.
https://doi.org/10.1007/978-3-030-31239-8_9

participating in. This oblivious reaction on the part of the public provides criminals a means to exploit opportunities and enable them to partake in their unlawful acts. Without a change in public perception and awareness, this will remain a guise for criminals.

While the purpose of this paper is not to promote potential criminal acts, we aim to highlight the risks to law enforcement and government agencies, as well as departments within a nation's critical infrastructure, in the hope that they can either design and/or enhance their existing defensive/mitigation measures. Specifically, we will be using the Routine Activity Theory (RAT) [2] to guide the discussion, i.e. identify the potential motivations behind each crime type, along with an explanation of the crime and any recorded events. RAT was a key theoretical approach in criminology in the late 1970's, which identified that people act in response to situations. This could determine the nature of the crime they may commit. The model identifies that a criminal act occurs as a result of the merging of a motivated offender, a suitable target, and a lack of guardianship.

Whilst this paper will primarily focus on the United Kingdom (UK), the scope of the research has been considered comprehensive enough to be utilised and/or adapted to be used by any agency within any country. It should also be noted that a large amount of the information provided in this paper is the result of primary research completed, and personal experience gained working within the field of UAV forensics, as well as academic research gathered from secondary sources. The combination of both experience and research has enabled the use of examples and reflection to past experiences in order to further the understanding of the reader. There has been limited work in this area, despite its relevance to national security. In 2017, for example, Kovar detailed the potential threat posed by UAVs [3]. He reviewed the threat of terrorist organisations using UAVs against nation states and comprehensively explained UAV neutralisation methods. In 2018, Roder, Choo and Le-Khac forensically studied DJI Phantom 3 UAV [4], and provided guidelines for the forensic examination of a UAV for law enforcement agencies and their partners. In 2014, a research team from Birmingham University [5] studied the potential for UAVs to be utilized in state-level weaponised activity.

2 UAVs Are the Target of a Crime/Nefarious Activity

With the average price of a DJI Phantom 4 Pro ranging between £1000–£1700 it is not surprising that UAVs can be the target during a burglary, robbery or theft from an individual. It should be noted that a suggested additional area of research would be to explore the trends in robberies and thefts where UAVs are the target. Additionally this research could highlight the frequency of assaults on UAV operators, since news articles have identified operators have been subject to abuse.

Should a UAV be used and seized during the course of committing a crime, one key factor for the suspect would be to prevent law enforcement agencies from using it to identify them. Seeking to address the gap, Roder, Choo and Le-Khac [4] presented a detailed UAV forensic guideline, which would assist the examiner in identifying the operator, along with other evidentially valuable data. Rather than the suspect risk identification by using their own UAV, it would be safer to obtain a UAV i.e. through theft in order to diminish any link back to themselves. There would also be the additional benefit in that law enforcement agencies may target an innocent party (The original owner) in the process.

Since the theft of UAVs seem to be the inevitable inference to the rise in criminal activity involving UAVs, law enforcement agencies will need to be creative in their approach to tackling the crime. The ideal scenario would be to have complete aerial coverage across the UK and/or US, providing monitoring of any aircraft or UAV within UK/US airspace. Further to aerial coverage would be a national database containing all registered UAVs. The result of this arrangement would be early response to any incursion into restricted airspace, along with evidence of illegal or nefarious use of a UAV, should a crime be committed. Any UAV, which was not registered, would be highlighted to those monitoring the airspace and action could be taken to ground the device or identify the operator. Goodwin [6] highlights the challenges 2019 will bring to those monitoring and protecting airspace, and concludes that increased collaboration will become fundamental to combat the increase in the volume of traffic. Whilst we believe that a system, similar to the one detailed above, will become necessary within the next 5–10 years, it is also understood that such a system may be too costly given the current threat level. As such, a simpler and cheaper method could be adopted in the short term. For example, in the short to medium term, a national static UAV detection system could be placed in every major city which would provide up to 40 km of cover [7]. In the case of London, this range would cover a significant distance as represented in Figs. 1 and 2.

 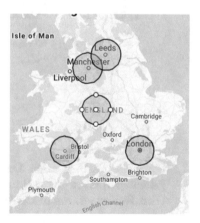

Fig. 1. 40 km radius of London, UK (Screen capture using www.freemaptools.com/radius-around-point.htm)

Fig. 2. 40 km radius of five of the UK's cities (Screen capture using www.freemap tools.com/radius-around-point.htm)

By placing a monitoring station in five (5) of the major cities, you can see from Fig. 2, that a large portion of the UK's population would have UAV coverage. Should this method of protection be adopted, rather than expect individual regions to monitor their own airspace, it would be more efficient for this role to be centrally managed, thereby maintaining continuity and increasing safety. Whilst the result of the implementation of monitoring stations in major cities, would be the reduction of UAVs becoming the target of criminal offences, the primary benefit would be safer airspace and a better response to criminal activity involving UAVs in those regions.

3 UAVs Are the Tool in the Commission of a Crime or Nefarious Activity

When considering UAVs used in criminal and nefarious acts, it must be taken into account the law which is being breached as well as the use of the UAV to commit the act in question. This section outlines the criminal offence, followed either by an offence reported in the media, and or a brief theoretical circumstance. To assist in assessing the threat level, we categorise the crimes into four levels of severity, with category one being the highest level – see also Appendix A.

Figure 3 provides a broad overview of the crime types, however this is subjective. Each act will have variations which will increase or decrease the threat level. As such Fig. 3 should be considered a guide to assist in determining the priority of a threat.

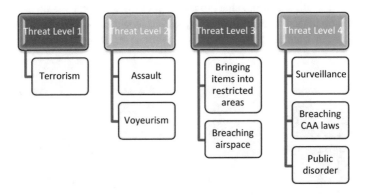

Fig. 3.   Crime types split into level of severity.

3.1 Breaching Civil Aviation Authority (CAA) Laws – Each Nation State Either Will Have Created or Will Have Considered Creating Their Own Rules or Laws to Control the Use of UAVs

A report by the National Air Traffic Service (NATS), estimated that over 2 million UAVs were now in circulation within UK. In October 2016, the UK MOD (Ministry of Defence) and CAA (In a joint enterprise) were authorised by Government ministers to conduct test crashes involving UAVs and Airliners. The tests were designed to simulate a collision and determine potential outcomes [9]. Whilst many breaches to the CAA laws (available on [8]) will be intentional, it seems far more likely that breaches of these rules will be due to miss-understanding or ignorance. These laws are designed to promote safe use of airspace, but in reality large sections of airspace are classed as

'Unrestricted'. Unrestricted airspace usually contains no air traffic control, and relies of pilots following the 'Rules of the Air' [10].

However, with the increase in UAV pilots it is unlikely that many will know or follows these rules, which increases the risk of collision. Operators who intend to use their UAV for commercial purposes currently require a PfCO (Permission for Commercial Operations). A PfCO requires that the operator obtain approved training and insurance to assure competence. Conversely, those who wish to fly recreationally do not require insurance or training, and would consider them to be too costly.

Breaching CAA guidelines is considered a low-level threat on the basis that breaching them would not present a threat to life. However, many of the crimes discussed further in this report could only be completed in the process of breaching the guidelines. This activity is classed as four (4) on the severity scale, which is the lowest level.

3.2 Breaching Airspace – UAV Flights into Civil and Military Airspace Have Become a Significant Risk to Flight Safety in Recent Years, with Multiple Reports of 'Near-Miss' Events Involving UAVs and Passenger Jets

A review of the Airprox board monthly summary for 2018, highlighted that multiple near-miss reports had been submitted by UAV operators and passenger jet pilots [13]. It has also been reported that the UK MoD (Ministry of Defence) and CAA (Civil Aviation Authority) are testing the effects of a drone strike on a plane [9]. In February 2018, a drone recorded video received global attention, when it showed a passenger plane passing by directly under the UAV [14]. More recently in September 2018 the University of Dayton published their work researching impact tests between large aircraft and small UAVs. The research concludes that an impact could cause significant damage to an aircraft [15, 16]. Recent information obtained from www.Airproxboard.org.uk (as of Feb 8, 19) confirm that the number of near-miss reports have increased year on year since 2014 [17]. The following extract (Table 1) has been taken from the Airprox reports listings [10].

The above report is one of 40 category A incidents recorded in 2018 and highlights the risk to aircraft and emphasises the likelihood that an impact may occur.

While organisations, agencies and authorities like the ICAO (International Civil Aviation Organisation), CAA (Civil Aviation Authority), FAA (Federal Aviation Authority) and EASA (European Aviation Safety Agency) are trying to provide guidance, and where necessary laws to make the airspace safe, ultimately there is no enforcement present to limit/prohibit an operator utilising their UAV within controlled airspace primarily used by commercial liners.

It can be determined that there is little control over much of the UK airspace, leading to UAV flights which can operate with almost complete anonymity. Furthermore, whilst key locations such are airports and military instillations may have some level of Aircraft Control Zone (ATZ), other key targets such as hospitals, power stations and other sections of national infrastructure will have no monitored ATZ, leading to an increased presence of viable targets. The motivation to strike any number of these locations could be criminal, but in all likelihood would be in the form of a

Table 1. Extract from Airprox reports for flight 2018225 (adapted from [10]).

Reference: 2018225	Date/Time: 20/08/18 15:47 h
Aircraft category: A319	Location: London CTA

Report:
The A319 pilot reports that on departure from Heathrow RW27R, passing 4DME and 3000 ft he saw a drone about 200–300 m ahead and 30 m right of the nose, about 200 ft above. The drone was black in colour, which contrasted with the white overcast cloud above, which is why it caught his attention. It appeared to be hovering just to the right of the departure track and as they climbed out they passed about 100 ft below it and 30 m horizontally. By the time he had seen it and communicated it to the Captain who was the handling pilot, they had passed it. It was only visible for 2–3 s. He opined that they were relatively heavy and therefore had a poor climb performance, if they had been lighter they would have passed much closer to the hovering altitude of the drone and risk of impact would have been very high. They reported it to ATC. The drone looked similar in size and design to the DJI Mavic Pro, a rectangular bodied quadcopter and they were between cloud layers, so it was unlikely to be flown by visual line of sight

Reported Separation: 100 ft V/30 mH Reported Risk of Collision: High	The NE Deps Controller reports that the A319 pilot reported a drone whilst on departure. He informed the following aircraft and the Heathrow Tower Controller. No other pilots reported seeing the drone

Cause: The drone was being flown above the maximum permitted height of 400 ft such that it was endangering other aircraft at that location. The Board agreed that the incident was therefore best described as the drone was flown into conflict with the A319

Risk: The Board considered that the pilot's overall account of the incident portrayed a situation where providence had played a major part in the incident and/or a definite risk of collision had existed

terrorist act. Breaching airspace is considered a medium-level threat in that the breach itself could present a threat to life. The intentional or accidently collision between a UAV and aircraft would cause damage to the craft and this may result in a catastrophic incident. This activity is classed as three (3) on the severity scale.

3.3 Public Disorder/Noise Disturbance/Harassment - This Could Be Encompassed in the Term 'Antisocial Behaviour'

Typically, 'off the shelf' UAVs generate noise, often described as being similar to a swarm of bees. This noise could be considered irritating if heard over a prolonged period of time. Additionally, people like the privacy of their own space, which would be infringed should a surveillance UAV fly overheard. Given the low level of crime this type of offence encompasses; it is understandable that few publicly reported events exist, however there are numerous news articles which highlight the frustration felt by the inappropriate use of UAVs in public. A 2017 news article, for example, stated that police reports for incidents logged involving 'Flying gadgets' had risen between 2014 and 2016, and raised concerns about privacy and the use of UAVs by criminals to preview properties prior to burglaries [18]. Another 2017 news article stated that between 2014 and 2016 complaints to police forces had risen by 1200% [19]. These articles highlight the predominant issues surrounding UAV related antisocial behaviour [18–20]. Complaints of this nature further stretch law enforcement agencies, who are often struggling to deal with traditional crimes.

Without any method of UAV registration, there is nothing in place to link a UAV to an owner, other than through investigation, which can be time consuming and may not achieve a satisfactory result. As such, UAVs can be flown with reasonable anonymity. Antisocial behaviour is not necessarily directed at an individual or group, rather is a lack of awareness or empathy towards those in the vicinity of the act.

Public disorder is considered a low-level threat on the basis that it would not present a threat to life. This activity is classed as four (4) on the severity scale.

3.4 Bringing Items into Restricted Areas – Whilst There Are Many Restricted Areas Within the UK, This Offence Appears to Primarily Focus on Flying Illegal or Banned Items into Prisons

This type of offence has obtained mainstream media attention, initially starting in around 2014. The threat was perceived to be so serious, that a dedicated task force was created to investigate the offences at a national level [21]. A 2016 news article highlighted that in 2013 there were no reported drone activities around prisons, in 2014 there were two reported drone activities, and in 2015 there are 33 reported drone activities. Items which were found included drugs, phones and USB thumb drives [22]. In December 2017, a news article published that an individual from Birmingham, UK had been convicted with 13 offences of conveying banned items into a prison. The charge, which was the first one brought by the West Midlands Police, involved three separate prisons [23]. Another news article in the same month confirmed that 10 people were sentenced for conveying banned items into a prison and conspiracy of conveying banned items into a prison, which involved multiple prison locations in the Midlands region and as far north as Scotland [24].

The authors' recent Freedom of Information Act request to the MoJ (Ministry of Justice) requested statistics for 2016–2018 was refused, with the MoJ stating 'the likely threat to the good order and security of prisons and the implications of this for prisoners and staff, favours non-disclosure of prison drone incidents'. The request is currently under appeal. A recent discussion with those working within the prison service has highlighted that UAV incursions into prisons only account for a small percentage of total incursions, however the increase noted between 2014–2016 must be taken into account.

Whilst there is a trial to incorporate UAV detection systems into prisons, a nationwide implementation is unlikely to occur for some time. Hence, there are several locations across the UK which have no defence against a UAV incursion. Even when a UAV is captured, it can take a significant amount of investigation time to potentially identify a suspect, which given the modification opportunities afforded to UAVs is not inevitable. For example, it was reported that a gang had used drones to fly over £500,000 worth of drugs into UK prisons [25], which highlighted the financial motivation to commit these types of crimes. Bringing items into restricted areas is considered a medium-level threat on the basis that those items may present a threat to life.

Whilst the primary offence appears to involve transporting drugs into prisons, there are several restricted areas in the UK, which include military installations, power stations and locations of importance. When determining the level of threat this offence poses, it cannot be negated that persons may be harmed in the process. This activity is classed as three (3) on the severity scale.

3.5   Voyeurism – Voyeurism Is the Act of Viewing Someone Conduct a Private Act Without Their Knowledge

UAVs predominantly carry a lot of sound and if within a close proximity of an individual they could be heard/recognised; with the advancements in the technology and quality of mounted recording devices, obtaining 4K resolution images and video from a distance means the requirement to be in close range of the target is not necessary.

By its very nature, voyeurism is an under reported crime, however the use of video capture on small mobile devices has increased this type of offence, including 'Up-skirting'. Whilst no offences of voyeurism using a drone have been recorded in the UK, in February 2017 a couple from Utah in the United States (US) were charged with voyeurism offences, after it was alleged that they spied on their neighbours through the bathroom and bedroom windows, using a drone [26]. Other incidents and concerns include those reported/raised in [27–29].

Since no registration is necessary to own or operate a UAV, any use to view or record a person or group in a private act can be completed with anonymity. The type of activity is dependent on the offender's sexual preference, as such the target opportunities are hard to calculate, but what can be determined is that it is unlikely there will be any preventative measures in place to prevent a UAV flying over a private address or location. Voyeurism is considered a high-level threat on the basis that this is a sexual offence. This activity is classed as two (2) on the severity scale.

3.6   Assault – An Assault Is Committed When a Person Intentionally or Recklessly Applies Unlawful Force to Another

Whilst not committed in the UK, in 2015 a male was convicted in Seattle, US, for reckless endangerment, after the UAV he was in control of crashed into a building and fell into a crowd, where it hit a female on the head and knocked her unconscious. Whilst this was clearly an unintentional act, it shows the risks associated with UAV flight, and the potential risks and harm to persons not in direct control of the UAV. For example, there have been numerous videos and images posted onto social media sites showing UAVs falling to the ground, often hitting people in the process. The videos, often referred to as 'Epic Drone Fails', highlight the risks posed to the public and anyone not in direct control of the UAV. For example, in July 2015 a video was uploaded onto Youtube by Austin Haughwout from Connecticut, USA [30]. The clip showed a firearm attached to a UAV, and appears to fire at least four rounds. The video went 'Viral' and attracted the attention of the FAA. Whilst no action was taken against the teenager, it highlighted the risks and potential uses of UAVs in illegal acts. Other videos involving concerning use of UAVs include [31, 32], and there are many other videos posted on social media sites claiming to have retrofitted UAVs with firearms, which are merely mimicking government agencies who are aiming to decrease their reliance on traditional 'manned' militaries. If UAVs are adapted for illegal acts, this type of weaponry could prove to be deadly.

Since no registration is necessary to own or operate a UAV, any use to attack a person or group can be completed with anonymity. With the UK population calculated at approximately 66 million, there are no shortage of potential targets, however the likely targets will either be political or high profile. An attack could be personally motivated, but given the current conflict with ICIS, the likely event will be terrorism related. Assault is considered a high-level threat on the basis that it could present a threat to life. This activity is classed as two (2) on the severity scale.

3.7 Surveillance/Counter-Surveillance - Obtaining Detailed and up to Date Information on National Assets and Infrastructure

Since the proliferation of internet mapping tools, including Google Earth and Google Streetview, the idea that sensitive locations could be protected from detailed mapping techniques has long since been forgotten. However, these tools usually have figures removed and can often be months or sometimes years out of date.

Criminal networks can now obtain detailed information prior to an act, which could include guard patrol routes, CCTV positions and angles, weaknesses and vulnerabilities, high value assets and a host of other critical information. In addition, surveillance/counter-surveillance has been conducted by nation states, who have become proficient in its implementation. Government buildings, national infrastructure, military bases and landmarks will likely have a level of defence against a UAV attack, however low level infrastructure i.e. police stations, hospitals and other similarly important institutions and buildings will likely be vulnerable to attack.

UAVs increase the ability to obtain valuable intelligence in the preliminary/planning stages prior to an attack. The benefit of their use, is the negligible risk to the operator and the vast amount of information that can be discreetly retrieved. Conversely, UAVs can be used as a counter-surveillance tool. For example, UAVs can be used to conduct perimeter checks at a military base, where a single operator could be tasked to view multiple autonomous UAVs programed to fly a pre-set route along the perimeter boundary. This use could reduce a manned personal presence.

Surveillance is considered a low-level threat on the basis that it would not present a threat to life. However, consideration must be given to the location under surveillance, along with the potential threat to life should a breach occur. This activity is classed as four (4) on the severity scale.

3.8 Terrorism

Scholars and practitioners, such as Kovar [3], have identified that terrorist groups have used UAVs in a surveillance and offensive capacity. For example, it was reported that over a two month period, Coalition forces observed almost one enemy UAV each day over Mosul [33]. Also in 2018, ISIS reportedly released an animated video showing drone strikes at football stadiums in Russia during the 2018 World Cup. It is uncertain if the threat was simply not realised or if it was prevented [34]. However, the video and other threats (e.g. [35]) are a chilling reminder that the terrorist organisation had considered the use of UAVs in this manner. We also remark that ISIS are not the only terrorist organisation, although they are the most prolific in recent history and have used social media to promote their ideals and campaign. As such it must be highlighted that any terrorist organisation could use UAVs to the same effectiveness. This information confirms the progressive use of UAVs by terrorist organisations and the escalation towards lethal intent.

Since no registration is necessary to own or operate a UAV, any use to strike a person or group can be completed with anonymity. As previously mentioned in this paper UK airspace is largely unrestricted, leading to a significant amount of potential targets, ranging from power stations, road networks, water supply facilities, hospitals and other locations which could cause a national emergency or significant loss of life. Terrorism is considered an extreme-level threat on the basis that it would almost certainly present a threat to life. This activity is classed as one (1) on the severity scale.

4   UAV Crime Prevention Strategies

4.1   UAV Pilots Licence

Whilst this concept may seem unachievable currently, it is obvious that the regulation of the use of UAVs will ultimately require implementation, i.e. through the use of licencing. This will promote the safe and lawful use of UAVS, whilst enabling a framework for executing penalties for illegal or dangerous behaviour.

When cars were first produced in 1892 there was no requirement for a licence, which was only introduced into the UK in 1903. Incidentally, driving tests were not started until 1935. The concept of introducing a licence was not popular, however government anticipated their use and decided to introduce the licence in an effort to identify vehicle uses. The driving test was later introduced to reduce road fatalities.

One must also assume that the use of UAVs, either through manual flight or automated flight paths, will become a standard sight in our skies, with companies such as Amazon researching automated package delivery through this medium [36].

UAV operators have the potential to become a recognised career path, much in the same way as taxi drivers are today. A method of identifying those who are trained and insured to carry out these tasks will become a necessity to maintain control and promote safe flight. By creating the legal requirement for all UAV pilots to hold a nationally recognised licence, will promote safe practice, whilst also supporting law enforcement agencies in identifying and prosecuting illegal activity.

4.2   UAV Insurance and Registration

Car insurance became mandatory under the Road Traffic Act 1930, which ensured all vehicle owners had to be insured for their liability for injury or death. Its enactment was brought about to reduce road fatalities through safer practices, whilst also linking an individual to a vehicle for the purpose of recompense to an injured party.

This legal requirement had the additional benefit of matching the owner to the vehicle, thereby putting the responsibility of the owner to account for the vehicles whereabouts at any given time. As such, it became less likely that the owner would commit crime in their own vehicle, which would be registered to them.

By creating mandatory UAV insurance, owners would be less likely to use UAVs registered to them to commit illegal or dangerous activities. Also, by necessitating that all UAVs are registered, the owner would become liable for its use, and law enforcement would have additional powers to seize UAVs that were not registered or insured. For example, Nijim and Mantrawadi [37] supported the assertion that UAV classification and identification is necessary and provided avenues for such a system.

Any positive introduction to regulate lawful use will inevitably breed negative connotations, such as the potential theft of UAVs for the participation in criminal activity. Since this practice of theft is likely in occurrence, registration and insurance would only help to support victims of crime. By creating a UAV pilots licence, with the supporting framework of insurance and registration, the opportunity to commit crime with relative anonymity would be greatly reduced.

4.3   Opportunities and Limitations

It cannot be denied that policing and military budgets within most countries are under strain in the current economic climate, as such any reference to the reduction of target vulnerabilities has to be given with due consideration. The simplest and most cost effective solution is for government to authorise research into a nation-wide detection system, which can be implemented by all law enforcement forces and military bases. The system would also have an oversight in the interest of national infrastructure. Such research into UAV detection technology would likely work with civilian agencies who have already created localised detection systems. This would reduce the initial cost, although likely increase the long term expenditure.

The system, once in place, would work alongside the UAV registration database with the aim of identifying UAVs that have not been registered, whilst also identifying registered UAV flights and tracking their activity. Any breach in lawful activity would be notified to local law enforcement to conduct enquiries and act as appropriate. The system could also be used by the military to identify breaches of restricted airspace, with sufficient warning for a proportionate response to be generated against the threat. By increasing the airspace monitoring and bolstering the defence of key military installations and locations or national infrastructure, the risk of an attack can be reduced.

5   Automated UAV Countermeasures

Although UAVs have proven to have a number of benefits, they are not without their limitations. To operate a UAV, there are two primary characteristics that must be present, the Unmanned Aerial Vehicle (UAV) and the Ground Control Station (GCS).

The devices are usually paired to each other and the link between the GCS and the UAV is maintained through a wireless interface (WiFi). Wireless connectivity is an established method of data communication but can open to manipulation.

5.1 WiFi Jamming

WiFi jammers are illegal within UK, US and a number of other countries; however, they could be a way to prevent a UAV from crashing into or flying close to a target. A WiFi jammer does not always have to be activated. However, should a UAV be spotted or intelligence obtained that a strike was imminent, the device could be activated, thereby negating that particular threat. There have also been counter-UAV studies, such as [38] that proposed detection and jamming systems.

5.2 UAV Interception

UAV interception and control systems appear to be in their infancy, with limited suppliers often charging significant amounts for their hardware. It is acknowledged that there are several methods to capture UAVs currently on the market, which range from net guns to trained birds; however, these methods require that a UAV has been spotted and there be sufficient personnel. With the declining or limited budgets available to government agencies, this paper focuses on autonomous equipment. This method is often considered not possible as the UAV and GCS will pair with each other to prevent such an event, but the WiFi connection is still open and visible and can be intercepted and manipulated. In recent years, there have been a number of commercial entities offering UAV interception services [7, 39–41]. It is clear that there is an evolving market for UAV detection and prevention systems, but their cost may prove to be prohibitive for anyone other than government agencies [42, 43].

6 Threat Analysis

UAVs continue to advance significantly, with enhanced capabilities in collision detection, stabilisation and extended flight time. For example, the DJI Phantom 4 UAV promotes a 28-min flight time, with a maximum speed of 20 m/s. This calculates to almost 70 km flight distance. The drone also reportedly has an 'Advanced Stereo Vision Positioning System (VPS)' giving precise hovering without the need for satellites. The Phantom 4 also boasts an 'Activetrack' feature, where a target can be recognised and followed and kept in frame. These features can be used in a host of creative environments with great effect, but they can also be abused.

The 2016 video [30] in which a pistol was attached to a UAV and remotely fired is just one example in the manipulation of features of the device, which ultimately can be used as a threat to life. If a pistol were to be attached to the Phantom 4 (which could be remotely fired), the features available would allow the operator to target an individual and follow them with minimal effort, with increased stabilisation and 'Activetrack' the operator would be able to accurately fire the pistol and hit the target. The 2018 research conducted by the University of Dayton [15] also highlights the risks to aircraft who fly in relatively narrow flight corridors. The increase in 'near-miss' reports with aircraft further compounds the realisation that a UAV striking an aircraft either intentionally or accidently is inevitable.

Terrorist organisations have used explosives to deadly effect for decades, however there were risks associated due to the act of placing an often volatile explosive device at the target location without being noticed. With the use of a UAV, this risk is negated. The Phantom 4 can fly up to 70 km, which is sufficient distance from the operator as to negate the risk of exposure whilst being relatively unstoppable.

Therefore, we posit that the probability of the following events has been deemed as high: a terrorist act within a first world country in 2019, a UAV related strike on an aircraft within 2019, and a UAV related criminal act within 2019.

Considering the abilities of UAV devices; the threat of a terrorist act within a first world country in 2019 can be deemed as high. ISIS have been driven out of Iraq and are unlikely to prevail in Syria. Whilst it will have some support in Afghanistan, the Taliban are much more entrenched in the region and cohabitation would be problematic. The conclusion is that ISIS will disperse and conduct hostilities against nation states which actively opposed them. Their efficiency in UAV strategy will undoubtedly be used within those nation states.

Their previous attacks highlight that a UAV related attack will likely involve either an explosive or chemical payload attack. Key locations would include government buildings, National infrastructure, military bases.

Similar to an act of terrorism, the likelihood of a UAV related strike on an aircraft within 2019 is considered to be high, as evidenced by the increase in the number of 'near-miss' reports and the reality that a small UAV (e.g. a DJI Phantom) could penetrate the wing (fuel tank location) of a passenger airliner [13].

The likelihood of a UAV related criminal act in the foreseeable future is real, also noted in the existing literature. For example, conveying an article into a restricted area (Prison) has already occurred in a number of countries and will likely continue through 2019. The more worrying trend is that associated with voyeurism and stalking. The main weakness in the successful investigation and prevention of this type of offence is that it will be almost impossible to prevent and difficult to prove unless the UAV is retrieved.

7   Conclusion and Future Work

It must be noted that the purpose of this paper is not to be perceived as 'anti-UAV'. UAVs have proven to be excellent when used as a proactive tool by emergency services and other law abiding agencies and individuals.

The increased use of UAVs is almost inevitable. Rather than try to prevent their use, it would be more productive to work with industry experts and enthusiasts to design a framework for lawful use which is robust and achievable for law enforcement agencies to act on. Any action by government should encourage proactive and lawful use of UAVs, which would promote 'self-policing' within the community.

Based on findings from this research, we posit the importance of undertaking further study to better understand the global crime and terrorist threats through the use of UAVs. Each country should commission research to identify vulnerable targets and design a strategy to limit their exposure to attack. In addition, countries such as UK do not have appear to have sufficient statistical crime analysis to accurately map the growth of UAV related criminal activity. Law enforcement agencies should consider adding the keyword 'UAV' or 'Drone' as a keyword indicator when recording crimes, which would assist in calculated growth trends.

A globally unified approach for intelligence gathering and sharing is also essential for each nation state, and necessary for accurate and intelligence lead dissemination of information. In reality, few agencies have the capacity to achieve this aim, with the exception of Interpol who encompass 192 countries.

We hope that governments continue the collaboration with their foreign counterparts in the sharing and dissemination of information. For example, global law enforcement agencies, such as Interpol and Europol, can commission a research project into creating an UAV intelligence database. This will assist in identifying crime trends, strategic weaknesses and support the identification and prosecution of offenders. Such a national/international database should include make, model, serial numbers (UAV, UAV battery, GCS etc), linked personal details (Email address, account info, etc), modifications (hardware, software and firmware), DNA, fingerprints, etc.

Acknowledgements. The assistance of Michaela Bradshaw in preparing this paper is acknowledged.

Appendix A: Crime Type Severity Guide

References

1. Duffy, R.: Understanding the public perception of drones (2018). https://www.nesta.org.uk/blog/public-perception-of-drones/. Accessed 07 Feb 2019
2. Cohen, L.E., Felson, M.: Social change and crime rate trends: a routine activity approach. Am. Sociol. Rev. **44**(4), 588–608 (1979)
3. Kovar, D.: UAV (aka drone) Forensics (2015). https://files.sans.org/summit/Digital_Forensics_and_Incident_Response_Summit_2015/PDFs/ForensicAnalysisofsUASakaDrones DavidKovar.pdf. Accessed 15 Mar 2017

4. Roder, A., Choo, R., Lekhac, A.: Unmanned Aerial Vehicle Forensic Investigation Process: DJI Phantom 3 drone as a case study (2018). https://arxiv.org/ftp/arxiv/papers/1804/1804.08649.pdf. Accessed 14 Dec 2018
5. Birmingham University: The security impact of drones: challenges and opportunities for the UK (2014). https://www.birmingham.ac.uk/Documents/research/policycommission/remote-warfare/final-report-october-2014.pdf. Accessed 06 Feb 2019
6. Goodwin, B.: Drones: how the world's airspace will change in 2019 (2018). https://www.aircargonews.net/news/technology/single-view/news/drones-how-the-worlds-airspace-will-change-in-2019.html. Accessed 24 Jan 2019
7. Coptrz: Real-time DJI Drone Detection System (2018). https://www.coptrz.com/dji-aeroscope/. Accessed 14 Dec 2018
8. CAA: Recreational drone flights. How the regulations apply to you (2015). https://www.caa.co.uk/Consumers/Unmanned-aircraft/Recreational-drones/Recreational-drone-flights/. Accessed 14 Dec 2018
9. uasvision: UK MoD and CAA to 'Test-Crash' Drones (2016). https://www.uasvision.com/2016/10/18/uk-mod-and-caa-to-test-crash-drones/. Accessed 14 Dec 2018
10. CAA: Standardised European Rules of the Air (2015). https://www.caa.co.uk/Commercial-industry/Airspace/Rules-of-the-air/Standardised-European-Rules-of-the-Air/. Accessed 14 Dec 2018
11. BBC: Drones and planes in mid-air near misses increase (2016). https://www.bbc.co.uk/news/uk-england-36734096. Accessed 14 Dec 2018
12. Davies, R.: Drone flew 'within wingspan' of plane approaching Heathrow (2017). https://www.theguardian.com/technology/2017/mar/31/drone-wingspan-plane-approaching-heathrow-near-misses. Accessed 14 Dec 2018
13. airproxboard: Airprox reports 2018 (2018). https://www.airproxboard.org.uk/Reports-and-analysis/Airprox-reports-2018/. Accessed 08 Feb 2019
14. RT: 'Reckless' drone flies dangerously close to landing plane in Vegas (VIDEO) (2018). https://www.rt.com/usa/417815-drone-plane-reckless-flying/. Accessed 14 Dec 2018
15. Gregg, P.: Risk in the Sky? (2018). https://udayton.edu/blogs/udri/18-09-13-risk-in-the-sky.php. Accessed 14 Dec 2018
16. uasvision: UDRI Tests Impact of Drone on Aircraft at High Speed (2018). https://www.uasvision.com/2018/10/09/udri-tests-impact-of-drone-on-aircraft-at-high-speed/. Accessed 14 Dec 2018
17. airproxboard: Drones (2018). https://www.airproxboard.org.uk/topical-issues-and-themes/drones/. Accessed 14 Dec 2018
18. Morley, K.: Rowing neighbours are reporting each other to the police over inappropriate use of drones (2017). https://www.telegraph.co.uk/news/2017/04/02/rowing-neighbours-reporting-police-inappropriate-use-drones/. Accessed 14 Dec 2018
19. TheSun: DRONE MOAN Complaints about nuisance drones up 1,200 per cent in just two years as cops deal with ten a day (2017). https://www.thesun.co.uk/news/3239924/complaints-about-nuisance-drones-up-1200-per-cent-in-just-two-years-as-cops-deal-with-ten-a-day/. Accessed 14 Dec 2018
20. Gilchrist, A.: Stop drones spoiling our wilderness (2017). https://www.theguardian.com/travel/2017/jul/24/flying-drones-in-the-uk-wild-places. Accessed 14 Dec 2018
21. Bulman, M.: Government plans to crack down on drones in prisons branded 'public relations exercise' by prison experts (2017). https://www.independent.co.uk/news/uk/home-news/drones-prisons-ministry-of-justice-jails-liz-truss-sam-gyimah-a7686561.html. Accessed 14 Dec 2018
22. BBC: Drones seized over HMP Pentonville carrying drugs and phones (2016). https://www.bbc.co.uk/news/uk-england-london-37152665. Accessed 14 Dec 2018

23. BBC: Man jailed for drone drug-drop into HMP Birmingham (2017). https://www.bbc.co. uk/news/uk-england-birmingham-42418064. Accessed 14 Dec 2018
24. BBC: Ten sentenced for smuggling drugs into prisons by drones (2017). https://www.bbc. co.uk/news/uk-42341416. Accessed 14 Dec 2018
25. BBC: Gang who flew drones carrying drugs into prisons jailed (2018). https://www.bbc.co. uk/news/uk-england-45980560. Accessed 26 Dec 2018
26. Papenfuss, M.: Utah Couple Arrested Over 'Peeping Tom' Drone (2017). https://www. huffingtonpost.co.uk/entry/peeping-tom-drone_us_58a6847fe4b045cd34c03e56? guccounter=2. Accessed 14 Dec 2018
27. Watson, L.: Woman claims she was sexually harassed by a DRONE after catching man flying remote-controlled plane at the beach that got uncomfortably close to female sunbathers (2014). https://www.dailymail.co.uk/news/article-2629459/Woman-claims-sexually-harassed-DRONE-catching-man-flying-remote-controlled-plane-beach-got-uncomfortable-close-female-sunbathers.html. Accessed 14 Dec 2018
28. Daily Mail: Voyeur uses drone to spy on nudists: furious naked sunbathers say they were buzzed by radio-controlled aircraft that had camera attached (2015). https://www.dailymail. co.uk/news/article-3175418/Voyeur-uses-drone-spy-nudists-Furious-naked-sunbathers-say-buzzed-radio-controlled-aircraft-camera-attached.html. Accessed 14 Dec 2018
29. Butler, D.: The dawn of the age of the drones: an Australian privacy law perspective (2014). https://heinonline.org/HOL/LandingPage?handle=hein.journals/swales37&div=23&id=&page=. Accessed 06 Feb 2019
30. Associated Press: Video of Gun-Firing Drone Spurs Investigation (2015). https://www. youtube.com/watch?v=FI–wFfipvA. Accessed 14 Dec 2018
31. Bieber, N.: Incredible flying flamethrower drones burn debris off of power lines in job too dangerous for humans (2017). https://www.mirror.co.uk/news/world-news/incredible-flying-flamethrower-drones-burn-9847548. Accessed 14 Dec 2018
32. ilipin: Paintball drone (2016). https://www.youtube.com/watch?v=8igjcW39T4k. Accessed 14 Dec 2018
33. Watson, B.: The Drones of ISIS (2017). https://www.defenseone.com/technology/2017/01/ drones-isis/134542/. Accessed 14 Dec 2018
34. Kitching, C.: World Cup stadium 'blown up' in new ISIS video depicting catastrophic attack against players and fans in Russia (2018). https://www.msn.com/en-gb/sports/football/ world-cup-stadium-blown-up-in-new-isis-video-depicting-catastrophic-attack-against-players-and-fans-in-russia/ar-AAyD8dA. Accessed 14 Dec 2018
35. Acton, M., White, D.: Tower drone attack as US security chiefs predict terror strike using a remote-controlled craft (2018). https://www.thesun.co.uk/news/7515439/isis-vows-eiffel-tower-drone-attack/. Accessed 14 Dec 2018
36. BBC: Amazon makes first drone delivery (2016). https://www.bbc.co.uk/news/technology-38320067. Accessed 07 Feb 2019
37. Nijim, M., Mantrawadi, N.: Drone classification and identification system by phenome analysis using data mining techniques (2016). https://ieeexplore.ieee.org/abstract/document/ 7568949. Accessed 06 Feb 2019
38. Shi, X., Yang, C., Xie, W., Liang, C., Shi, Z., Chen, J.: Anti-drone system with multiple surveillance technologies: architecture, implementation, and challenges (2018). https:// ieeexplore.ieee.org/abstract/document/8337899/authors#authors. Accessed 03 May 2019
39. droneshield: Complete Drone Security (2018). https://www.droneshield.com/. Accessed 14 Dec 2018
40. Telegraph: British prison is first to use 'disruptor' to create drone-proof 'shield' around jail (2017). https://www.telegraph.co.uk/news/2017/05/16/british-prison-first-use-disruptor-create-drone-proof-shield/. Accessed 14 Dec 2018

41. BBC: Southend Airport trials detect 'rogue' drone operators (2018). https://www.bbc.co.uk/news/uk-england-essex-44099674. Accessed 14 Dec 2018
42. Humphreys, T.: Statement on the vulnerability of civil unmanned aerial vehicles and other systems to civil GPS spoofing (2012). http://rnl.ae.utexas.edu/images/stories/files/papers/Testimony-Humphreys.pdf. Accessed 03 May 2019
43. BBC: Amazon testing drones for deliveries (2013). https://www.bbc.co.uk/news/technology-25180906. Accessed 07 Feb 2019

Using Modeled Cyber-Physical Systems for Independent Review of Intrusion Detection Systems

SueAnne Griffith[✉] and Thomas H. Morris

University of Alabama in Huntsville, Huntsville, AL 35899, USA
{s.griffith, tommy.morris}@uah.edu

Abstract. In this paper, the author proposes a methodology to perform comparison and validation of proposed intrusion detection and prevention systems (IDS/IPSs) designed for cyber-physical systems (CPSs). This approach consists of a software model of a CPS, as well as a variety of sample cyber attacks and a metric for comparing IDS/IPS performance. Securing critical infrastructure from cyber attack is an important step in reducing the likelihood of a system failure and the resulting losses of property and human life. Independent review is necessary in the scientific research process to determine the viability of proposed solutions, their reproducibility, and their usefulness when compared to other potential defenses. The design of the model and test attacks are complex enough to show their impacts, yet simplistic enough to allow researchers to easily reproduce them and to focus instead on the results of their testing.

Keywords: Cyber physical system · Virtual testbed · Independent review · Intrusion detection

1 Overview

As technology advances, with new and innovative ways to automate activities and process information, security for these systems must be reassessed and redesigned in order to keep pace with the growing number of threats [1]. As hackers' techniques improve and new methods are developed to compromise computer systems, new defenses must also be created and implemented in what could be described as a game of catch-up.

One such area of technology in need of stronger defense is that of cyber-physical systems (CPSs), including supervisory control and data acquisition (SCADA) systems. From car manufacturers and weapons systems to oil pipelines and hydroelectric dams, CPSs have become an integral part of society. A CPS is defined as a system in which changes in a digital component have an impact in the physical world. Many aspects of daily life, such as city utilities, rely heavily on digital systems to monitor and control mechanical functions. In manufacturing plants, these systems include conveyor belts and the heavy machinery used to lift, cut, and assemble products; in gas and oil pipelines, as well as hydroelectric dams and other water retention systems, the flow and pressure of the fluids contained is monitored by a computerized control system. In all

K.-K. R. Choo et al. (Eds.): NCS 2019, AISC 1055, pp. 116–125, 2020.
https://doi.org/10.1007/978-3-030-31239-8_10

instances, a failure of the CPS can potentially lead to extensive property damage, economic tolls, and loss of life.

Modern CPSs are becoming more integrated with high-speed networks, many of which use off-the-shelf technology that was not designed for such high-stakes systems as critical infrastructure [2]. Therefore, these systems must now be protected by retroactively adding intrusion detection systems or intrusion prevention systems (IDS/IPSs), often to the PLC used to control the physical portions of the system. A variety of issues may arise when implementing an IDS/IPS. Along with the difficulties in accurately determining what incoming data is and is not malicious, the memory space and processing power required for many IDSs can be burdensome to less powerful systems such as PLCs. In the specific case of CPS controllers, limited resources mean that host IDSs cannot be installed [3].

The hypothesis of the research presented in this paper is that a software model of a CPS can be useful in cybersecurity testing and protection for such systems and that the developed methodologies for testing can be applied to other ICSs to allow for a fair comparison across systems and devices. Peer reviewing IDS/IPSs by implementing them on this model will allow the researcher to verify these systems' functionality on a level playing field while discovering vulnerabilities and exploring the viabilities of these defenses on a dam.

In the papers found in a literature search [1, 2, 4–11], each proposed IDS/IPS was tested against a set of attacks, but there was no cohesive approach to use to compare these systems against one another. A series of attacks, based upon those found in the literature review and designed to encompass three types of network attacks (reconnaissance, man-in-the-middle (MitM), and denial of service (DoS)), will be created to encompass numerous likely scenarios faced by a CPS. This entire series of attacks should be used against each of the selected IDS/IPSs. By using the same set of attacks on all tested systems, this will further create an environment for fair comparison of security solutions.

Lastly, this research contributes a furthering of the importance of independent review in the field of cybersecurity. Papers proposing innovative solutions to CPS defense and protection are presented and published frequently, but these potential solutions are of little use in the real world if they are not reproducible. If other engineers and system designers cannot recreate the successes boasted by these publications, the research into protecting devices and networks does not "advance the field". In some cases, these published solutions may only function well under select circumstances. While each proposed security solution, in this case IDS/IPSs, may be successfully validated using attack scenarios chosen by the researcher(s) who developed the solution, independent review of these systems is necessary in order to compare them and determine which system is most statistically effective for real world use in defending against a wide range of plausible cyber attacks.

This paper is arranged as follows. Section 2 describes the structure of the modeled CPS to be used as a testbed. Section 3 discusses the attacks chosen for the independent study and the metrics by which the IDS/IPSs will be compared. Section 4 proposes future related research in this field.

2   The Testbed

CPSs can be broken down into five primary components: the physical system, the network connecting this system to a programmable logic controller (PLC), the PLC, the network between the PLC and the control center, and the human-machine interface (HMI) in the control center [12]. These five components are depicted in Fig. 1.

Fig. 1. The five components of a CPS

Because of the high stakes associated with hacking a CPS, which can be costly to repair and cause harm to life and property when malfunctioning, researchers often use small scale or software models to perform cybersecurity testing of CPS defenses [13].

The physical system, the component labeled with a 1 in Fig. 1, is the portion of the CPS with moving, mechanical parts, generally including valves, fans, and hydraulic pumps. Additionally, any sensors used to read data from and actuators used to control the mechanical components are considered part of the physical system [14]. In the case of a spillway, the portion of a dam which holds back the reservoir water, the movable gates used to regulate the flow of water can be controlled via a computer, making them a cyber-physical component.

The physical system connects to the PLC via a wired or wireless connection used to send and receive data. The "supervisory control" part of the SCADA acronym refers to actuator data, or commands, sent to cyber-physical components, while the sensor data, such as readings from thermometers and flow and depth sensors, comprises the "data acquisition" portion. The protocols used in CPS communication include Distributed Network Protocol (DNP3), Modicon Bus Protocol (Modbus), Inter-Control Center Communication Protocol (ICCP), Process Field Net (Profinet), and many proprietary protocols [14, 15]. The most commonly used of these is Modbus, due to its simplicity and availability as an open-source protocol without licensing fees [15]. The Modbus protocol will be used for the testbed discussed in this paper.

PLCs are used to control the physical system, as well as relay information to the human operators. These microcontrollers commonly have a limited memory and processing abilities, with just enough power and storage to perform the task they were initially programmed to do [3]. Register values in the PLC will be modified and the program logic installed on the PLC will be executed based on the sensor data received from the physical system and the commands received from the human operator. PLCs are programmed by firmware, which is usually written in one of the five languages defined in the IEC 6D31-3 standard [14, 16]. For this testbed, the open-source OpenPLC software will be run on a Linux operating system and used as the PLC platform, with the firmware written in ladder logic in the PLCOpen editor and uploaded to the PLC using the structured text (ST) format [15].

CPS networks differ from "conventional" enterprise networks in that they are real-time systems in which availability and timeliness cannot be sacrificed [17]. Delays in the sending of CPS sensor data or commands could result in issues if the received data is no longer current and valid or if the control signals arrive later than intended. Because CPSs impact the physical world, these issues could result in safety concerns. The actual type of network over which the communication is sent varies by system, but examples include both wired and wireless networks, as well as public switched telephone networks [17]. The network between the PLC and the system operation software will use uses TCP Modbus protocol on a simulated wired network.

The human operators of the CPS often control and monitor the system(s) under their supervision remotely, rather than being physically near the controlled system at all times. Human-machine interfaces (HMIs) are used to display data and provide input sources; this is often accomplished through software, with the operator viewing information on a graphical user interface (GUI). HMIs may use computer screens, with graphs and clickable control buttons, or may have a hard control panel with dials, seven-segment displays, and other physical components. These HMI softwares are sometimes run on commercially available systems such as non-industry-specific desktop computers, leading to increased vulnerability [2]. ScadaBR, an open-source HMI software, should be implemented on the simulation's host computer and viewed using an internet browser [18].

3 Attacks

3.1 Categories and Designs of Attacks

A contribution of this paper is the development of a cohesive set of attacks to use in IDS/IPS tests for CPSs. The sixteen attacks designed to test these systems will fit within the following three attack categories: reconnaissance, MitM (including injection, replay, and alteration), and DoS. The planned attacks were chosen based upon the frequency with which they were found in the chosen publications detailing IDS/IPSs. These attacks were all either used by the researchers to validate their research or mentioned as a potential threat. Though publications on newly developed IDS/IPSs may only discuss one type of attack, to assess an IDS/IPS's usefulness in the field, a variety of attack types must be considered as any of them could be faced by the system while it is in operation. This assessment, performed by an outside researcher, is the independent review.

In reconnaissance attacks, the intruder does not modify or disrupt the system in any way; the aim is only to gain information. This is usually done by eavesdropping on network traffic or sending information queries to the device, though side-channel analysis may also be used [7, 10, 19]. Three of the papers describing IDS/IPSs made reference to reconnaissance attacks, with two of these papers offering a solution [7, 9, 10]. The reconnaissance attack to be used in this research is an address scan, as described in [10], to determine what addresses on the PLC are in use. If a system using Modbus protocol is addressed by a query, it will send a response code back if that address exists; if no response is sent, that indicates that the address is not in use. This reconnaissance is useful for planning attacks in which data is to be injected into the system. The address scan used in testing will be implemented with a script that cycles through all legal Modbus addresses and sends a query to each, first in order and then in a second test with only a few selected addresses queried.

In MitM attacks, the attacker uses their access to the CPS network to interfere with the sending and receiving of legitimate packets; attack types include injection, replay, and alteration, which will be further described in the following paragraphs [20]. In a protocol that does not have authentication or encryption methods in use, these MitM attacks are easier to implement because the PLC does not have a way to confirm that the messages received are from legitimate sources [21].

Injection attacks are performed when data is sent from the intruder's device, masquerading as a legitimate device, to an actual legitimate device on the network. This includes command injection, in which the attacker tells the system to perform an action that the legitimate controller or operator did not request [9]. An example of command injection is an attacker broadcasting the command to close all breakers on an electrical circuit while spoofing the IP address of the HMI; the attacker's goal is to make it appear as though the HMI sent this request. Another type of injection attack is response injection, where false data is sent by the attack [9]. Having a rogue device send false pressure readings in a pipeline to the PLC is a type of response injection, for example.

Of the ten IDS/IPS papers studied in the literature review, seven mentioned command injection and four mentioned response injection [2, 4, 5, 7–11]. Both types of injection will be implemented in separate tests during the independent review stage of this research. The command injections will all consist of the attack device spoofing the IP address of the HMI using a script to send packets to the PLC. The response injection attacks will send false sensor readings to the PLC; there will be three different implementation of this attack. In the first, the attack program will randomize what sensor values to spoof and what values to choose for them and send. In the second, spoofed values chosen by the researcher will be sent; these will be designed to seem like viable, yet incorrect, responses. In the third, values that could potentially yield a PLC error, such as negative or potentially overflowing values, will be sent. This third set of attacks will test the IDS/IPSs' ability to identify illegitimate responses while the PLC is processing data that could result in an error.

Two other types of MitM attacks will be implemented in testing: replay and alteration. Ettercap will be used to implement the replay and alteration MitM attacks used in this research [22]. It is a useful tool in performing MitM attacks, as it has the capability to perform Address Resolution Protocol (ARP) poisoning. This is a method of attack which can be implemented against Ethernet based systems, such as those using Modbus [20].

Networks using Ethernet use ARP to match Media Access Control (MAC) addresses with IP addresses [23]. Senders broadcast an ARP request in order to find the MAC address of the intended recipient; ideally, this request message will bounce around the network until it reaches the legitimate recipient. This device will send an ARP reply back, listing its MAC address. Once the sender receives and saves the MAC address that matches the recipient's IP address, it can transmit packets to that specific device. This protocol has no authentication method, and therefore no built-in way to address spoofing [21]. In ARP poisoning, the attacker pretends to be the intended recipient. When the attacker's device sees the ARP request, it replies with its own MAC address, thereby telling the sender that it is the intended recipient of messages to the original recipient's IP address. Because the sender will cache this MAC and IP combination, messages will be directed from the sender to the attacker's device, allowing the attacker to intercept packets. This method is used in both replay and alteration MitM attacks.

In a replay attack, the attacker eavesdrops on the network to record packets, then plays them back at a later time [19, 24]. This type of attack is different than injections in that the commands and responses being sent were originally sent over the network legitimately then resent by the attacker, rather than being simply created by the attacker. Replay attacks can be thwarted by the use of timestamps or sequence numbers in packet headers, but the Modbus protocol does not use these methods [4, 25]. For the tests to be performed for this research, an Ettercap program will be set to record packets with a predetermined type and play them back after a random amount of time. There will be two versions of replay attack: one recording and re-sending sensor data from the physical system and one doing the same with commands from the HMI.

Alteration attacks, the second type of MitM attack to be implemented on the testbed while comparing IDS/IPSs, involve intercepting data sent over the CPS network and modifying it prior to sending it to its originally intended destination [4]. By altering legitimate commands instead of manufacturing commands from scratch, as seen in injection attacks, the attacker can ensure that the addresses, signature, and header portions of the packet are correct. Four types of alteration MitM attacks will be implemented against the tested IDS/IPSs; both command and sensor response packets will be recorded separately and altered both randomly and in a set, sensical way as determined by the researcher. This will test the IDS/IPSs' ability to recognize commands and responses that are both wildly different from and within the bounds of reasonable.

In a DoS attack, the attacker attempts to render the targeted device unreachable by other devices on the network [26]. Two different methods will be used in these tests to attempt to achieve this goal. In the first, an open-source program called Low Orbit Ion Cannon (LOIC) will be used to rapidly send packets to the PLC in the hope of overwhelming the device by providing too much network traffic for it to process. This is intended to cause the PLC to lock up, rendering it unresponsive and useless [15, 27]. This method is sometimes referred to as "flooding" [15].

The second method to make the PLC unreachable over the network is a combination of DoS and MitM in which packets sent to or from the PLC from the HMI will be intercepted by the attack computer using Ettercap. These packets will be subsequently dropped, meaning that no traffic to or from the PLC will reach its intended destination. By doing this, the PLC will be cut off completely from the rest of the CPS. It is worth noting that five of the ten papers mentioned in the literature review also discussed insider threat as a valid security concern in IDSs. Insider threats occur when someone who already has access to a system, such as an employee or contractor, attempts to cause harm [9]. As previously mentioned, this is the scenario under which the Maroochy attacks were implemented. For this reason, privilege escalation will not be addressed by any of the chosen IDS/IPSs; it will be assumed for the sake of testing that the attacker already has access to the CPS's network. Table 1 summarizes the attacks proposed for use in testing the selected IDS/IPSs.

Table 1. Listing of proposed attacks to implement in the testing of IDS/IPSs.

	Category	Description
1	Recon	Query all addresses to find which are in use
2	Recon	Query select few addresses to find which are in use
3	Injection (MitM)	Inject random commands
4	Injection (MitM)	Inject sensical commands chosen by researcher
5	Injection (MitM)	Inject random response values
6	Injection (MitM)	Inject sensical response values
7	Injection (MitM)	Inject out of bounds response values
8	Replay (MitM)	Record and re-send sensor readings
9	Replay (MitM)	Record and re-send commands from HMI
10	Alteration (MitM)	Record, change payload value randomly, re-send
11	Alteration (MitM)	Record, change payload by set amount, re-send
12	Alteration (MitM)	Record, change command randomly, re-send
13	Alteration (MitM)	Record, change command to pre-chosen one, re-send
14	DoS	Flood with nonsensical packets
15	DoS	Flood with valid packets
16	DoS/MitM	Intercept and drop all packets

The previously described attacks will be used against the modeled system when it is unprotected as well as when defended by the chosen IDS/IPSs.

3.2 Comparison Metrics

All sixteen of the attacks listed previously should be implemented against each of the chosen IDS/IPSs on the testbed. Each system will be installed, one at a time, on the PLC(s) in use. Then, traffic containing the chosen types of attacks, intermingled with valid traffic, will be passed through the CPS network and the results will be compared for each combination of protection and threat. Metrics used for comparison will be the number of detected attacks, false positives, and false negatives, as these statistics determine the IDS/IPSs' accuracy and reliability. Additionally, the time taken to detect the attack and the state of the system following the attack will be considered. Because PLCs often have very little space memory space, the memory consumed by the IDS/IPS will be evaluated, as well. An example of a "scorecard" to compare the performance of each IDS/IPS is shown in Fig. 2. This will allow for the proposed security solutions to be compared in an even and fair environment.

IDS/IPS NAME HERE	Detected Attacks	False Positives	False Negatives	Speed	Size of IDS/IPS	System functioning?
Query 1						
Query 2						
Com. Inj. 1						
Com. Inj. 2						
Resp. Inj. 1						
Resp. Inj. 2						
Resp. Inj. 3						
MitM Replay 1						
MitM Replay 2						
MitM Alt. 1						
MitM Alt. 2						
MitM Alt. 3						
MitM Alt. 4						
DoS Flood 1						
DoS Flood 2						
DoS/MitM						

Fig. 2. Scorecard for use in comparing each IDS/IPS's performance.

The definition of a "functioning system" may be further defined at the time of research and should be held consistent across all tests. This definition may vary with the criticality of the system being defended.

4   Conclusion and Future Work

This paper has described the need for independent recreation and review of proposed IDS/IPSs for CPSs, including a cohesive testing methodology in order to compare them fairly and determine the most successful system for defending against numerous attacks. After reviewing types of threats against CPSs and various proposed solutions in academic literature, a collection of attacks and IDS/IPSs were chosen for this comparison.

The author also introduced a proposed testbed consisting of constructing a virtual model of a dam, including controlled and uncontrolled spillways, navigational locks, and a hydroelectric power production plant. This testbed will utilize Matlab Simulink, UDP, VirtualBox, OpenPLC, TCP/Modbus, and ScadaBR to emulate the five parts of a CPS: the physical components, the wired connection, the PLC, the network connection, and the HMI. The design of this testbed was chosen based upon a literature review of common components of dams, their characteristics, and potential vulnerabilities to these systems.

Future work on this project will include the continued construction of the dam model testbed, implementation of the chosen attacks, and a comparison of the selected IDS/IPSs.

References

1. Barbará, D., Wu, N., Jajodia, S.: Detecting novel network intrusions using Bayes estimators. In: Proceedings of Siam Conference on Data Mining (2001)
2. Pan, S., Morris, T., Adhikari, U.: A specification-based intrusion detection framework for cyber-physical environment in electric power system. Int. J. Netw. Secur. 17(2), 174–188 (2015)
3. Garitano, I., et al.: A review of SCADA anomaly detection systems. In: 6th International Conference SOCO - Soft Computing Models in Industrial and Environmental Applications, pp. 357–366 (2011)
4. Fovino, I.N., et al.: Modbus/DNP3 state-based intrusion detection system. In: 24th IEEE International Conference on Advanced Information Networking and Applications, pp. 729–736 (2010)
5. Adhikari, U., Morris, T., Pan, S.: Applying non-nested generalized exemplars classification for cyber-power event and intrusion detection. IEEE Trans. Smart Grid 9(5), 3928–3941 (2018)
6. Yang, D., Usynin, A., Hines, J.W.: Anomaly-based intrusion detection for SCADA systems. In: Proceedings of the 5th International Topical Meeting on Nuclear Plant Instrumentation Controls, and Human Machine Interface Technology, pp. 797–803, 12–16 November 2006
7. Alves, T., Morris, T.: OpenPLC: an IEC 61131-3 compliant open source industrial controller for cyber security research. Comput. Secur. 78, 364–379 (2018)
8. Düssel, P., et al.: Cyber-critical infrastructure protection using real-time payload-based anomaly detection. In: Rome, E., Bloomfield, R. (eds.) Critical Information Infrastructures Security (CRITIS). LNCS, vol. 6027. Springer, Heidelberg (2009)
9. Richey, D.J.: Leveraging PLC ladder logic for signature based IDS rule generation. MS thesis, Mississippi State University, Starkville (2016)
10. Gao, W.: Cyberthreats, attacks and intrusion detection in supervisory control and data acquisition networks. Ph.D. dissertation, Mississippi State University, Starkville (2013)
11. Denning, D.: An intrusion-detection model. IEEE Trans. Softw. Eng. SE-13(2), 222–232 (1987)
12. Igure, V., Laughter, S., Williams, R.: Security issues in SCADA networks. Comput. Secur. 25, 498–506 (2006)
13. Alves, T., Das, R., Morris, T.: Virtualization of industrial control system testbeds for cybersecurity. Presented at ICSS 2016, Los Angeles, CA, USA, 06 December 2016 (2016)
14. Morris, T., et al.: A control system testbed to validate critical infrastructure protection concepts. Int. J. Crit. Infrastruct. Prot. 4, 88–103 (2011)
15. Alves, T.: OpenPLC: towards a fully open and secure programmable logic controller. Ph.D. dissertation, ECE, UAH, Huntsville (2019)
16. John, K., Tiegelkamp, M.: IEC 61131-3: Programming Industrial Automation Systems. Springer, Heidelberg (1993)

17. Zhu, B., Sastry, S.: SCADA-specific intrusion detection/prevention systems: a survey and taxonomy. In: Proceedings of Workshop on Secure Control System (2010)
18. ScadaBR: Principle Functionalities (in Portuguese). http://www.scadabr.com.br/. Accessed 5 Mar 2019
19. Zhu, B.: A taxonomy of cyber attacks on SCADA systems. In: Proceedings of the 2011 International Conference on Internet of Things and 4th International Conference on Cyber, Physical and Social Computing, pp. 380–388 (2011)
20. Mitchell, R., Chen, I.: A survey of intrusion detection techniques for cyber-physical systems. ACM Comput. Surv. (CSUR) **46**(4), Article no. 55 (2014)
21. Nayak, G., Samaddar, S.: Different flavours of man-in-the-middle attack, consequences and feasible solutions. In: 2010 3rd International Conference on Computer Science and Information Technology, Chengdu, China, 9–11 July 2010 (2010)
22. Ettercap: Homepage. https://www.ettercap-project.org/. Accessed 10 Mar 2018
23. Plummer, D.: An ethernet address resolution protocol. Network Working Group Request For Comments: 826, November 1982
24. Papp, D., et al.: Embedded systems security: threats, vulnerabilities, and attack taxonomy. In: 2015 13th Annual Conference on Privacy, Security and Trust (PST) (2015)
25. Reaves, B., Morris, T.: Analysis and mitigation of vulnerabilities in short-range wireless communications for industrial control systems. Int. J. Crit. Infrastruct. Prot. **5**(3–4), 154–174 (2012)
26. Alves, T., Das, R., Morris, T.: Embedding encryption and machine learning intrusion prevention systems on programmable logic controllers. IEEE Embedded Syst. Lett. **10**(3), 99–102 (2018)
27. abatishkev: LOIC. SourceForge: https://sourceforge.net/projects/loic/. Accessed 24 Apr 2019

An Efficient Profiling-Based Side-Channel Attack on Graphics Processing Units

Xin Wang and Wei Zhang[(✉)]

Virginia Commonwealth University, Richmond, VA 23284, USA
{wangx44,wzhang4}@vcu.edu

Abstract. The encryption/decryption algorithms have been ported to GPU platforms to take advantage of the GPUs' high-throughput computing capability. The downside of moving the cryptographic algorithms onto GPUs, however, is that the vulnerability of side-channel attacks for GPUs has not been well studied and the confidential information may be under a great risk by processing encryption on GPUs. In this paper, we proposed to leverage a profiling-based side-channel attack (SCA) to expose GPUs' side-channel vulnerability and the weakness of security services provided by GPUs. Our results show that GPUs are particularly vulnerable to profiling-based side-channel attacks and need to be protected against side-channel threats. Especially, for AES-128, the proposed method can recover all key bytes in less than 1 min, outperforming all prior SCAs we know.

Keywords: Security · Side-channel attack ·
Graphics Processing Units (GPU) · Performance profiling

1 Introduction

GPUs which served as the graphic-oriented computation platform are increasingly used to host more general-purpose applications such as compute-intensive and data-parallel scientific computing programs. In particular, taking advantage of the high computation throughput offered by GPUs, the security services such as cryptographic application have been growingly deployed on GPUs to enjoy remarkable performance boosting [1–5]. Unfortunately primary existing studies concentrate on taking care of identifying vulnerabilities based on mainstream security computing platforms such as CPUs and FPGAs, while the vulnerabilities exposed on the promising GPU-based security systems have not been thoroughly explored leaving severe vulnerabilities for an adversary to pull down the whole security system.

In this paper, we propose a novel profiling-based side-channel attack that information leaked from GPU performance profiling can be extracted when executing on an SIMT-based GPU to fully recover the encryption secret key. Unlike power and timing side-channel attacks, instead of bridging the number of unique

© Springer Nature Switzerland AG 2020
K.-K. R. Choo et al. (Eds.): NCS 2019, AISC 1055, pp. 126–139, 2020.
https://doi.org/10.1007/978-3-030-31239-8_11

cache line requests with the key via a differential of power consumption or execution time, our profiling-based strategy accomplish straightforward key recovery procedure by sampling the exact number of unique cache line requests during the run-time and simply checking all 256 possibilities for each byte of all 16-byte AES key to determine the correct answer. The number of samples and time needed for recovering the key are dramatically reduced while the accuracy can be always guaranteed. Moreover, the profiling-based side-channel attack has great scalability, as the profiling time only increases slightly with no additional samples required when the length of the AES key scales up. Compared to the prior work in [13] of GPU side-channel attack, for which only GPUs supporting the execution of multiple kernels from different programs are vulnerable since the spy program needs to implant malicious kernels to probe victim program; creating contention between spy and victim programs is unnecessary for the attack proposed in this work, which can be more generally applied to GPUs. Furthermore, instead of profiling coarse-grained information or collecting only one sample for each kernel executed, we propose a fine-grained profiling strategy which can break the limitation of sample frequency set up by the profiling APIs and get more details during the kernel execution for an efficient and accurate key recovery. We have implemented a run-time profiling tool, which can sample two performance matrices, the number of the memory load and memory store requests. Based on the two performance matrices, the number of unique memory load requests in the last round encryption for each byte except the first byte then can be determined. The 16-bytes AES key can be recovered byte by byte. By calculating the number of unique memory load requests with 256 possibilities of a single byte of the AES key and comparing to the profiled number, the number of possibilities of corresponding byte key can be narrowed down. The number of possibilities can be further decreased to 1 by repeating the same procedure with different input data for several times.

2   Background

In this section, the Fermi GPU architecture for profiling-based side-channel attack evaluation is introduced. We also describe the CUDA programming model and the GPU implementation of AES algorithm.

2.1   GPU Architecture

In this study, we demonstrate a profiling-based side-channel attack on two NVIDIA GPUs, including an NVIDIA Tesla C2075 which has 14 streaming multiprocessors (SMs) and 448 CUDA cores in total with 32 cores per SM, and an NVIDIA Quadro 2000 which has different parameters on the number of CUDA cores, the size of on-chip memories, and on-chip interconnection bandwidth and etc. Although the hardware resources for two GPU cards are different, they both follow the Fermi architecture [15]. Figure 1 shows the overview of a typical Fermi GPU architecture. It consists of 15 GPU cores called Streaming Multiprocessors

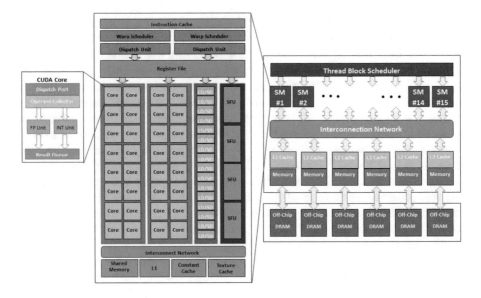

Fig. 1. Baseline GPU architecture

(SMs) where each SM core contains 32 single instruction multiple data execution units, 16 load/store units and 4 special function units (SFUs). Each SM core owns two warp schedulers and two instruction dispatch units, enabling to issue two independent instructions from two different warps. The 32 execution units are assigned to two shader processors (SP) each containing 16 execution lanes called SIMT lanes. Since the execution units are operating at double clock frequency of the SMs, 32 threads can then be running upon a single SP concurrently.

All SMs have their own private L1 data cache, read-only texture cache, constant cache and software-managed shared memory(scratchpad memory). Total size of L1 cache and shard memory is 64 KB and they can be configure to 16 KB L1 cache and 48 KB shared memory and vice verse through software. The read-only texture cache loads data from off-chip texture memory and is designed to identify the spatial locality of memory access patterns. The read-only constant cache works with the off-chip constant memory and is utilized to optimize the data sharing among all the threads in a warp. All SMs share an on-chip unified L2 cache partitioned into 6 tiles and an off-chip global memory. The SMs and the shared L2 cache are connected via an on-chip network. The private L1 cache per SM and shared L2 cache and global memory cooperate to perform fast memory accesses. Although GPU architecture inherits similar memory hierarchy from CPU architecture, GPU caches result in much smaller size and much higher bandwidth that caches in CPU. Moreover, Unlike CPUs which primarily count on caches to reduce memory latency, GPUs rely on massive TLP to hide memory latency and increase throughput. To support fast and low-cost context switching for massive concurrently running threads, a large number of registers are necessary in GPUs. The massive TLP supported by a large size of register files contribute most performance enhancement and achieve high throughput. On the other hand, the functionality of GPU caches on boosting performance is more like a complement instead of an essential.

2.2 CUDA Programming Model

CUDA programming language allows the programmer to define C functions as several kernels which consist of thousands of threads that are executing in parallel [14]. Each thread within a CUDA kernel is marked with a assigned unique thread ID which is accessible through the built-in threadIdx variable. GPGPU applications always contain multiple kernels that contain a group of thread blocks. A thread block is formed by one-dimensional, two-dimensional or three-dimensional thread index allowing a vector, matrix or volume computation domain. Every 32 Threads within the same thread block with consecutive thread IDs are grouped as a warp. A warp is executed in a single instruction multiple-threads (SIMT) way and has only one PC (Program Counter). The warps in a thread blocks allow GPU to overlap long latency by conserving stalled warps context and switching a oldest ready warp. The NVIDA officially set limitation for the number of threads running concurrently on a SM. For Fermi architecture, up to 48 active warps or 8 thread blocks can be hosted per SM, this is in total 1536 active threads per SM. On the other hand, each thread consumes a portion of hardware resources such as register files and shared memory to support the execution environment and conserve execution data and the available hardware resources of a SM also limit the number of concurrent threads per SM.

Although a single warp only maintains one PC, threads in the same warp can access different memory address or follow different control flow paths. To take care of all 32 memory accesses initiated by threads in the same warp which may point to different addresses, a coalescing unit accepts all these memory requests and assign them to different cache lines according to their address. Memory requests with addresses located at the same cache line then can be merged generating unique cache line requests which then are forwarded to L1 cache controller for further processing. Indeed, the SIMT execution pattern and the memory coalescing mechanism can achieve high throughput and efficient memory bandwidth utilization respectively, but also they conspire to cause a secure deficiency which can help adversaries to access confidential information and will be detailed in Sect. 3.

2.3 AES GPU Implementation

In the AES algorithm, the size of the minimum encryption element is fixed to 128 bits. The encryption key length, however, can be varied from 128 to 196 and 256 bits, and the total encryption round is 10, 12, and 14 respectively. We focus on 128 bit AES in this work which contains 10 encryption rounds in total. Before the first encryption round starts, the original 16-byte key expands itself into a key schedule consisting of 10 more 16-byte keys for the following 10 rounds

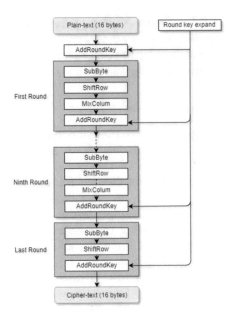

Fig. 2. The operations of encryption rounds in AES

of processing and XORs itself with the 16-byte plain-text to form the initial state. All encryption rounds except the last round execute the same operations composed of SubByte, ShiftRow, MixColumn and AddRoundKey. The 10th (i.e. last) round which only contains SubByte, ShiftRow and AddRoundKey operations skips MixColumn. Figure 2 shows the operations of encryption rounds in AES.

The LUT (Look Up Table) based AES is one of the most classical implementations on many general-purpose computing platforms. In the LUT based AES implementation, the SubByte, ShiftRow, MixColumn operations are equalized to four table lookups. For each round in the 1st to the 9th round, four table lookups accessing four different tables are performed. Since the last round does not involve the MixColumn operation, only one table is accessed, which is different from the four tables used by the previous 9 rounds. To complete the encryption round, the AddRoundKey operation is XORed with the corresponding round key at the end of each round after table lookups. As shown in Fig. 3, our GPU implementation of AES arranges each GPU thread to work on one block of the plain-text of 16 bytes independently and a warp composed of 32 threads can then process 32 blocks in parallel.

3   Profiling-Based Side-Channel Attack

The scenarios of this attack and the threat model are the follows:

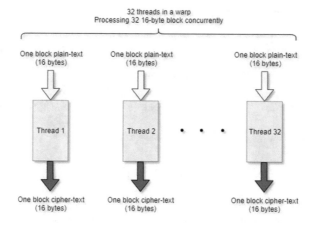

Fig. 3. The GPU based AES implementation

1. The attack owns the local computer and thus has the authority to launch a profiling tool which can monitor the GPU performance counters. The spy program contains a profiling tool which is purely constructed by profiling APIs with no kernels.
2. The victim program communicates with the outside world through a secure channel that is protected by the GPU AES encryption and the AES key is hidden from the attack.
3. The victim program is always running and waiting for data from outside to be encrypted. The spy program keeps sending data to victim program for encryption and meanwhile launches the GPU profiling tool to get samples.

We build a profiling strategy which is able to accurately distinguish the number of unique memory load requests for each of the 16 load instructions in the last round based on the three following findings.

First, we find that due to no MixColumn operation, the last round can be simply performed by one table lookup (i.e., load instructions), XORing the round key (i.e., logic instructions) and result write-back (i.e., store instructions). Among these instructions, we only concern about load and store instructions, as they are relevant to information leaking. In particular, the execution order and data dependencies of load and store instructions provide the basis for our profiling based attack. To generate and preserve one cipher-text byte, one load and one store instruction are required to read the table and write back the byte generated. To generate the entire 16 bytes block cipher-text, 16 load and 16 store instructions are executed alternately and hence the store instruction is dependent on the previous load instruction because it cannot be executed to store the cipher-text byte which has not yet been generated by executing the load instruction preceding it.

Second, the memory request coalescing which is considered as an effective method for GPU to reduce the number of memory accesses and enhance the

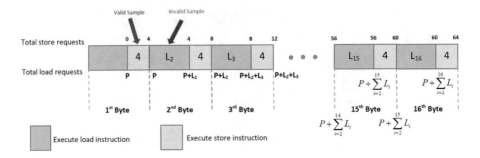

Fig. 4. The overview of the profiling strategy.

throughput can be a channel to leak key related information to attackers. In the GPU-based AES, a warp with 32 threads can process 32 16-byte blocks simultaneously and a table lookup initiating a memory load instruction generates 32 memory requests to fetch the data from different addresses within the memory space where the table is stored. The size of the table in the last round is 1024 bytes with 256 4-byte elements. A cache line of the L1 cache in the GPU is typically 64 bytes and therefore 32 memory requests generated by one load instruction can result in memory accesses varying from 1 (if all 32 threads request data locating in the same cache line) up to 16 (i.e., 16 consecutive cache lines totally are necessary to serve the entire table) after coalescing. The number of unique memory requests of one load instruction of 32 threads in a warp is highly dependent on the table indexes which can be calculated with cipher-text byte and corresponding round key using an inverse lookup table.

Third, we find that there are in total 66 memory store transactions when 32 threads in a warp process 32 16-byte blocks. Specifically, 2 of 66 memory store transactions are caused by system calls at the end of the program. The remaining 64 transactions are used to store $32*16$ bytes data for 32 threads and storing $32*1$ bytes costs 4 transactions. Therefore, the number of unique memory store request that can be profiled ranges from 0 to 66.

We use the CUDA Profiling Tools Interface (CUPTI) to implement our own profiling tool. The Event API of CUPIT allows to access multiple performance counters at the same time and therefore at one profiling point we can get two parameters (the number of the memory load requests and the number of the memory store requests) forming a pair. Due to the poor profiling rate, only one profiling point can be found for a single run of the kernel. The profiling point may happen anywhere and offers both valid and invalid information. To increase the profiling resolution, we keep sending the same plain-text to the victim program so that it is able to profile the same case under the same circumstance multiple times until the details to generate a sample are all captured.

Figure 4 describes the scenario to determine the validity of a profiling point. P denotes the total number of unique memory load requests before generating the second cipher-text byte in the last round. $\{L_2, L_3, ..., L_{16}\}$ denotes a sample composed of the numbers of unique load requests for $2nd$ to $16th$ load instructions in

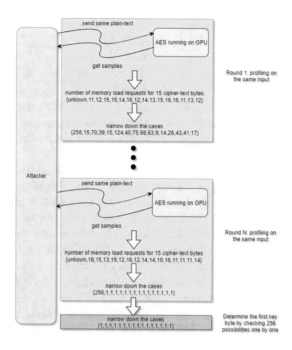

Fig. 5. Procedures to recover the entire 16-byte AES key

the last encryption round. The profiling point that happens during the execution of the store instruction (i.e., the gray regions) is valid, providing the total number of unique memory load requests of previous load instructions. In the gray regions, the number of memory store requests may change due to the ongoing store instruction. However, the number of unique memory load requests remain constant because there is no load instruction executing simultaneously. This is because the load instruction before the store instruction has already been finished and the next load instruction should not start before the completion of the current store instruction. For example, profiling in the first gray region pointed by the red arrow in Fig. 4 can get a performance counter pair (3200, 2), in which 3200 is the total number of unique memory load requests generated by the load instructions in previous encryption rounds and the first load instruction in the last round whereas 2 is the number of unique memory store requests initiated by the first store instruction. As aforementioned, each store instruction costs 4 transactions and therefore, if only 1, 2 or 3 of 4 have been completed, the store instruction is still under execution. In this case, P equals to 3200. To summarize, for a performance counter pair (num_load, num_store), if $num\_store = a*4+b$ and $b = 1, 2 \ or \ 3$, $P + \sum_{i=2}^{a+1} L_i = num\_load$. By comparison, the profiling point locating in the blue region is invalid since a load instruction is during execution. We then can filter the performance counter pair (num_load, num_store), if $num\_store = 4*a$. For two invalid performance counter pair, even they have the same num_store, num_load may vary according to the profiling timing during the execution of the same load instruction. By collecting all 16 valid numbers at the right profiling points (i.e., $\{P, P+L_2, P+L_2+L_3, ..., P+\sum_{i=2}^{15} L_i, P+\sum_{i=2}^{16} L_i\}$), the numbers of unique memory load requests for all 16 load instructions except for the first one then can be recovered by calculating the differences between consecutive numbers in $\{P, P+L_2, P+L_2+L_3, ..., P+\sum_{i=2}^{15} L_i, P+\sum_{i=2}^{16} L_i\}$. For example, $L_{16} = (P + \sum_{i=2}^{16} L_i) - (P + \sum_{i=2}^{15} L_i)$. However, the number of unique memory load requests for the first cipher-text byte cannot be calculated directly based on profiling because there is no store instruction preceding it as the boundary.

Figure 5 depicts the procedure to recover the entire 16-byte AES key. The exact number of unique memory load requests in the last round encryption for generating every cipher-text byte except the first byte then is already known by the attacker, based on which the corresponding AES key bytes can be recovered one by one. To recover one key byte, using an inverse lookup table, the attacker checks all 256 guesses of a single byte of the AES key by calculating and comparing to the profiled number. The number of possibilities of the corresponding key byte can be narrowed down. The number of possibilities can be further decreased to 1 case by repeating the same procedure with different input data for several times, which must be the true key byte. The profiling-based method can only recover the last 15 bytes of the AES key. However, the attack can reveal the first key byte by checking at most 256 cases with the same input if the later 15 key bytes are already known from profiling.

Table 1. Experimental results

	Tesla C2075	Quadro 2000
Key recovery done	Success: 100	Success: 100
Recovery success rate	100%	100%
Profiling done	Success: 676	Success: 656, Fail: 1
Profiling accuracy	100%	99.85%
The average profiling time	13814.99 ms	29837.02 ms
The average recovery time	93389.37 ms	196029.20 ms
The average inputs profiled	6.76	6.57

4   Experimental Results

We implement a python script to control and complete the AES key cracking procedure automatically and repeatedly which can also monitor the procedure and finally output the results. Using this script, we have done systematic experiments and verification. To simulate a attack as real as possible, we let the AES program keep running and waiting data from outside. The adversary program

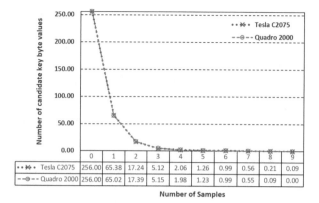

	0	1	2	3	4	5	6	7	8	9
Tesla C2075	256.00	65.38	17.24	5.12	2.06	1.26	0.99	0.56	0.21	0.09
Quadro 2000	256.00	65.02	17.39	5.15	1.98	1.23	0.99	0.55	0.09	0.00

Number of Samples

Fig. 6. The convergence speed of the number of possible key byte cases with the increasing number of samples.

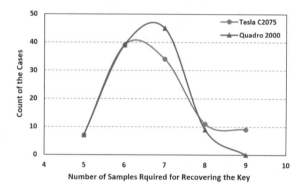

Fig. 7. The distribution of the number of samples required to recover an entire AES key.

keep sending data to AES program for encryption and meanwhile launches the GPU profiling tool to get samples. We have evaluated the profiling-based AES key recover on two GPU platforms: NVIDIA Quadro 2000 and NVIDIA Tesla C2075.

To test the accuracy of the profiling tool and success rate of key recovering, we recover the key 100 with different inputs. Table 1 shows the experimental results for both GPU cards. Both recovery success rate and profiling accuracy are 100% for NVIDIA Tesla C2075 while NVIDIA Quadro 2000 fails to get the correct number of memory load requests only once. However, the success on the rest of 656 inputs profiled results in 99.85% profiling accuracy. Although the profiling accuracy is not 100%, the recovery success rate can still achieve 100% since the inaccurate samples can always be filtered by the profiling tool noticing the abnormal pattern. The reason for the profiling failure on NVIDIA Quadro 2000 is that this GPU card serves for both general-purpose and graphics computing.

Consequently, the profiling tool may be misled by graphic computation tasks. This profiling failure can be easily avoided by prohibiting other tasks running on GPU when initiating the profiling based side-channel attack. NVIDIA Tesla C2075, which is dedicated to general-purpose computing, can always achieve 100% on both recovery success rate and profiling accuracy.

Table 2. Comparison of existing GPU side-channel attacks

SCA	Correlation	Number of samples
Profiling	No correlation analysis is required by this profiling-based SCA	**Less than 9** samples is required to shrink down 256 possibilities to a single case for all 16 key bytes values
Timing [6]	Execution time relies on the number of unique memory requests after coalescing	The collected samples contain 32-block encrypted messages and associated timings. The correlation coefficient which represents the strength of this relationship is calculated by a correlation analysis model for all 256 possible cases of a single key byte. The value with the maximum correlation coefficient is the right key. The necessary amount of the samples varies for different timing SCAs due to their differences on Signal to Noise Ratios (SNRs). In particular, SCAs in [6,7] and [8] need 10^6, 10^5 and 10^7 samples, respectively.
Timing [7]	Execution time relies on cache access conflicts decided by the accessed address of each thread	
Timing [8]	Execution time relies on shared memory conflicts decided by the accessed address of each thread	
Timing [9]	Power consumption depends on the Hamming distance which is calculated by the power model with the corresponding key value	With 10,000 power consumption traces (each contains the encryption of 49,152 blocks of plain-text), the correct key byte value stands out with the maximum correlation coefficient [9]. To extract the whole last round key, the same procedure needs to be repeated 16 times and **160,000** traces are necessary in total

Figure 6 shows the convergence speed of the number of possible key byte cases with respect to the increasing number of samples. Both GPU cards present the same convergence trend that the number of cases drops from 256 to around 65 after profiling one input and the convergence speed slows down afterward. Figure 7 presents the distribution of the number of samples required to recover the entire AES key for 100 times. In most cases, information retrieved from 6 or 7 samples is enough to recover all the key bytes on both GPU cards. On Tesla C2075, the largest number of samples is 9 for a complete key recovery, while it is 8 on Quadro 2000. The average number of samples is 6.76 and 6.57 for Tesla C2075 and Quadro 2000 respectively, which is much smaller than the amounts needed by existing GPU side-channel attacks [6–9] as shown in Table 2.

We also evaluate the performance of the proposed attack as the length of the AES key is increased to 192-bit and 256-bit as shown in Table 1. The extended AES key does not increase the number of samples for the key recovery; however, the time to profile one sample increases a little bit since it takes longer time to execute the encryption kernel due to the additional encryption rounds. The scalability of the profiling-based AES key recovery method is very good as the profiling time appear to increase linearly with respect to extending the length of the AES key (Table 3).

Table 3. Scalability with longer AES keys

	Profiling one sample	Average samples
Tesla 128-bit	13814.99 ms	6.76
Tesla 192-bit	15011.62 ms	6.68
Tesla 256-bit	16825.12 ms	6.71
Quadro 128-bit	29837.02 ms	6.57
Quadro 192-bit	32521.45 ms	6.6
Quadro 256-bit	34651.12 ms	6.57

5   Countermeasures

A straightforward way to completely hide the variation on the number of memory load requests and close the opportunity for the proposed attack to thieve the secure key is to disable the memory coalescing mechanism, which however, may result in substantial performance degradation. Kadam et al. [12] propose the Rcoal techniques which can offer the security at the cost of the tuned performance overhead. The Rcoal techniques randomize the coalescing logic such that it is hard for the attacker to guess the correct number of coalesced accesses. Specifically, the size of the subwarp and the threads assigned to each subwarp are randomized. Accordingly, they propose three coalescing randomization strategies, fixed sized subwarp (FSS), random-sized subwarp (RSS), and random threaded subwarp (RTS). Moreover, the combinations (FSS+RTS and RSS+RTS) are demonstrated to enable 24- to 961-times improvement in the security against the correlation timing attacks with 5 to 28% [12] performance degradation. However,

1. This technique is implemented by additional micro-architecture in GPUs leading to hardware overhead;
2. Moreover, since it is hardware-based, the randomized memory coalescing has to be effective for all load instructions and may consequently result in significant performance overhead;
3. Furthermore, the randomized coalescing technique can only improve GPU security instead of completely eliminating the threat of side-channel attacks.

Our future work will focus on developing more effective countermeasures that introduce as low as possible hardware and performance overheads and as well are able to offer flexible deployment and firm security.

6   Related Work

Side-channel attacks have recently been demonstrated to be feasible on GPUs for the AES key recovery [6–11]. Luo et al. [9] propose a power side-channel attack on GPU to crack AES key. Jiang et al. [6–8] proposed several GPU side-channel attacks based on correlation analysis. By observing that the kernel execution time is linearly proportional to the number of unique cache line requests generated during a kernel execution which is dependent on the input data and encryption key, [6] successfully establishes the relationship between the execution time of an encryption kernel and the encryption key. [7,8] implement side-channel timing attacks using similar scenarios with [6], which however is relying on different relationships describing the time of execution differentiated by the number of L1 cache bank conflicts [7] and shared memory bank conflicts [8] respectively. However, all these methods require much more samples and time than the profiling-based SCA.

7   Conclusion

In this paper, we propose a novel profiling-based side-channel attack on GPUs which leaks critical information to an adversary to fully recover the encryption secret key. The profiling-based side-channel attack accomplishes the high resolution profiling and is well scalable. The number of samples for recovering the key is dramatically reduced and much smaller than all the existing GPU side-channel attacks. We demonstrate this profiling-based side-channel attack on two Nvidia GPUs to recover 16-byte AES keys in as short as 30 s with very high accuracy (approaching 100%).

References

1. Biagio, A.D., Barenghi, A., Agosta, G., Pelosi, G.: Design of a parallel AES for graphics hardware using the CUDA framework. In: Proceedings of the 2009 IEEE International Symposium on Parallel and Distributed Processing, pp. 1–8. IEEE Computer Society (2009)
2. Cohen, A.E., Parhi, K.K.: GPU accelerated elliptic curve cryptography in GF(2m). In: 2010 53rd IEEE International Midwest Symposium on Circuits and Systems (MWSCAS), pp. 57–60. IEEE (2010)
3. Iwai, K., Kurokawa, T., Nisikawa, N.: AES encryption implementation on CUDA GPU and its analysis. In: 2010 First International Conference on Networking and Computing, pp. 209–214. IEEE (2010)
4. Le, D., Chang, J., Gou, X., Zhang, A., Lu, C.: Parallel AES algorithm for fast data encryption on GPU. In: 2010 2nd International Conference on Computer Engineering and Technology (ICCET), vol. 6, p. V6-1. IEEE (2010)
5. Manavski, S.A., et al.: CUDA compatible GPU as an efficient hardware accelerator for AES cryptography. In: Signal Processing and Communications 2007 (2007)
6. Jiang, Z.H., Fei, Y., Kaeli, D.: A complete key recovery timing attack on a GPU. In: 2016 IEEE International Symposium on High Performance Computer Architecture (HPCA), pp. 394–405. IEEE (2016)
7. Jiang, Z.H., Fei, Y.: A novel cache bank timing attack. In: Proceedings of the 36th International Conference on Computer-Aided Design, pp. 139–146. IEEE Press (2017)
8. Jiang, Z.H., Fei, Y., Kaeli, D.: A novel side-channel timing attack on GPUs. In: Proceedings of the Great Lakes Symposium on VLSI 2017, pp. 167–172. ACM (2017)
9. Luo, C., et al.: Power analysis attack of an AES GPU implementation. J. Hardw. Syst. Secur. **2**(1), 69–82 (2018)
10. Luo, C., Fei, Y., Luo, P., Mukherjee, S., Kaeli, D.: Side-channel power analysis of a GPU AES implementation. In: 2015 33rd IEEE International Conference on Computer Design (ICCD), pp. 281–288. IEEE (2015)
11. Gao, Y., Cheng, W., Zhang, H., Zhou, Y.: Cache-collision attacks on GPU-based AES implementation with electro-magnetic leakages. In: 2018 17th IEEE International Conference on Trust, Security and Privacy in Computing and Communications/12th IEEE International Conference on Big Data Science and Engineering (TrustCom/BigDataSE), pp. 300–306. IEEE (2018)
12. Kadam, G., Zhang, D., Jog, A.: RCoal: mitigating GPU timing attack via subwarp-based randomized coalescing techniques. In: 2018 IEEE International Symposium on High Performance Computer Architecture (HPCA), pp. 156–167. IEEE (2018)
13. Naghibijouybari, H., Neupane, A., Qian, Z., Abu-Ghazaleh, N.: Rendered insecure: GPU side channel attacks are practical. In: Proceedings of the 2018 ACM SIGSAC Conference on Computer and Communications Security, pp. 2139–2153. ACM (2018)
14. Nvidia, C.: Programming Guide (2008)
15. Nvidia, C.: Nvidia's next generation CUDA compute architecture: Fermi. Comput. Syst. **26**, 63–72 (2009)

Enforcing Secure Coding Rules
for the C Programming Language Using
the Eclipse Development Environment

Victor Melnik[1], Jun Dai[1(✉)], Cui Zhang[1], and Benjamin White[1,2]

[1] Computer Science, California State University, Sacramento, CA 95819, USA
jun.dai@csus.edu
[2] Mother Lode Holding Company, Roseville, CA 95747, USA

Abstract. Creating secure software is challenging, but necessary due to the prevalence of large data breaches that have occurred for organizations such as Equifax, Uber, and U.S. Securities and Exchange Commission. Many static analysis tools are available that can identify vulnerable code, however many are proprietary, do not disclose their rule set or do not integrate with development environments. One open source tool that integrates well with the Eclipse development environment is the *Secure Coding Assistant* that was developed at California State University, Sacramento (CSUS), which is featured by early error detection. The tool provides support for secure coding rules for the Java programming language that were developed at the CERT division of the Software Engineering Institute at Carnegie Mellon University. The tool also provides error correction and contract programming support. To provide secure coding assistance in C programming, we further extend the tool to support the C programming language by semi-automating a subset of the CERT secure coding rules for C. The tool detects rule violations for the C programming language in the Eclipse development environment and provides feedback to aid and educate developers in secure coding practices. The tool is *open source* to the community and maintained at *GitHub* (http://benw408701.github.io/SecureCodingAssistant/).

Keywords: Secure coding · Software security · C programming

1 Introduction

Developing software using secure coding practices is becoming increasingly important as the frequency and severity of data breaches continue to rise. According to the Identity Theft Resource Center, 2017 set a record of the highest number of data breaches in the United States of America, with an increase of 44.7% compared to the previous year [1]. In 2017 the world also observed some of the largest data breaches to date. For instance, in the beginning of 2017, Uber disclosed that 57 million Uber users and driver's information was stolen, which included "names, email addresses, phone numbers, driver's license numbers", and other personal information [2]. Later that year the largest data breach to date occurred at Equifax, a consumer credit reporting agency. Hackers were able to steal "145.5 million records containing social security numbers, names, addresses, credit card numbers and other personal information" [7]. Lastly, the

© Springer Nature Switzerland AG 2020
K.-K. R. Choo et al. (Eds.): NCS 2019, AISC 1055, pp. 140–152, 2020.
https://doi.org/10.1007/978-3-030-31239-8_12

U.S. Securities and Exchange Commission's Electronic Data Gathering, Analysis, and Retrieval (EDGAR) system was infiltrated and information regarding mergers, acquisitions and other company data was exfiltrated [3]. The severity of this data breach is difficult to assess, because the data retrieved could be used in the future to make millions to billions of dollars for criminal organizations. Many of these attacks could have been mitigated or prevented if the organizations enforced more stringent coding practices.

There are many vulnerabilities that are reported and published on the Common Vulnerability Enumeration (CVE) website. It would take a good deal of effort to keep up with ever newly published vulnerability. In 2017 alone, 14,712 CVEs were published [12]. This was an unprecedented spike in code vulnerabilities compared to 2016, where only 6,447 CVEs were published [12]. According to IEEE Senior Member Gary McGraw, "there has been too much focus on common bugs and not enough on secure design and avoidance of flaws" [13].

To stay ahead of the curve of newly published vulnerabilities, various tools were developed to provide code weakness detection and secure coding assistance. Our tool named *Secure Coding Assistant* is one of these efforts, which is open source and implements the CERT secure coding rules for Java programming language [7, 18, 19]. It is a static analysis tool that was developed in 2016 [18, 19] and later enhance in 2017 at [8]. The tool, featured by early detection, provides support for the CERT secure coding rules for the Java language. It also provides error correction and contract programming for the Java language. The rules were developed at the CERT division of the Software Engineering Institute at Carnegie Mellon University. By enforcing the rules throughout coding, newly developed software can avoid common security pitfalls.

This paper is focused on the enhancement of the tool by semi-automating the secure coding rules for C programming language. To achieve this goal, a subset of the CERT secure coding rules for C will be carefully selected and implemented. Specifically, the tool will flag unsecure code segments similar to problem markers generated during the compilation process. These markers will provide the developer with the name of the violated rule and information on how to remediate the vulnerable code. These problem markers will help educate software developers on secure coding principles.

Throughout this paper, the enhancement to provide support for the C language to the Secure Coding Assistant will be referred to as the *Secure Coding Assistant for C. Secure Coding Assistant for Java* will be used to refer to the original software that was developed for the Java language. The Secure Coding Assistant for C and Secure Coding Assistant for Java are integrated as part of the same tool but are mutually exclusive components within the tool, due to their inherent difference in programming language.

2 Related Work

There are currently many static analysis tools that are available to aid developers in making secure software. Table 1 provides a list of some of these available tools. The first five are commercial tools while the rest are open source ones.

All the tools that are closed source do not disclose the rule set or the methodologies that are used to detect vulnerabilities in the developer's source code. The first four open source tools, scan source code for vulnerabilities but do not disclose which rule set the tool is based on. Also, two of these open source tools have not been updated for a few years. *VisualCodeGrepper* has not been updated in the past two years, while *PreFast* has not been updated since 2005. The tool that is most closely related to our tool in *Flawfinder*. *Flawfinder* is an open source tool that is available for download on GitHub. *Flawfinder* is based on the Common Weaknesses Enumeration (CWE) database and detects vulnerable code segments by matching code against a database of C/C++ functions with known problems. Unlike *Flawfinder*, Secure Coding Assistant is based on an established secure coding rule set and does not rely on new vulnerabilities to be published to update the tool. Secure Coding Assistant will be maintained and further developed by the Department of Computer Science at CSUS.

Table 1. Current secure code analysis tools.

Company	Tools	Rule Set	Open/Closed
Synopsys	Coverity Static Analysis Tool	Proprietary	Closed
Veracode	Static Analysis SAST	Proprietary	Closed
Rouge Wave Software	KlocWork	Proprietary	Closed
Viva64	PVS-Studio Analyzer	Proprietary	Closed
Micro Focus	Fortify Static Code Analyzer	Proprietary	Closed
Microsoft	PreFast	Custom	Open
NCC Group	Visual Code Grepper	Custom	Open
Michael Scovetta	Yasca	Custom	Open
Daniel Marjamäki	CPPCheck	Custom	Open
David Wheeler	Flawfinder	CWE	Open

3 Design

3.1 Goals

There are two goals that are expected by enhancing the Secure Coding Assistant. The first goal is to provide developers with feedback when compiling their source code. This will be similar to warnings and error problem reports that are generated during the compilation process. This feedback will allow developers to mitigate security vulnerabilities during the development of their software.

The second goal is to educate developers on secure coding practices for the C language. This goal will be accomplished by providing developers with problem alerts that provide a clear message that specifies the violated rule and guideline on how to remediate the unsecure code segment. These two implemented goals will create a learning environment that will educate software developers on the secure coding practices for the C language.

3.2   Architecture

The Secure Coding Assistant for C runs when the *build* command in Eclipse is called. The *build* command is used to compile all the C source code files within an open project. Eclipse refers to source code files that are inputted into a compiler as translation units. As the *build* command runs, all the nodes in the translation unit are analyzed to determine if any rules are violated. Figure 1 shows the high-level flow on the overall design for the Secure Coding Assistant for C. When the *build* command is called all the pre-existing markers in the source code are cleared, and the first node within the first translation unit is visited. If a rule is violated in the node, a *marker* is generated with the name of the rule violated and its remediation information. Then the next node in the translation unit is visited. This process continues until all the nodes in the translation unit have been visited and analyzed. If there are more translation units in that need to be compiled, the next translation unit is visited, and all its node are subsequently analyzed. Once all the translation units within the project are visited and analyzed, the Secure Coding Assistant for C displays all the *markers* that have been created during the *build* processes. The Secure Coding Assistant for C will run and display all the problem markers in the project's translation units, even if the *build* fails to compile the project successfully.

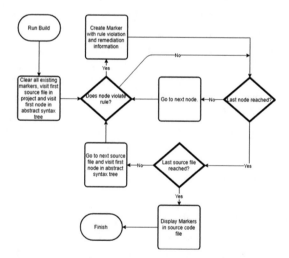

Fig. 1.   Secure coding assistant for C high-level flow chart.

4   Implementation

The idea to use the Eclipse Development Environment as the common platform decides that the Secure Coding Assistant for C and the Secure Coding Assistant for Java could share methodologies for implementation. The difference between the two analyzers is mainly that they utilize a different Eclipse tooling library. Specifically, the Secure Coding Assistant for C utilizes the Eclipse C/C++ Development Tooling (CDT) library, while the Secure Coding Assistant for Java utilizes the Eclipse Java Development Tooling (JDT) library.

4.1   Rule Selection

The CERT secure coding standard provides a total of 120 rules for C which are divided into 17 specific categories. To determine which rules are to be incorporated into the Secure Coding Assistant for C, the rules are first divided into two categories: rule that could be automated and rules that could not be automated. For a majority of the C secure coding rules, the CERT website provides information on whether the rule can be automated or not.

An example of a rule that could not be automated is the *FIO32-C* rule, which states to not perform file operations on devices that are only appropriate for files [15]. In the UNIX and Windows operating systems, special files are used to represent devices. To determine if this rule was violated, the tool would require a mechanism of identifying each file as it was inputted into a file operation function. Since this information could only be gathered during runtime, this rule could not be automated in a static analysis tool.

Additionally, the CERT secure coding standard for C contained three rule categories that did not contain any rules that could be automated. One of these rule categories is the *Preprocessor* category. The *Preprocessor* rule category could not be automated due to the limitation of the Eclipse CDT library. The library did not provide a method to analyze preprocessor code segments in a translation unit. This limitation prevented the tool from being able to automate any of the rules within this rule category.

From the 120 CERT rules for C, 38 were determined to be automatable. From the 38 rules that were determined to be automatable 20 rules were selected to be automated in the tool. The 20 rules that were selected for this tool were determined based on their severity, and the likelihood that the rule violation would occur. The CERT website provided the classification for each rule. Additionally, rules were also selected to represent all the 17 rule categories that did contain automatable rules.

4.2   Plugin Implementation

To develop the Secure Coding Assistant for C, the Eclipse Plugin Development Environment (PDE) was utilized. The Eclipse PDE provides developers with extension points that can be used to improve and customize the existing development environment. Extension points are a combination of XML mark-up language and a Java interface, that allow for one plugin to extend and customize the functionality of another plugin [4].

The Secure Coding Assistant for C extends one extension point. The extension point is *org.eclipse.cdt.core.ErrorParser*. This extension point allows the plugin to fulfil two functions. First, it allows the plugin to interact with the C *build* process. *Build* is used to compile and link the source files in an open project. Second, it allows for the generation of problem markers. Problem markers are used to mark the segment of code that contains a rule violation and provide a tool-tip that contains information on the violated rule and how to remediate the unsecure code.

4.3   Abstract Syntax Tree

Each translation unit in a C project is represented as an *Abstract Syntax Tree* (AST). An AST is a tree model that is used to represent the structure of a programming language's source code file. An AST can be traversed depth-first from top to bottom or bottom to top.

The Eclipse CDT library provides a mechanism to examine the AST through the *org.eclipse.cdt.core.dom.ast* package. To traverse the AST, the *org.eclipse.cdt.core.-dom.ast* package provides the class *ASTVisitor*. *ASTVisitor* provides a *visit()* method for each of the different types of nodes (variable declaration, expression statement, function parameters, etc.). The *visit()* method allows for each node within a translation unit to be visited and examined.

The Secure Coding Assistant for C has two classes that extend the *ASTVisitor* class: *SecureCodingNodeVisitor_C* and *ASTNodeProcessor*. *SecureCodingNodeVisitor_C* class is used to access the AST during the build process. *ASTNodeProcessor* class is used by the *Utility_C* library to aid in the detection of rule violations.

4.4   Rule Detection

The Secure Coding Assistant for C uses two Java classes to fulfil the task of detecting rule violations: *ASTNodeProcessor_C*, and *Utility_C*.

ASTNodeProcessor_C is at the heart of rule detection. *ASTNodeProcessor_C* traverses the AST of a translation unit a second time and creates collections of various node types such as variable declarations, function definitions, assignment statements, etc. *ASTNodeProcessor_C* also assigns a numerical value to each node to keep track of the order in which the nodes appear in the source code. These collections of nodes allowed for easy retrieval of nodes that were called before and after the node being currently analyzed.

Table 2. *Utility_C* library.

Utility	Method
Get scope of node	getScope(IASTNode)
Determine if inner node is contained within outer node	isEmbedded(IASTNode, IASTNode)
Get list of all variables in the same scope as the node	allVarNameType()
Get list of function call parameter	getFunctionParamaterVarName()
Get list of function call parameters for printf functions	getFunctionParameterVarNamePrintf()

Utility_C library is a collection of methods that are used by more than one rule. Since many of the CERT rules share common rule detection logic, *Utility_C* library was used to simplify the logic for each rule. This library created a list of methods that could be used by future developers to expand the tool. The list of methods in the *Utility_C*, along with the purpose they serve is show in Table 2. The *Utility_C* library was expanded during the development of the Secure Coding Assistant for C tool. A new method was added when more than one rule was determined to share similar rule detection logic. Using both the *ASTNodeProcessor_C* class and the *Utility_C* library simplified the rule logic for each rule and allows for code reusability.

4.5   Rule Interface

Each rule implements the *SecureCodingRule_C* interface. The interface provides methods for detecting a rule violation and for provide feedback to the user of the tool. Table 3 provides the methods contained in the *SecureCodingRule_C* interface.

Table 3. *SecureCodingRule_C* interface [18].

Method signature	Description
Boolean violated_CDT(IASTNode)	Checks to see if a rule has been violated for a node
String getRuleText()	The description of the violated rule
String getRuleName()	The description of the violated rule
String getRuleID()	The ID of the violated rule
String getRuleRecommendation()	Suggestions to remediate the insecure node
Int securityLevel()	The security level of the violated rule: HIGH, MEDIUM, LOW
String getRuleURL()	The URL to the rule on the CERT website

This interface is borrowed from the Secure Coding Assistant for Java developed by [18, 19]. However, since both tools use different Eclipse development libraries, the *SecureCodingRule_C.violated()* function is modified to accommodate the difference.

The *SecureCodingRule_C.violated()* method takes one parameter, i.e. the node that is currently being processed by the *SecureCodingNodeVisitor_C*. The node is analyzed by the method and returns true if the rule has been violated. This method made the code required for running each rule against all the nodes in a translation unit simple. Figure 2 displays the rule traversal logic used in *SecureCodingNodeVisitor_C*.

5   Evaluation

5.1   Accuracy

5.1.1   CERT Validation

The CERT website provides a list of example code as well as the definitions for each of the CERT rules. Each rule contains a pair of code samples: one with a rule violation and one with the rule violation remediated. Some of the rules contained more than one pair of code examples. To initially develop the Secure Coding Assistant for C, the tool focused on detecting the rule violation in the unsecure code segments. It also made sure that any false positives were remediated during this process. Once the Secure Coding Assistant for C was able to detect all the rule violation in the CERT's rule sample code, the rule logic was considered to be complete.

5.1.2   False Positive

```
public void traverseRule(IASTNode checkNode)
{
    for (IRule_C rule : c_rules)
    {
        if(rule.violate_CDT(checkNode))
        {
            Globals.insecureGlobalNode = checkNode;
            Globals.cdt_InsecureCodeSegments.add(
                    new InsecureCodeSegment_C(checkNode,rule, localITU));
        }
    }
}
```

Fig. 2. Rule detection logic in *SecureCodingNodeVisitor_C*.

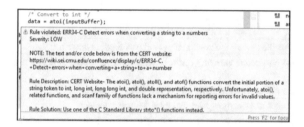

Fig. 3. *ERR34-C* rule violation from Juliet Test Suite for C/C++ detected by Secure Coding Assistant [11].

The *Juliet Test Suite* for C/C++ developed by the NSA Center for Assured Software was used to conduct a false positive study [11]. This test suite consists of 64,099 C/C++ source code files which are categorized under 118 different CWEs. Each source code file contains an unsecure code example paired with a secure code correction. The authors of the files provide comments within each file to identify the code segments that contain weaknesses. Many of the weaknesses that were documented in the Juliet Test Suite for C/C++ were not detected by the Secure Coding Assistant for C because most CWEs do not directly translate over to any CERT rules. For example, CERT does not include any rules for code weaknesses such as unchecked return values or unreachable code segments.

The Secure Coding Assistant for C generated 11,021 secure coding warning which are shown in Table 4. Ten of the 20 rules that were implemented in the tool detected rule violations. The top two rules that were detected are the *ERR34-C* and *MEM31-C* rules, which collectively account for 68% of all the rule violations. The *ERR34-C* rule states to detect errors when converting strings to a number [5]. This rule detects rule violations when using *string* to *integer* conversion functions that lack error reporting mechanism such as *atoi*, *atoll*, and *atoll* [5]. Figure 3 shows an example of a rule violation for the *ERR34-C* rule with its accompanied problem alert window. The rule *MEM31-C* states that dynamically allocated memory should be freed once it is no longer needed by the program [16]. This rule was detected, since many CWEs are associated with memory leakage and corrupt memory pointers.

Table 4. Juliet Test Suite for C/C++ results.

Level	Rule name	Total	Percent
L3	ERR34-C. Detect errors when converting a string to a number	3784	34.3%
L2	MEM31-C. Free dynamically allocated memory when no longer needed	3750	34.0%
L2	INT33-C. Ensure that division and remainder operations do not result in divide-by-zero errors	2010	18.2%
L2	MSC30-C. Do not use the rand() function for generating pseudorandom numbers	812	7.4%
L1	ENV33-C. Do not call system()	557	5.0%
L2	ARR36-C. Do not subtract or compare two pointers that do not refer to the same array	36	0.3%
L2	FIO45-C. Avoid TOCTOU race conditions while accessing files	36	0.3%
L1	SIG30-C. Call only asynchronous-safe functions within signal handlers	18	0.2%
L2	SIG31-C. Do not access shared objects in signal handlers	18	0.2%
L2	FIO47-C. Use valid format strings	18	0.2%
L2	DCL36-C. Do not declare an identifier with conflicting linkage classifications	0	0.0%
L3	DCL38-C. Use the correct syntax when declaring a flexible array member	0	0.0%
L3	DCL41-C. Do not declare variables inside a switch statement before the first case label	0	0.0%
L2	EXP32-C. Do not access a volatile object through a nonvolatile reference	0	0.0%
L2	FLP30-C. Do not use floating-point variables as loop counters	0	0.0%
L2	STR34-C. Cast characters to unsigned char before converting to larger integer sizes	0	0.0%
L3	STR37-C. Arguments to character-handling functions must be representable as an unsigned char	0	0.0%
L1	STR38-C. Do not confuse narrow and wide character strings and functions	0	0.0%
L2	FIO47-C. Use valid format strings	0	0.0%
L2	CON40-C. Do not refer to an atomic variable twice in an expression	0	0.0%
L2	POS33-C. Do not use vfork()	0	0.0%
	Total	11,021	

Each rule detection in Table 4 was manually inspected to determine if the alert was a true positive or false positive. Table 5 displays the false positives that were identified. False positives accounted for 25% of all of the rule detections. Only two rules were determined to have false positive detections: the *INT33-C* and the *MSC30-C* rules.

The highest false positive result was attributed to the *INT33-C* rule. This rule states that "division and modular operations should not result in a divide-by-zero error" [14]. These false positives stem from floating point division, where a conditional statement checks to see if the divisor is greater than the value of .00001 before performing division. The rule logic in the tool is structured to check if the divisor is greater than zero, greater than or equal to one, or not equal zero. It would be difficult to account for the different variations of conditional statements that can be satisfied to check if a floating-point number is not equal to zero. This makes avoiding false positives for this rule difficult. This rule highlights that the rule detection logic for this rule should be revisited.

The second highest false positive result is attributed to the *MSC30-C* rule. This rule states to not use the function *rand()* to generate pseudorandom numbers for application that have a strong pseudorandom number requirement [9]. The false positive results found were in source files that were using *rand()* for purposes that did not need strong pseudorandom values. It would be difficult to fix the false positives that were generated by this rule, because it requires context into how these random number will be used in an application. Future release of the Secure Coding Assistant for C could provide the option to hide a secure coding rule violation if there is disagreement with the tool. This would help minimize the number of false positive detections.

Table 5. False positive results.

Rule	Total count	True Pos. count	True Pos. (%)	False Pos. count	False Pos. (%)
INT33	2,010	1,519	75.57	491	24.43
MSC30	812	505	62.19	307	37.81
Total	2,822	2,024	71.72	798	28.28

5.1.3 False Negative

To conduct a false negative study on the Secure Coding Assistant for C, the Juliet Test Suite for C/C++ [11] and the CWE website database [10] were used. These sources were used because they contained code segments that provided documented vulnerabilities. The false negative study was done by looking through both sources and determining if the documented vulnerability should have been picked up by the tool. The tool failed to detect rule violations for the *FIO45-C* and *STR34-C* rules.

The false negative instance for the *FIO45-C* rule was found in the Juliet Test Suite for C/C++. The *FIO45-C* rule states that a TOCTOU (time-of-check, time-of-use) race conditions should be avoided when more than one concurrent process is operating on a shared file system [17]. The code segment that should have been picked up by the tool is shown in Fig. 4. The Secure Coding Assistant for C did not flag this code segment as a vulnerability because the *#define* preprocessor directive was used to rename the file operations *stat* and *open* to *STAT* and *OPEN*, respectively.

```
if (STAT(filename, &statBuffer) == -1)
{
    exit(1);
}
fileDesc = OPEN(filename, O_RDWR);
if (fileDesc == -1)
{
    exit(1);
}
```

Fig. 4. Code segment from Juliet Test Suite for C/C++ [11].

The false negative instance for the *STR34-C* rule was discovered on the CWE website under *CWE-843*: Access of Resource Using Incompatible Type [10]. *CWE-843* does not relate to the CERT rule *STR34-C*, however the CWE code example contained a segment of code that violated the *STR34-C* rule. The *STR34-C* rule states that *char* should be cast to an *unsigned char* before converting the value to a larger integer size [6]. Figure 5 displays the code segment from *CWE-843* that should have been detected as a rule violation under the *STR34-C* rule. The character variable *defaultMessage* is cast to the integer *buf.nameID* without first casting the *char* to an *unsigned char*. Custom code was written to identify the variable being accessed using the member access operator for variables declared within complex data structures such as *union* and *struct*. This code was written since the Eclipse CDT library lacked this mechanism. The logic failed to consider a complex data structure being nested within another complex data structure. This case was not considered because none of the CERT examples provided code segments where this case occurred. This is a limitation of the tool that will be addresses in future developments.

```
#define NAME_TYPE 1

struct MessageBuffer
{
int msgType;
union {
char *name;
int nameID;

};

};

int main (int argc, char **argv) {
struct MessageBuffer buf;
char *defaultMessage = "Hello World";

buf.msgType = NAME_TYPE;
buf.name = defaultMessage;
printf("Pointer of buf.name is %p\n", buf.name);

buf.nameID = (int)(defaultMessage + 1);
printf("Pointer of buf.name is now %p\n", buf.name);
}

}
```

Fig. 5. *CWE-843* code segment from CWE website [10].

5.2 Accuracy

The tool's efficiency was measured by running the *build* command against test suites from [11] and test files that were generate from the CERT website examples to initially test this tool. Each project was built 3 times with and without the tool enabled to gather the average build time. After each build, the *clean* command was called to delete all the generate binaries. The Secure Coding Assistant for C efficiency result are shown in Table 6. The second to last column in Table 6 shows the increase in time to build the binaries for a project. The time it takes to build a project appears to be correlated with the number of files in a project, as well as the number of detected violations. There is an average 4.45% increase in *build* time with the tool enabled.

Table 6. Efficiency test results.

Project	Files	Alerts	Time increase (s)	Increase (%)
CERT	20	50	1.21	5.74
Test 45	66	13	3.48	15.84
Test 46	64	18	4.75	21.65
Test 101	58	29	5.42	9.04
Test 106	247	113	14.71	12.44

6   Limitations, Conclusion and Future Work

The enhancement to the Secure Coding Assistant for C programming language has proven to be pragmatic, efficient and accurate. The future developments will focus on improving the efficiency of the tool by fine tuning the rule logic and by minimizing the false positive and false negative rates. There will also be a focus on adding additional features such as providing the user the ability to hide problem markers if they disagree with the tool and by providing support for the C++ language. Additionally, the rest for the CERT rules for C that were identified as automatable will be implemented.

There are many static analysis tools that provide secure code analysis that are available for developers. However, none of these tools implement the CERT secure coding rules for the C programming language. This paper provides C programmers with an educational development tool that enforce secure coding standards. This tool is open source and will continue to be maintained by the Department of Computer Science at CSUS. The tool is available on the project website at GitHub (http://benw408701.github.io/SecureCodingAssistant/).

This project was conducted when Victor Melnik was a student in MS Computer Science program at California State University, Sacramento. More implementation details can be found in his Master Project Report [20], that is an extended version of this paper.

Acknowledgements. Acknowledgements and attributions are given to Carnegie Mellon University and its Software Engineering Institute, as this publication incorporates portions of the "SEI CERT C Coding Standard" (c) 2017 Carnegie Mellon University, with special permission from its Software Engineering Institute. Any material of Carnegie Mellon University and/or its software engineering institute contained herein is furnished on an "as-is" basis. Carnegie Mellon University makes no warranties of any kind, either expressed or implied, as to any matter including, but not limited to, warranty of fitness for purpose or merchantability, exclusivity, or results obtained from use of the material, Carnegie Mellon University does not make any warranty of any kind with respect to freedom from patent, trademark, or copyright infringement. This publication has not been reviewed nor is it endorsed by Carnegie Mellon University or its Software Engineering Institute. CERT and CERT Coordination Center are registered trademarks of Carnegie Mellon University.

References

1. 2017 Annual Data Breach Year-End Review (2018). https://www.idtheftcenter.org/images/breach/2017Breaches/2017AnnualDataBreachYearEndReview.pdf. Accessed 27 Feb 2019
2. Bearak, S.: Uber Data Breach Affects 57 Million: It is Time to Own Our Identities (2017). https://www.identityforce.com/business-blog/ubers-data-breach-affects-57-million-its-time-to-own-our-identities. Accessed 27 Feb 2019
3. Cimpanu, C.: SEC Says Hackers Breached its System, Might Have Stolen Data for Insider Trading (2017). https://www.bleepingcomputer.com/news/security/sec-says-hackers-breached-its-system-might-have-used-stolen-data-for-insider-trading/. Accessed 27 Feb 2019
4. Eclipse: Extensions and Extension Points (2018). http://help.eclipse.org/luna/index.jsp?topic=%2Forg.eclipse.pde.doc.user%2Fconcepts%2Fextension.htm. Accessed 27 Feb 2019
5. Hicken, A.: ERR34-C. Detect errors when converting a string to a number (2018). https://wiki.sei.cmu.edu. Accessed 27 Feb 2019
6. Hicken, A., Seacord, R.: STR34-C. Cast characters to unsigned char before converting to larger integer sizes (2018). https://wiki.sei.cmu.edu. Accessed 27 Feb 2019
7. Leary, J.: Equifax Breach Impacts 147.9 Million: Steps to Keep Your Identity Protected (2018). https://www.identityforce.com/business-blog/equifax-breach-impacts-143-million-steps-to-keep-your-identity-protected. Accessed 27 Feb 2019
8. Li, C., White, B., Dai, J., Zhang, C.: Enhancing secure coding assistant with error correction and contract programming. In: Proceeding of National Cyber Summit 2017, Huntsville, AL, 6–8 June 2017 (2017)
9. Long, F., Hicken, A.: MSC30-C. Do not use the rand() function for generating pseudorandom numbers (2018). https://wiki.sei.cmu.edu. Accessed 27 Feb 2019
10. MITRE: CWE-843: Access of Resource Using Incompatible Type ('Type Confusion'). Common Weakness Enumeration (2018)
11. NIST: Test Suites, 4.9 (2017). NIST Samate: https://samate.nist.gov/SARD/testsuite.php. Accessed 27 Feb 2019
12. Ozkan, S.: Browse Vulnerabilities by Date (2018). https://www.cvedetails.com/browse-by-date.php. Accessed 27 Feb 2019
13. Pretz, K.: 10 Recommendation for Avoiding Software Security Design Flaws (2014). http://theinstitute.ieee.org/special-reports/special-reports/10-recommendations-for-avoiding-software-security-design-flaws. Accessed 27 Feb 2019
14. Razmyslov, S.: INT33-C. Ensure that division and remainder operations do not result in divide-by-zero errors (2018). https://wiki.sei.cmu.edu. Accessed 27 Feb 2019
15. Seacord, R., Flynn, L.: FIO32-C. Do not perform operations on devices that are only appropriate for files (2018). https://wiki.sei.cmu.edu. Accessed 27 Feb 2019
16. Seacord, R., Hicken, A.: MEM31-C. Free dynamically allocated memory when no longer needed (2018). https://wiki.sei.cmu.edu. Accessed 27 Feb 2019
17. Svoboda, D., Snavely, W.: FIO45-C. Avoid TOCTOU race conditions while accessing files (2017). https://wiki.sei.cmu.edu. Accessed 27 Feb 2019
18. White, B., Dai, J., Zhang, C.: Secure coding assistant: enforcing secure coding practices using the eclipse development environment. In: Proceeding of National Cyber Summit 2016, Huntsville, AL, 8–9 June 2016 (2016)
19. White, B., Dai, J., Zhang, C.: An early detection tool in eclipse to enforce secure coding practices. Int. J. Inf. Priv. Secur. Integr. (IJIPSI) 3, 284–309 (2018)
20. Melnik, V.V.: Enhancing secure coding assistant: enforcing secure coding rules for C programming language. Master report at California State University, Sacramento (2018)

A Specification-Based Intrusion Prevention System for Malicious Payloads

Aaron Werth$^{(\boxtimes)}$ and Thomas H. Morris

University of Alabama in Huntsville, Huntsville, USA
aww0001@uah.edu.com, tommy.morris@uah.edu

Abstract. In this work, a control/command analysis-based intrusion prevention system (IPS) is proposed. This IPS will examine incoming command packets and programs that are destined for a PLC interacting with a physical process. The IPS consists of a module that examines the packets that would alter settings or actuators and incorporates a model of the physical process to aid in predicting the effect of processing the command and specifically whether a safety violation would occur for critical variables in the physical system. Essentially, a simulation of both the model of the physical system and a process running a copy of the ladder logic of the real PLC is performed in the module. Also, uploaded programs will be evaluated to determine whether the programs would cause a safety violation. Previous research has studied making predictions based on the payloads of packets where cumbersome specifications must be developed by a human expert for the model of the physical system and safety conditions. This work seeks to eliminate or minimize the amount of specifications to be developed by a human through system identification and machine learning to allow the IPS to be more generic and deployable. Another contribution of this work is a broader and more generic understanding of the threat model that causes unsafe or inefficient consequences. The accuracy in prediction and latency in analysis are metrics used when evaluating the results in this work.

Keywords: SCADA · Cybersecurity · Industrial control systems · Intrusion detection · Semantic-based · Specification-based · Malware

1 Introduction

In recent times, highly notable cyberattacks have targeted critical infrastructure with the goal of affecting physical processes to cause harm. These processes are under the control of Programmable Logic Controllers (PLCs) and are managed and monitored by an overall Supervisory Control and Data Acquisition (SCADA) system. Specific examples of these attacks include Stuxnet [28], the attack on the water treatment plant in Maroochi [19], and the attack on the Ukrainian power system [29]. These attacks have caused destruction, inefficiencies, and sabotage of cyber-physical systems, a generic term encompassing SCADA systems and emphasizing computer (cyber) systems that have an impact in the physical world [27]. Such cyber-attacks have come to the attention of governments and research organizations, who seek to better understand and address these threats to protect SCADA systems, which manage critical

© Springer Nature Switzerland AG 2020
K.-K. R. Choo et al. (Eds.): NCS 2019, AISC 1055, pp. 153–168, 2020.
https://doi.org/10.1007/978-3-030-31239-8_13

infrastructure. If these systems fail, potentially catastrophic consequences may result to society. SCADA systems consist of a network and devices, such as PLCs, servers, and other computers which make up the nodes of the network. This makes them like typical general IT (Information Technology) networks, but SCADA systems differ in that they interface with the physical world through actuators and sensors of PLCs and other IEDs (Intelligent Electronic Devices). Research in mitigations against cyberattacks have been quite extensive in the literature for SCADA systems and include firewalls, encryption, authentication and other techniques and technologies [22, 23]. One type of mitigation involves intrusion detection systems (IDSs) to detect malicious behavior on networks. IDSs can be used in general IT networks. However, the aforementioned examples of cyberattacks were largely able to perform their destructive acts before being discovered. These cyberattacks were only discovered after the destruction occurred. There is a concern in the research community of zero-day attacks, where exploits and vulnerabilities are unknown officially and certainly not yet published. As a result, the proper way to mitigate against these exploits has not been determined. Therefore, it is necessary to understand how to detect and mitigate against attacks even if unknown exploits are performed.

What the above examples have in common, regardless of the method used to infiltrate the main network of the SCADA system or the trusted computer nodes in the network, is that they each involve a malicious command or program that places the physical system in an undesired condition – whether unsafe or inefficient. Therefore, the threat model under consideration consists of attacks where the payload sent by an adversary has a malicious objective to cause one of those states. Commands can be considered packets that direct the PLC to change a setting associated with it. A program for a PLC can be considered as a sequence of commands to the actuators affecting the physical process that are usually dependent on the sensor data, internal states, or timers. The threat in the case of this work may appear in every way to be normal from the perspective of the network so that the threat can be hidden from certain types of IDSs that are focused on behavior of the network.

1.1 Understanding ICS to Mitigate the Threat

To address threats like these examples, governments and research organizations have worked to develop mitigations. Methods of mitigation include those that are standard to general IT systems. However, ICSs have several major differences that distinguish them from general IT systems: (1) ICSs have regular patterns of behavior and typically a static set of nodes, whereas General IT systems possess more random behavior as nodes are regularly added or removed and programs are run at various random times. (2) ICSs interact directly with physical processes in the real world. (3) ICSs traditionally have less computational processing capability and more limited memory than the computers of IT systems do. (4) ICSs tend to have strict real-time requirements or at least more so than General IT systems.

Therefore, is it necessary to respond to these threats by taking advantage of these unique aspects of SCADA systems. This work focuses on the fact that SCADA systems impact the physical world. This impact is brought about through commands and programs that direct the behavior of the PLC's actuators. If the impact of these commands and programs can be predicted before being processed, then unsafe and inefficient states of the ICS can be prevented.

1.2 Summary of Proposed Approach

The approach and contribution of this work is to create a proof-of-concept Intrusion Prevention System (IPS) that will monitor commands and programs delivered to the PLC in the form of packets before these packets are processed by the PLCs and other devices, which interact directly with a physical process through sensors and actuators. As these packets are examined, a prediction is made on what effect the packet would have on the system. The process of doing so is summarized as follows: A simulation is run in an IPS module. The simulation incorporates a modified OpenPLC process and a model of the physical process controlled by the PLC. The command is executed on the OpenPLC process and applied to the model. This simulation is useful as a means of prediction because it allows for the dynamic testing of programs. To create the model of the physical process, training must be performed where inputs and outputs of the physical system are collected and parameters of models are tuned to cause the model to fit the observed behavior. Techniques in system identification and machine learning are used.

The remaining sections of this work are as follows: Sect. 2 is a literature review. Section 3 is the research objectives and approach. Section 4 consists of the Preliminary Results. Several testbeds are studied as case studies with experiments of the proposed methodology. Section 5 is the conclusion.

2 State of the Art in Research Literature

Several notable surveys discuss intrusion detection/prevention for cyber-physical systems. It should be noted that intrusion preventions systems (IPS) extend IDS by also responding to detected threats [14–16]. Major types of IDS/IPS are discussed including their advantages and disadvantages. Some types are optimized for detecting certain attacks but not for other attacks.

Broadly speaking, there are several major types of intrusion detection systems according to work [15]: (1) Knowledge-based or misuse-based IDS/IPS involve signatures of known attacks. If the IDS/IPS encounters an intrusion with a signature stored in its database, the IDS/IPS is very effective at detecting it. However, if the intrusion has never had the signature of an attack catalogued in its database, the attack can go undetected as in the case of Zero-day attacks (attacks that exploits vulnerabilities that are yet unknown to the vendors of the hardware or software involved.) (2) Behavioral-based attacks have a concept of normal behavior determined through machine learning algorithms and other methods. (3) Behavior-Specification-Based Intrusion Detection, which is a subcategory of the second, consists of well-defined conditions for normal behavior that make up the specification. This type has the advantage of being highly accurate in determining violations of normal behavior but may be very cumbersome. Generating specifications requires a human domain expert.

On the other hand, the work [16], has a different, but at the same time useful, way to organize IPS/IDS, especially for IPSs/IDSs intended for cyber-physical systems. That survey uses the following categories: (1) Protocol analysis–based IDS/IPS, (2) Traffic mining–based IDS/IPS, (3) Control process analysis–based IDS/IPS. The last category places an emphasis on the process under consideration and semantics of the data and signals that have physical meanings (i.e. temperature, voltage, pressure, etc.). The focus of this work falls under the control/command analysis-based subcategory of control process analysis–based IDS/IPS. [16]. Using the previous taxonomy of IDS/IPS for cyber physical systems, this work also may be considered a variant of Specification-based ID/PS since it incorporates a model of the physical system and defined safety and efficiency requirements, which together form the specification. The latter survey [16] only describes two examples of the control command analysis-based type. What follows is a more expanded review and discussion of the relevant literature for this work.

Several recent works incorporate models of power systems, where incoming commands that are sent to a PLC or RTU (Remote Terminal Unit) are evaluated with the model in a simulation or simply a calculation to determine what the effect would be before the command is processed [1–5]. This contrasts with other works that attempt to detect abnormal behavior once the cyber-attack has already begun to have an impact on the physical process [13, 25]. If the effect is a safety violation, the IDS/IPS will be able to alert the operator of the power system and potentially drop or delay the command. In power systems, safety requirements would specify certain ranges of acceptable voltages for buses and currents for transmission lines ($I < I_{max}$). The IDS/IPS in [1] uses this approach, even though a real testbed appears not to be used. Instead, calculations to determine if safety violations occur are performed for various scenarios even though an actual IDS/IPS is not implemented. In [3], the same authors propose a similar IDS/IPS building on their previous work, which involves creating a local IDS/IPS as opposed to a central IDS/IPS since modern smart grids are better suited for distributed control to achieve stability and balanced operation. One advantage of the localized approach to the centralized IDS/IPS is that the centralized IDS is likely to be affected by Man in the middle attacks (MitM), where a threat actor can intercept communication from the PLC to the centralized IDS/IPS to deceive the IDS/IPS from being aware of the true condition of cyber-physical system. Also, a Denial of Service (DoS) attack can prevent situational awareness as well for the centralized IDS. Like the previously mentioned works, the authors in [4] also use a power grid. But in their work, an implementation of the power system and IDS/IPS is developed and tested. A "look ahead" analysis is performed before execution of the commands that would affect the breakers of the power grid. These commands would be considered legitimate in their structure; however, their contents may be harmful in terms of what actions would be taken on the power grid. That work explicitly refers to itself as a specification-based IDS/IPS. Matpower, a toolbox provided in MATLAB, is used to simulate the power grid. Several methods of power flow analysis are used that include DC analysis and Newton Raphson for AC analysis. Therefore, this work allows for a dynamic simulation of the power system as opposed to a simpler steady-state analysis where transient behaviors of the power systems are neglected. In [6], authors develop a similar system in principle to the previous works with a centralized system that serves as a form of access control. Certain users are permitted to control only certain variables associated with the cyber-physical system and see only certain variables based on their assigned roles and privileges.

All these works make meaningful contributions to IDS/IPS for cyber-physical systems. However, they are limited in that they focus on power systems only. They also assume that the topology of the power systems and other information such as branch impedance, transformers, generation and loads are readily available since they would be for fault diagnosis and analysis. However, such is not the case for all types of cyber-physical systems. Another problem with constraining the physical system to only these power systems is that these systems will have an almost immediate effect once a change in breaker status occurs, etc. Such behavior is not representative of all systems or processes in other domains of engineering, which have slower response times, meaning that the safety violation will not occur as quickly.

There is a work, however, which does not assume a model is known a priori [7]. That work uses autoregression (AR) to make predictions of general trends for process variables. AR is a method that will be extended and described later in this work, to create a basic model of the physical system. However, in that work, the model is not used to predict the behavior of a command before it is processed. It is used instead to predict the expected behavior of sensor data according to the model based on previous readings of the sensor. The IDS compares this expected behavior as determined by an AR model of the process with the current reading of the sensor data. If there is a significant deviation, that deviation is considered an abnormality and a potential sign of a cyberattack against the integrity of sensor data, such as a MitM attack. Data is gathered concerning process variables from the packets collected by sensors associated with network-based IDS to create time series of data of the physical sensors. This notion of memory map is important because the current proposed work also uses a memory map. The work [12] is also of interest to this paper. The work involves critical state analysis and a distance metric. Critical states are finite in number and known in advance. At a certain distance from the critical state, a warning can be issued. Detecting when variables enter critical states is useful for IDS. But that work does not make a prediction that the cyber-physical system is headed toward a critical state. The work only indicates if the system is in a critical state or what is the current distance between the state of the system and the critical state. Other relevant research includes works that evaluate the PLC code and every path of execution to determine if some defined set of undesired states is reached. However, without a knowledge or a model of the dynamics of the physical system, those methods are very limited in being able to detect malicious code [8–10].

3   Research Objectives and Approach

Unlike the previous works, this proposal does not assume that a model of the system is readily available. Instead this work assumes that time series data of the physical model is readily available or can be collected from sensors or actuators. These time series data include inputs of the physical process (Actuator signals/Data) and outputs of the physical process (Sensor signal/Data) and are used for the purpose of training a model of the physical system.

3.1   Problem Formulation

It is necessary to understand a basic configuration of the ICS/SCADA system for this work. In this work, it is assumed that there are a set of PLCs, each that manage some aspect of a physical process. The physical process can be said to have a single input and a single output (SISO) or multiple inputs and multiple outputs (MIMO). The conceptual SCADA System that is used and studied in this work is the one described by [9], where the following is assumed: (1) A PLC or multiple PLCs directly interface with the physical process or system through wired connections that allows for transmission of signals to actuators and from sensors connected to the process. (2) A network facilitates communication between the PLCs and the HMI (Human Machine Interface). More specifically, the SCADA System consists of five major components: (1) physical system/process, (2) cyber-physical link (3) PLC, (4) Network, (5) Remote Monitoring and Control, which generally consists of an HMI. What is also important for this work are the loops that illustrate the flow of information as seen in the following Fig. 1:

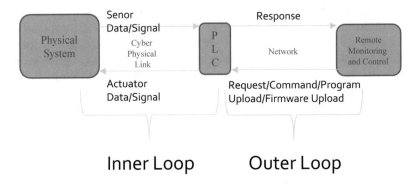

Fig. 1. SCADA system decomposed into major components.

The focus of the problem is on the outer loop, or the wide area control loop, which consists of the network that facilitates communication between the PLC and HMI. The inner loop, or local control loop, is assumed to have virtually no attack surface since it consists of wires.

3.2   A Closer Look at the Threat Model

The basic threat model of this work includes malicious command packets and ladder logic programs that are sent to PLCs to affect the physical system. In the case of this work, harm or maliciousness is defined as causing the cyber-physical system to enter into an undesired state, whether the state be unsafe or inefficient [24]. The following is assumed about these malicious attacks regarding their origin in this work: (1) They may originate from a foreign node that is able to connect to the ICS network. (2) They may also originate from a trusted system in the network, such as an HMI, or a trusted server, where the trusted system has become compromised. The word trust in cybersecurity implies that a system is considered safe or acceptable to use or transact with whether in communication or in transferring data/instructions and is not compromised [20]. Often, trusted systems are whitelisted by the operator of the SCADA system so that other systems that are associated with or depend on the trusted system will be able to identify (Perhaps by IP address) the trusted system by the fact that it can be found in the white list. Note, this simple approach is not totally effective since it allows the possibility for spoofing attacks. The Moroochi attack is an example of the first case of having a foreign node. Stuxnet and the Ukrainian attack are examples of the latter case since trusted nodes on the network became compromised. In the case of Stuxnet, the trusted PC networked to the PLC became compromised. With the Ukrainian attack, trusted computers that human operators used in the control room were infected by remote malware.

Threat Model from Perspective of Kill Chain. Also, the threat model should be understood in terms of the ICS kill chain. The threat is assumed to have reached an advanced stage of the ICS Cyber kill chain [26]. In this work, the cyberattack has reached Stage 2 and specifically the deliver phase before the phases (1) Install/Modify and (2) Execute ICS Attack. These advanced stages are worth studying since the attacker cannot hide its true intentions based on what is in the payload to be installed or executed in terms of the effects or consequences that can potentially be predicted before the payload is fully processed by the PLC's or the other hardware that is directly interfacing the physical system. The threat model must also be understood in terms its consequences. The threat model of this work is like another threat model which consists of attacks that cause safety violations and inefficiencies for the SCADA system.

Threat Model from Perspective of PLC Software. The threat model that can induce the above consequences consist of Layer 2 - commands born of the SCADA network to PLC that direct the PLC to change settings, states, or output of the PLC, Layer 1 - Programmation/Ladder Logic Updates, and Layer 0 - Firmware Updates. Note that the firmware updates are out of scope of this work but still worthwhile to investigate. Malicious firmware updates may require a full virtual machine to investigate changes of underlying software. Virtual machines have much greater complexity and fidelity to evaluate such treats.

Layer 0. The firmware is the underlying software of the PLC. It consists of drivers for the hardware that serve as the interface for the ladder logic to retrieve input values and to set output for the hardware. This firmware may also have an operating system, such as Linux or an RTOS (Real Time Operating System). Even if the PLC software is considered to be running on "bare metal" (without an operating system), there is still likely to be underlying code that serves as the hardware driver.

Layer 1. The programmation consists of the ladder logic program, which serves as the software to control the PLC. ladder logic is characterized by a scan cycle, in which input to the PLC presumably coming from sensors is read and new outputs of the PLC are computed based on the internal logic of the program. This scan cycle is repeated over many iterations during the operation of the industrial plant that the PLC controls.

Layer 2. The programming or ladder logic will have internal states or settings associated that may be modified by the operator through Modbus commands [17]. Specifically the Modbus packets with "write" function codes (Table 1):

Table 1. Modbus "Write" function codes

Function code	Function name
5	Write Single Coil
15	Write Multiple Coils
6	Write Single Holding Register
16	Write Multiple Holding Registers

These settings exist as registers in the memory of a PLC. Examples of these settings may be setpoints, such as the set point for a PID controller, high low and low set points for an on/off controller. Other settings are the gains of the PID controller or other control algorithm. The output is important to consider since it directly affects the physical plant. It should be noted that it is essentially affected indirectly by changes in layers 1–2, and even layer 0 if the driver software is modified. For a PLC running a simple program, it is conceivable that in typical operation of a SCADA system that oversees the PLC an operator may wish to change the output of the PLC, such as in the case of a breaker in a power grid. Note that with more complex ladder logic, the ladder logic may be such that the output is given an updated value every scan cycle, which would override any attempt by an external request to change the value.

What is of interest in this work is malware that can change layers 1–2. For malware to make changes to these layers, commands are sent to the PLC to change the values of registers, which may be outputs or internal settings (Layer 2). To change the ladder logic, the ladder logic upload request is sent to the PLC (Layer 1).

3.3   Proposed Intrusion Detection System and Methods

For the proposed IDS, there will be a software-based module that either exists as a separate process within the PLC or is located in a hardware device that exists between the PLC and the main SCADA network. The IPS Module's design can be described as follows (See illustration in Fig. 2): It relays the request packets coming from the HMI to the PLC (Flow of packets depicted as arrows moving toward PLC). It receives the response packets from the PLC and stores the data from the packets into its own shadow memory map similar to [7]. Note that in the Modbus protocol and other SCADA protocols, a node on a network that receives request packets for information contained in its registers will respond with such information to the requesting node. Not only will the most current sensor data be stored in the memory map but also the data over time will be logged in another data structure or to a file. The module then relays response packets originating from the PLC to the HMI. The module can be considered as an analysis engine like many of the previous papers. When a packet is sent to the PLC that contains a command (the packet has a function code indicating a write command), several steps are taken: (1) The model of physical system receives state variable information from the real system (e.g. Water tank level.), which is found in the memory map in the module. This is important because the model must have initial conditions similar to the current state of the true physical process as indicated by the sensor; (2) The command to be evaluated is applied to ModOpenPLC; (3) The simulation of the Physical System Model combined with the ModOpenPLC is run; (4) The simulation is evaluated to determine if any "critical variables" veer outside of a safe region, which is defined by a human expert or is determined by the module during training by examining the region of operation. (5) A decision is made by the IPS depending on whether the potential action by the command will cause harm (unsafe, inefficient). If the command is determined to be safe, the command is processed. Otherwise, the IPS module will take one or more of the following options: (1) alert the SCADA operator, (2) delay execution of the command to allow time for further investigation of the situation. (3) Drop the command altogether (4) Process the command despite unsafe predictions (This may be reasonable in cases where the critical variable does not become unsafe immediately). The following image depicts this module:

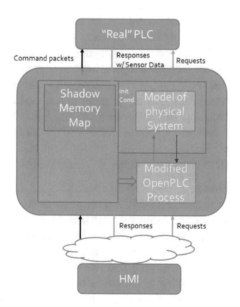

Fig. 2. Proposed IPS in a Nutshell

Operation of IPS Module. There are two main phases for the operation of the proposed IDS module. These are training (Phase 1) and testing (Phase 2). Essentially, in this work, a model is chosen to represent the physical system. The model will be used to predict if an action will cause the physical system to enter either into an unsafe state or an inefficient state. Previous work has also dealt with unsafe states. Phase 1 is made up Steps 1–3 and involves training the IDS: (1) Collection, (2) Parameter Estimation of the Models, and (3) Validation and selection of Model. For this first phase, it is assumed that the ICS is not under any cyberattack. Therefore, there are no attacks on integrity of sensor data that are used in the training data as an example. This assumption can especially hold true if the placement of the IPS is in the PLC or between the PLC and the network. In those placement scenarios, an integrity attack affecting the network would not adversely affect the training. However, if the placement is in a centralized node in the network away from the PLCs, it is conceivable that a cyberattack originating from a foreign node and connected to the SCADA network can use MitM attacks to modify sensor data in packets and therefore compromise integrity. If such is the case, certain mitigations that include encryption and authentication may be employed along with the proposed IPS. In the case that the cyberattack is originating from a trusted node, the following mitigations and setup must be used: Steps 4–5 comprise the testing phase: (5) Monitoring and (6) response. The following paragraphs will describe each of the phases and steps in details.

Phase I – Training

Step 1 - Collection: This step consists of collecting input and output data over time from local, distributed PLCs. This input data is defined as input into the physical systems over time. This input data would also be considered actuator commands. The output data is defined as output from the physical system over time. This consists of discrete-time data from the sensors in the physical system. The time series data is collected by sampling the sensor data at regular intervals. There are two main ways that data can be collected. One way is to have the control system in the PLC operate using a typical program for normal operation of the plant which would be used by the plant personnel. As PLC is running and controlling the plant, the input or sensor data is collected by the module and the output commands or data that is applied to the physical system is also recorded. The other approach would be to not run a specific program considered "normal" but to choose inputs and experimentally observe the outputs [11].

Step 2 - Parameter Estimation: This step involves using the collected time series data to train a set of trial mathematical models and select parameters. System Identification techniques and machine learning techniques, such as gradient descent will be used in this step in order to tune the parameters of the models. System Identification takes the physical system as a black box, where sample inputs and outputs are given and allows for the generations of models. This proposed work will use Auto-Regressive Moving Average (ARMA), along with other techniques, since both inputs and outputs are to be incorporated into the model, not just simply the states or outputs of the physical system as in the case of [7]. These inputs are essentially actuator signals that impact the model.

$$X[n] = a_1X[n-1] + a_2X[n-2] + \ldots + b_0U[n] + b_1U[n-1] + \ldots e,$$

Where X is the output and U is the input and e is an error term.

The parameters $(a_1, a_2\ldots,b_1, b_2)$ must be tuned so as to enable the model to fit the observed data as closely as possible.

Step 3 - Validation: This step involves validating which model fits the data the best and selecting the best model for the IPS to be used for the Training phase. To choose the correct structure for the model, several such structures must be trained. That is, parameters associated with them must be tuned. Then, other time series data other than the data that was used to train them must be used. A metric is then used to determine how well the models fit the validation data. The purpose of the validation is to prevent overfitting of the data.

Phase II - Testing

Step 4 - Monitoring and Examination of Incoming Commands and Packets: This involves using the model to evaluate incoming packets and ladder logic. The approach to evaluation is as follows. A command packet that is sent to a PLC will be checked by a special software-based module that uses the model determined earlier to predict whether the command if applied as an input to the model would cause a violation to the safety conditions for the system. This module at minimum contains a copy of the compiled ladder logic code such that it is interfaced with the model inputs and outputs. The goal of being able to run a scenario with the changed setting is to be able to detect if the change would be disastrous to the physical system. This is determined by a given metric concerning a safety violation or inefficiency in the system's operation.

Step 5 - Response: For this, Possible actions include the following: (1) Process packet or load ladder logic, (2) drop packet. (3) Delay the processing of the packet of the loading of the new ladder logic. (4) Alert the user.

3.4   Hypotheses to Be Evaluated in This Work

Using System Identification and machine learning techniques, accurate models of the physical system or process can be generated that will allow the IPS to predict the effects of an action by the PLC before it occurs. To test this hypothesis, experiments will be conducted. A command injection attack will be performed on the physical system while recording the input and output data. Also, an experiment with a model that has previously been generated based on steps 1–3 of the training phase will predict the same attack.

Physical-based metrics can be used to determine how well the behaviors for physical variables for the virtual testbed match those of the actual testbed. It is necessary to compare time-series data from the actual and predicted behavior. One metric to determine accuracy is Mean Absolute Percentage Error (MAPE) defined by the following equation:

$$M = \frac{100\%}{n} \sum_{t=1}^{n} \left| \frac{A_t - F_t}{A_t} \right|$$

Where At is the actual value and Ft is the forecast value. n is the number of samples taken over time for the time series. Historically, this has been a useful metric for this type of data. However, this metric has limitations when samples for At are close to zero and when there is a substantial difference between At and Ft. metrics such as SMAPE (Symmetric Mean Absolute Percentage Error) were created to help with this issue.

SMAPE is calculated as follows:

$$\text{SMAPE} = \frac{100\%}{n} \sum_{t=1}^{n} \frac{|A_t - F_t|}{(|A_t| + |F_t|)/2}$$

Where A is the actual value and F is the forecast value.

SMAPE mitigates the problem of division by a number close to zero, which would skew the results as can be seen in MAPE. SMAPE does so by taking the average of the actual and the forecast value for the denominator. The smaller the resulting value for SMAPE, the more accurate the virtual testbed is.

4   Preliminary Results

4.1   Proof-of-Concept Experiments

The experiments include running attack scenarios that represent the previously described threat model on the water testbed. Later work as part of this proposed research will involve more complex testbeds and more sophisticated attacks from the threat model. For these experiments, a virtual testbed is setup that includes the major components of SCADA systems as described in [18]. The latency for the IPS to make its prediction is determined. Also, the behavior of the model is compared with the real behavior of the system to determine in the accuracy in how well the model represented the real event.

Preliminary Experiments with Python Implementations of Basic IPS

In this work, several experiments were performed illustrating stage II of the IPS, where commands packets that change the state of the actuator or a parameter used by the latter logic program. It has been assumed that stage I has already been performed such that the values for the ARMA Model have been determined for the physical process. A MATLAB script was used for Stage I.

For the following experiments, latency and accuracy in the prediction were evaluated for an implementation of the IPS running on a (1) virtual machine assigned to core of i7 Intel Processor in a personal computer and (2) a Raspberry Pi. The results for these two platforms are side by side for each experiment to allow for ease of comparison. Presumably the Raspberry Pi will have poorer results. However, the Raspberry Pi may serve as a useful and cheap implementation of the proposed IPS where the IDS Device is placed between PLC and Network. Such a device with that placement may be especially necessary when the PLC is proprietary in its hardware and software and having software processes in the PLC is not possible for the user to setup.

Manual Mode Test with Pump Turned on

An experiment is performed for predicting the behavior of the physical system when manual mode in the water tank latter logic is set and the pump is turned on. The IPS will predict when the critical state is reached, which is defined as 50%. The accuracy in the overall graph of the prediction is given in the table below in terms of MASE and MATLAB's fit metric (Fig. 3).

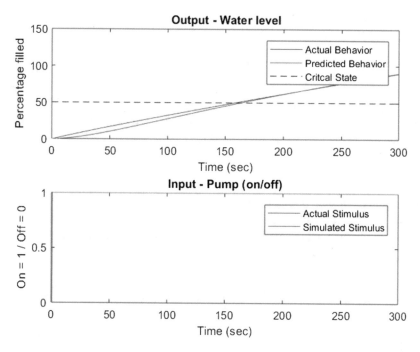

Fig. 3. Input and output behavior for water tank with pump on

Also, the following results were compiled for the VM assigned to a single core of the processor and the Raspberry Pi for the cases when the data is saved to a file as the simulation is run and when the data is saved to a data structure to be stored to a file after the simulation is run. The Latency to predict the next 10000 scan cycles, where a scan cycle lasts for 100 ms, is shown, as well as the final simulation time. The latency at the point that the Criticality is reached in the simulation is also given. The estimated time to criticality is a prediction that the IPS makes for when the behavior of the physical process will reach the critical state. This is a useful metric that allows a user or the IPS to know how soon danger may occur that is associated with a particular action (Tables 2 and 3).

Table 2. Accuracy of prediction compared with actual observed behavior

100% – SMAPE (Accuracy)	81.91%
MATLAB's fit Metric	86.26%

Automatic Mode Test with Different Parameters for Set Points
Additional experiments were performed with the water tank placed in automatic mode, which is defined as an on-off controller with low and high set points. If the water level is about to go below the low set point the pump is turned on. If the water level is about to go above the pump is turned. The following graphs show the water tank controllers

Table 3. Results on latency and predictions for criticality in seconds

Platform	Saving data to file		Storing to data structure	
	VM on single core	Raspberry Pi	VM on single core	Raspberry Pi
Latency	0.100765	0.361271	0.093953	0.29277
Final simulation time	1000.00	1000.00	1000.0	1000.0
Latency at criticality	0.016553	0.058519	0.013437	0.046684
Estimated time to criticality	163.9	163.9	163.9	163.9

in automatic mode with different set points for the high set point (70% on left and 40% on the right). The IPS will predict that reaching the critical state is imminent when 70% is chosen as the set point and will predict that the process will not reach the critical state when the set point is at 40% (Fig. 4).

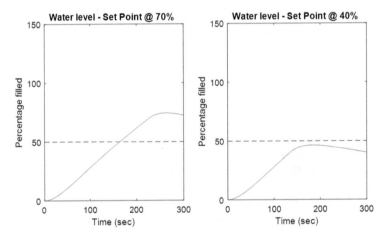

Fig. 4. Behavior of water tank in automatic mode with high set point set to 70% (Left) and 40% (Right)

5   Conclusion

In this work, methods were used to detect whether a malicious attack was occurring based on predicting what the effect of the attack would be on the system with the aid of modeling and simulation before allowing the PLC to carry out the command or to run the program. The attack takes the form of a command or a program with a malicious objective. In the case of SCADA systems, where physical processes are involved, it is important to take into account that certain actions can perturb these physical processes in undesired and harmful ways. General IT mitigations may be effective to some degree in detecting and combating attacks. However, these methods may fail because they fail to take into account the effect of a command packet or program that have physical consequences. Therefore, a malicious command or program may appear as normal to common IDSs and may have behavior from the network perspective that would not flag these IDSs. The results of this work serve as a preliminary proof of concept.

Future work will be to use more types of physical processes and those of greater complexity. Also, experiments will be performed with ladder logic malware that include both time bombs and logic bombs [21]. Methods of dynamically and statically analyzing the ladder logic to determine whether it is malicious will be explored. Also, experiments will be performed on responding to threats.

References

1. Chromik, J.J., Remke, A., Haverkort, B.R.: What's under the hood?
2. Improving SCADA security with process awareness. In: JointWorkshop on Cyber-Physical Security and Resilience in Smart Grids (CPSR-SG), pp. 1–6. IEEE (2016)
3. Chromik, J.J., Remke, A., Haverkort, B.R.: Improving SCADA security of a local process with a power grid model. In: ICS-CSR (2016)
4. Lin, H., Slagell, A., Kalbarczyk, Z., Sauer, P.W., Iyer, R.K.: Semantic security analysis of SCADA networks to detect malicious control commands in power grids. In: Proceedings of the First ACM Workshop on Smart Energy Grid Security, pp. 29–34. ACM, (2013)
5. Lin, H., Slagell, A., Kalbarczyk, Z., Sauer, P., Iyer, R.: Runtime semantic security analysis to detect and mitigate control-related attacks in power grids. IEEE Trans. Smart Grid **9**, 163–178 (2016)
6. Etigowni, S., Tian, D.J., Hernandez, G., Zonouz, S., Butler, K.: CPAC: securing critical infrastructure with cyber-physical access control. In: Proceedings of the 32nd Annual Conference on Computer Security Applications, pp. 139–152. ACM (2016)
7. Hadžiosmanović, D., Sommer, R., Zambon, E., Hartel, P.H.: Through the eye of the PLC: semantic security monitoring for industrial processes. In: Proceedings of the 30th Annual Computer Security Applications Conference, pp. 126–135. ACM (2014)
8. McLaughlin, S.: On dynamic malware payloads aimed at programmable logic controllers. In: Proceedings of the 6th USENIX Conference on Hot Topics in Security, HotSec 2011, Berkeley, CA, USA, p. 10. USENIX Association (2011)
9. McLaughlin, S.: Cps: stateful policy enforcement for control system device usage. In: Proceedings of the 29th Annual Computer Security Applications Conference, ACSAC 2013, pp. 109–118. ACM (2013)
10. McLaughlin, S., McDaniel, P.: SABOT: specification-based payload generation for programmable logic controllers. In: ACM Conference on Computer and Communications Security, pp. 439–449. ACM (2012)
11. Ljung, L.: System Identification: Theory for the User. Prentice-Hall, Upper Saddle River (1987)
12. Carcano, A., Fovino, I.N., Masera, M., Trombetta, A.: State-based network intrusion detection systems for SCADA protocols: a proof of concept. In: International Workshop on Critical Information Infrastructures Security, pp. 138–150. Springer, Heidelberg (2009)
13. Cárdenas, A.A., Amin, S., Lin, Z.-S., Huang, Y.-L., Huang, C.-Y., Sastry, S.: Attacks against process control systems: risk assessment, detection, and response. In: Proceedings of the 6th ACM Symposium on Information, Computer and Communications Security, pp. 355–366. ACM (2011)

14. Zhu, B., Sastry, S.: SCADA-specific intrusion detection/prevention systems: a survey and taxonomy. In: Proceedings of the 1st Workshop on Secure Control Systems (SCS), vol. 11, p. 7 (2010)
15. Mitchell, R., Chen, I.-R.: A survey of intrusion detection techniques for cyber-physical systems. ACM Comput. Surv. (CSUR) **46**(4), 55 (2014)
16. Ding, D., Han, Q.-L., Xiang, Y., Ge, X., Zhang, X.-M.: A survey on security control and attack detection for industrial cyber-physical systems. Neurocomputing **275**, 1674–1683 (2018)
17. Gao, W., Morris, T.H.: On cyber attacks and signature based intrusion detection for modbus based industrial control systems. J. Digit. Forensics Secur. Law **9**(1), 3 (2014)
18. Alves, T., Das, R., Werth, A., Morris, T.: Virtualization of SCADA testbeds for cybersecurity research: a modular approach. Comput. Secur. **77**, 531–546 (2018)
19. Abrams, M., Weiss, J.: Malicious Control System Cyber Security Attack Case Study–Maroochy Water Services, Australia. The MITRE Corporation, McLean (2008)
20. Taipale, K., Cybenko, G., Yen, J., Rosenzweig, P., Sweeney, L., Popp, R.: Homeland security. IEEE Intell. Syst. **20**(5), 76–86 (2005)
21. Govil, N., Agrawal, A., Tippenhauer, N.O.: On ladder logic bombs in industrial control systems. In: Computer Security, pp. 110–126. Springer, Cham (2017)
22. Dempsey, K.L., Witte, G.A., Rike, D.: Summary of NIST SP 800-53, Revision 4: Security and Privacy Controls for Federal Information Systems and Organizations. No. Computer Security Resource Center (2014)
23. Giraldo, J., Sarkar, E., Cardenas, A.A., Maniatakos, M., Kantarcioglu, M.: Security and privacy in cyber-physical systems: a survey of surveys. IEEE Design Test **34**(4), 7–17 (2017)
24. Huang, Y.-L., Cárdenas, A.A., Amin, S., Lin, Z.-S., Tsai, H.-Y., Sastry, S.: Understanding the physical and economic consequences of attacks on control systems. Int. J. Crit. Infrastruct. Prot. **2**(3), 73–83 (2009)
25. Giraldo, J., Urbina, D., Cardenas, A., Valente, J., Faisal, M., Ruths, J., Tippenhauer, N.O., Sandberg, H., Candell, R.: A survey of physics-based attack detection in cyber-physical systems. ACM Comput. Surv. (CSUR) **51**(4), 76 (2018)
26. The industrial control system cyber kill chain. SAN Institute
27. Lee, E.A.: Cyber-physical systems-are computing foundations adequate. In: Position Paper for NSF Workshop on Cyber-Physical Systems: Research Motivation, Techniques and Roadmap, vol. 2, pp. 1–9. Citeseer (2006)
28. Langner, R.: Stuxnet: dissecting a cyberwarfare weapon. IEEE Secur. Priv. **9**(3), 49–51 (2011)
29. Alert (IR-ALERT-H-16-056-01). ICS. https://ics-cert.us-cert.gov/alerts/IR-ALERT-H-16-056-01. Accessed 29 May 2019

Analyzing the Vulnerabilities Introduced by DDoS Mitigation Techniques for Software-Defined Networks

Rajendra V. Boppana$^{(\boxtimes)}$, Rajasekhar Chaganti, and Vasudha Vedula

Department of Computer Science, University of Texas at San Antonio,
San Antonio, TX, USA
`rajendra.boppana@utsa.edu`

Abstract. Software defined networking facilitates better network management by decoupling the data and control planes of legacy routers and switches and is widely adopted in data center and production networks. The decoupling of control and data planes facilitates more optimal network management and deployment of elaborate security mechanisms, but also introduces new vulnerabilities which could be exploited using distributed denial of service (DDoS) attacks. In this paper, we identify several protocol vulnerabilities and resource limitations that are exploited by DDoS attacks. We also analyze the vulnerabilities introduced by several DDoS mitigation techniques, discuss attacks that exploit them, and quantify their impact on the network performance using experiments on a 5-node testbed. We show an approach to mitigate such vulnerabilities while minimizing the introduction of any exploitable vulnerabilities.

Keywords: Software-defined networks ·
Distributed denial of service attacks · TCP SYN attacks ·
TCP ACK attacks

1 Introduction

Software defined networking facilitates better network management by decoupling the data and control planes of legacy routers and switches [1,2]. The data plane consists of simplified switch elements that keep a table of flow processing rules and process packets according to the rules. The control plane, running on one or more computers or virtual machines but logically centralized, specifies the flow table entries and packet handling policies to be used by the switches. The control plane has a global view of the network status and helps manage, control, and secure the network and automate network tasks by facilitating implementation of various network applications on top of it (see Fig. 1). Software-define networks (SDNs) are widely adopted in data center networks and production (Internet core and campus/organization) networks. While offering significant security and performance management advantages, SDNs introduce new vulnerabilities. For example, the entire organizational network can be brought down

© Springer Nature Switzerland AG 2020
K.-K. R. Choo et al. (Eds.): NCS 2019, AISC 1055, pp. 169–184, 2020.
https://doi.org/10.1007/978-3-030-31239-8_14

by congesting the channels (Southbound API in Fig. 1) between the data and control planes in an SDN.

Distributed denial of service (DDoS) attacks are powerful attacks that bring down portions of the Internet and render popular Internet-based services and commerce unavailable for significant periods of time [3]. In recent DDoS attacks, the attack traffic directed toward the victims was as high as 1 trillion bits/second [4]. DDoS attacks are hard to mitigate since the attack traffic can be distributed over millions of bots (computers, internet of things, or networking devices compromised by the attacker) [4–8].

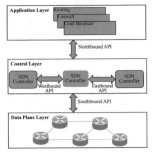

DDoS attacks are known to be particularly harmful to SDNs [9–13]. DDoS attacks on SDNs

Fig. 1. Software-defined network architecture.

can be launched by exploiting the commonly used protocols and networking standards at the network, transport, or application levels. For example, an attacker can send multiple TCP (transmission control protocol [14]) SYN requests from several distinct IP addresses to saturate the data plane, control plane, or both. The legacy network defenses such as firewalls and intrusion detection systems, residing on top of the controller, are too far removed to prevent or respond quickly to Southbound API congestion.

Several solutions to the DDoS problem in the context of SDNs have been proposed [13, 15–19]. The limitations and new vulnerabilities introduced by the detection and mitigation techniques are often left uninvestigated. For example, Shin et al. [13] extend the functionality of switches in the data plane by having them send SYN-ACK responses to new TCP SYN requests so that the impact of the attack in which the attacker sends a large volume of SYNs but ignores SYN-ACKs is mitigated. However, Ambrosin et al. [15] show that this technique introduces new vulnerabilities, which can be exploited. The issue of new vulnerabilities is all too common with the current mitigation techniques.

In this paper, we investigate the vulnerabilities introduced by DDoS mitigation techniques for SDNs. More specifically, we consider TCP SYN flooding attacks and the solutions that use simple thresholds to infer attacks and protocol and architectural enhancements to mitigate them. We make the following contributions.

- We enumerate several vulnerabilities and resource limitations that are often introduced by DDoS mitigation schemes.
- We investigate some of the well-known DDoS mitigation solutions for SDNs, identify the vulnerabilities introduced by them, and discuss potential new attacks which can exploit them.
- We quantify, using experiments on a testbed, the impact of the new attacks.
- We present a solution to diffuse ACK flooding attacks, which exploit a vulnerability introduced by one of the mitigation techniques for SYN flooding, experimentally demonstrate its effectiveness, and analyze its vulnerabilities and limitations.

The rest of the paper organized as follows. Section 2 discusses various vulnerabilities and resource limitations that are typically exploited in DDoS attacks and analyzes several current DDoS mitigation techniques and their vulnerabilities. Section 3 presents an experimental study of vulnerabilities of some of the well-known DDoS mitigation techniques. Section 4 presents modifications to make a previously vulnerable mitigation technique resistant to further exploits. Section 5 discusses the related work. Section 6 concludes the paper.

2   Limitations and Vulnerabilities of Mitigation Techniques

In this section, we discuss various protocol vulnerabilities and resource limitations that are exploited to launch DDoS attacks. Using these vulnerabilities and limitations, we show how to identify the new security vulnerabilities introduced by mitigation techniques. First, we briefly describe SDN packet handling and TCP SYN flooding attacks on SDNs to provide the context for the discussion on vulnerabilities and their exploits.

2.1   SDN Vulnerabilities and SYN Flooding Attack Mitigation

We assume that the OpenFlow protocol [20] is used for the control-data plane interface, and denote the data plane switch elements as OpenFlow switches and the controller as the OpenFlow controller. Each new packet received by an Open-Flow switch (OFS) signifies a potential new flow and can trigger a packet_in message from the switch to its OpenFlow controller (OFC). The latter responds, if the new flow is admissible, with a flow_mod message to add a flow entry in the switch. If new packets arrive at a high rate, then the communication channel(s) between the data plane and the controller (Southbound API) will be congested, and the controller will be overwhelmed by the volume of flow requests. (Some of the recent SDN switches cannot handle more than a few hundred packets/second on the Southbound API.) Furthermore, if the Flow Table (FT), implemented using an expensive but fast ternary content-addressable memory (TCAM), is filled with FT entries (FTEs) for the attack traffic, then new packets (flows) are denied of service. Therefore, an SDN architecture exposes new attack surfaces, such as the Southbound API, controller, and TCAM. TCP SYN floods, in which a large number of SYN (TCP connection request) packets are sent, are particularly harmful to SDNs.

Shin et al. [13] proposed Avant-Guard (AG) to mitigate saturation of the Southbound API and the control plane due to TCP SYN flooding by tweaking the TCP. AG specifies a few changes in OFS functionality. When a TCP SYN packet without a matching FT entry is received, the OFS will act as the intended server and sends a SYN-ACK response packet using the SYN proxy technique with a SYN cookie, a cryptographically strong hash value generated based on the received SYN packet, local time, and other data, as its sequence number [21,22]. If the SYN is from an attacker who ignores the SYN-ACK responses it receives,

the SYN flood's impact is limited to the switches that receive it. If the SYN
is from a legitimate client, the client replies with an ACK packet, completing
the 3-way handshake for the TCP connection. At that point, the OFS acts as
the client and sets up a second TCP connection with the intended server and
modifies the packet headers flowing in each direction suitably to simulate the
desired client-to-server connection. This is called connection migration. (Shin et
al. also describe a delayed connection migration option, which requires the client
to send a data packet to trigger the connection migration activity by the switch.
However, the primary vulnerabilities introduced by AG are unchanged.)

The connection migration technique is interesting because it aims to mitigate
SYN flooding attacks by enabling OF switches to respond without impacting the
Southbound API or controller processing resources. Several recent SDN products
from Cisco, Juniper and other vendors implemented the SYN proxy mechanism
to address this attack [23–26].

2.2 Frequently Exploited Limitations and Vulnerabilities

Based on our literature review, we identified the following list of most common
vulnerabilities and limitations that are exploited by DDoS attacks and, sometimes, introduced by DDoS mitigation techniques.

VL1 Vulnerabilities inherent in the design or modification of the protocols used,
VL2 Memory or processing resource requirement that grows linearly or faster
   with the number of flows,
VL3 Significantly large-size or multi-packet responses to a single query/request
   packet,
VL4 Accepting received information without verifying sender's authenticity,
VL5 Using simplistic indicators such as minimal followup of requests (for exam-
   ple, an attacker replying to SYN-ACKs with ACKs) or simple rate thresh-
   olds to determine how new packets are handled,
VL6 Rejecting new flow requests or packets based on a blacklist of sender IDs or
   other indicators created using unverified, potentially spoofed, information,
VL7 Using whitelists to accept new flow requests without further checks,
VL8 Throttling OpenFlow messages, sent over the Southbound API to the controller, beyond set thresholds, and
VL9 Responses that reveal configuration, security posture, and timing or functional characteristics.

We identified the above vulnerabilities based on the functionality and architecture of SDNs, the known DDoS attacks on them, and the mitigation techniques proposed for the same in literature. Most of these vulnerabilities are not
listed in public CVE and NIST vulnerability databases [27, 28], which cover primarily the core OpenFlow switch software vulnerabilities owing to code bugs
and improper implementation.

In the remainder of the section, we discuss a few vulnerabilities introduced
by the recent SYN flooding mitigation techniques.

2.3 Exploiting Limitations of New Resources and Lists

TCP SYN flooding exploits a TCP connection setup vulnerability (VL1). To mitigate this, AG enhances the OFS to send SYN proxies, serve as an invisible intermediary, and modify the TCP packet headers for the lifetime of the connection. This enhancement requires the OFS to use a TCP ephemeral port and store a small amount of data (for packet header translation) for each connection, which leads to the vulnerability (VL2). Ambrosin et al. [15] show that AG is vulnerable to saturation attacks on the header translation data store and exhaustion of ephemeral ports available in the switch.

To address AG's vulnerabilities, Ambrosin et al. proposed Lineswitch (LS), which uses a probabilistic approach to selectively use Avant-Guard's SYN proxy for a small percentage of SYN requests and blacklisting the senders who do not respond to them. However, LS itself introduces the vulnerability (VL6): the blacklisting feature of LS can be exploited by sending TCP SYNs with spoofed source IP address to blacklist a victim. This is even more debilitating than the SYN-flood attack.

2.4 Exploiting Connection Migration Vulnerabilities

The connection migration technique used by AG and LS can be easily defeated. Using a few probing packets, an attacker can easily distinguish SYN-ACKs generated by an OFS from those by the targeted server based on the response times or the values used in various TCP header and option fields (exploitation of the vulnerability VL9). Following the approach by Bifulco et al. [29] to fingerprint SDNs, we verified that the origin of SYN-ACKs can be easily determined based on the RTTs observed. Also, the SYN-ACKs sent using SYN proxy mechanism by an OFS will need to keep track of the options such as window scaling and maximum segment size (MSS) indicated in the SYN packet; since the SYN proxy mechanism does not save any state information from SYNs, to avoid resource depletion during a flooding attack, some of the options information in the SYN must be coded in the SYN-ACK's sequence number field or its options. Packet content analysis allows an attacker to identify SYN-ACKs from a switch.

Therefore, an attacker can successfully launch SYN flooding attacks, even with LS or AG extensions to OFS, by responding only to SYN-ACKs from switches. This attack exploits the simplistic test (VL5)—replies to SYN-ACKs received or not—used by AG and LS to determine the attack traffic.

Chen et al. [16] proposed SDNShield, which uses statistical methods to determine which TCP SYN requests are to be handled using the connection migration technique is susceptible to selective ACK responses by the attacker.

Another vulnerability introduced by AG (and LS) stems from its use of SYN proxy and not maintaining any state for SYN-ACKs sent or ACKs received by OFS. So when an ACK is received and a matching FT entry does not exist, the ACK field of the received packet is used to verify that it was in response to a SYN-ACK it sent earlier. Upon successful verification, the OFS sends a packet_in message to OFC for processing and possible FT entry. This is a form of simplistic test/threshold vulnerability, (VL5). A resourceful attacker can exploit this by sending multiple ACKs to cause congestion on the Southbound API and overwhelm the controller processing capacity.

2.5 Exploiting Simplistic Indicators

Mohammadi et al. [17] propose SLICOTS, a two-part solution to mitigate DDoS attacks. A TCP SYN packet sent by a new client is processed by installing a temporary FT entry which will expire in 3 s. A SYN-ACK reply by the server results in another temporary FT entry in the opposite direction. If the 3-way handshake is completed within that time, the 3-second expiration on both FT entries is removed. The second part of the solution keeps a list of received SYNs from the client and their status based on the packets seen for this connection request: only SYN seen, or matching SYN-ACK, RST, or ACK is seen. If a matching ACK or RST is seen, the SYN is removed from the list; otherwise, the outstanding SYN count of the sender is incremented and, if it is over a specified threshold, the sender is blacklisted. This solution suffers from the simplistic response/threshold (VL5) and blacklisting (VL6) vulnerabilities, which can be exploited as follows. First, the SLICOTS logic for changing the temporary rules to permanent is triggered by and ACK from the sender of SYN. So an attacker can send a SYN and follow it up with a fictitious ACK that matches the original SYN, making the FT entry permanent. Repeating this, TCAM can be filled with bogus FT entries. A resourceful adversary can also overwhelm SLICOTS logic by sending SYNs from many bots and fill TCAM with too many temporary FT entries, although this attack is not as effective as sending matching ACKs, which creates permanent FT entries for a longer-lasting impact. Sending SYNs with spoofed MAC addresses, which SLICOTS uses instead of IP addresses to keep track of SYN counts, leads to blacklisting of a victim.

Fichera et al. [18] propose OPERETTA, which determines clients who complete a connection request as legitimate and do not have to go through SYN proxy for future connection requests. It also keeps a count of outstanding SYNs from each client and blacklists those with SYN count over a specified threshold. A successful 3-way handshake puts the sender in a whitelist. This mitigation technique introduces three vulnerabilities: blacklists (VL6), whitelists without additional checks (VL7), and simplistic response/rate threshold (VL5). An attacker can exploit these by completing the first connection to the server to get into a whitelist and launch the SYN flooding attack. Even if the unsuccessful SYN count is kept for whitelist nodes (not indicated in OPERETTA logic), an attacker can overcome this by completing one connection in every k SYNs, where k is the SYN count threshold to blacklist a node. Sending SYNs with spoofed MAC addresses can lead to blacklisting of a victim.

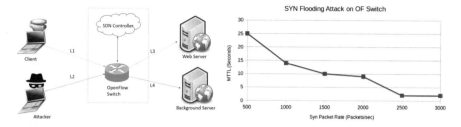

Fig. 2. Hardware testbed used for experiments.

Fig. 3. MTTL of client-server connectivity during SYN attacks on the OF switch.

3 Impact of the Vulnerabilities Introduced by DDoS Mitigation Techniques

In this section, we evaluate impact of some of the vulnerabilities introduced by TCP SYN flood mitigation techniques. More specifically, we show attacks that exploit the connection migration and blacklisting features of AG and LS and quantify their impact.

We use a hardware testbed of five Linux workstations (Fig. 2). One workstation runs the OF v1.0 protocol with OF vSwitch as OFS and pox controller as OFC; the Southbound interface is a 1-Gbps TCP channel between them via loopback interface. Two client machines, one of which is used as a normal client and another as an attacker, and two web servers are connected to the OF switch with dedicated Ethernet cables, creating a private subnet among the five machines.

The SDN switch and controller are run on a workstation with Ubuntu 16.04 LTS operating system (OS), Linux 4.15.0-34-generic kernel, Intel core i7-7700 processor at 3.69 GHz clock rate, and 32 GB, 2400 MHz DRAM memory, and a 4-port gigabit Ethernet interface card. Each of the two servers is run on a workstation with Ubuntu 16.04 LTS, Linux 4.4.34abc, Apache 2.4.18 (Ubuntu), Intel Core2 Quad processor Q9550 at 2.83 GHz, 8 GB synchronous 800 MHz DRAM and a gigabit Ethernet interface card. The normal client is run on a workstation with Ubuntu 16.04 LTS, Linux 4.4.0-138 kernel, Intel Core2 Quad processor Q9550 at 2.83 GHz, 8 GB synchronous 800 MHz DRAM and a gigabit Ethernet card. The attacker is run on a workstation Ubuntu 16.04 LTS OS, Linux 4.4.0-134 kernel, Intel Core i7-2600 processor at 3.4 GHz, and 16 GB DRAM at 1333 MHz, and a gigabit Ethernet card. We experimented with different link speeds and link delays. All results we report in this paper are based on link settings with 1 Gb/s capacity and no additional link delay introduced.

The client uses cURL [30], a tool to download web pages, to download 1 KB files from Server 1 (Web Server in Fig. 2) back to back throughout each experimental run. The attacker uses BoNeSi [31], a DDoS botnet simulator, to launch desired SYN floods or other attacks at the specified rates on Server 2 (Background Server in Fig. 2). The attack and normal traffic compete for only one resource: the SDN switch. We implemented AG and LS extensions to OF vSwitch, pox controller, and the OF protocol. For ease of presentation, we denote the standard OF vSwitch as OF switch, and the one with AG extensions as AG switch and the one with LS extensions as LS switch.

We use two metrics: the mean time to loss (MTTL) of client-server connectivity, which is the average time it takes the client to experience first loss of connection, and the average client download (DL) time of a file from Server 1.

Before we show the impact of attacks on vulnerabilities introduced by DDoS mitigation schemes, we present in Fig. 3 the MTTL of client-server connectivity under SYN flooding attack. This serves as a reminder of the need for AG or some other SYN flood mitigation technique.

3.1   ACK Flooding Attack

The ACK flooding attack exploits AG's connection migration technique in which an ACK sent by the client in response to SYN-ACK proxy results in a packet_in message from the switch to the controller and the controller processing the connection request. However, since no state is maintained for the received ACKs, an attacker can simply send more than one ACK to effectively launch an ACK flooding attack.

We conducted several experiments to quantify the impact of this attack. The attacker uses the BoNeSi tool to send SYN requests to the Background Server at the rate of 1 to 10 per second (denoted as connections/second or CPS). For each SYN-ACK received, the attacker sends multiple ACKs so that the volume (number) of ACKs sent per second is 250, 500, 750, or 1000. The client uses cURL to request a 1 KB file from the Web Server back to back until the experiment is terminated or the Web Server is unresponsive. In each experiment, the client starts at time 0 s and the attacker starts at 30 s; the total experiment time is 2 min: 30 s without attack and 90 s with attack. We ran experiments for AG, LS, and OF switches. For each scenario, we ran the experiment 20 times, dropped two best and worst results and averaged the remaining 16 results.

Figure 4 shows the MTTL of client-server connectivity from the start of the attack. The maximum possible time is 90 s if the client is able to download files throughout the experiment. As the ACK volume and CPS increase, the downloads stop before the end of the experiment. With the original OF switch, the MTTL is 90 s in these experiments (not shown in the figure).

The average download times seen by the client during the attack increase with CPS and ACK volume. Figure 5

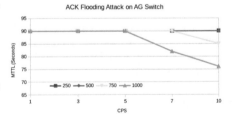

Fig. 4. MTTL of client-server connectivity during ACK attacks on the AG switch.

presents these results for AG and OF switches. With the OF switch, the download times increase because the duplicate ACKs require FT entry lookup forwarding them to the server. This results in steady but manageable increase in download times seen by the client. On the other hand, with the AG switch, the download times increase more rapidly because duplicate ACKs received by the switch result in additional packet_in messages to the controller until it receives a FT entry. Figure 6 shows that the controller's response time to packet_in messages by the AG switch increases significantly during the ACK attack.

The LS switch performs only slightly worse than the OF switch since most of the SYN requests are handled normally; only a small percentage (5%) or the SYNs are handled using AG's connection migration technique. However, a different attack, in which the attacker sends SYNs at a high rate, ignores SYN-ACKs from the server, but sends multiple ACKs in response to SYN-ACKs from the switch will successfully result in reducing the MTTL to a few seconds similar to that seen for SYN floods on the OF switch (Fig. 3). Given the more debilitating impact of SYN floods, it is safer to use AG's connection migration for all SYNs from external clients when SYN flooding attack is suspected.

3.2 Blacklisting Vulnerability

In another set of experiments, we ran attacks on the LS switch to exploit its blacklist of nodes that send SYNs but do not respond to SYN-ACKs. The attacker spoofs client's IP address to send SYN requests at various rates (1 to 7 per second). We measured the MTTL of client-server connectivity during the attack. Even with a low rate of SYNs sent by the attacker, the MTTL is 4 s or less as shown in Fig. 7.

4 ACK Flooding Attack Mitigation

In this section, we present a simple technique to mitigate ACK flooding attacks on the AG switch and discuss its vulnerabilities and merits. The ACK flooding attack on an AG switch works because the switch does not keep track of the ACKs received in response to its SYN proxy (vulnerability VL5). To address this shortcoming, we use an ACK cache, which keeps track of the recently used SYN cookies in the SYN-ACK responses sent by the AG switch. Let k be a positive integer. We implement an ACK cache to keep track of, up to, k SYN cookies based on the ACKs received and verified to be responses to the SYN-ACKs the switch generated recently. The ACK cache works as follows.

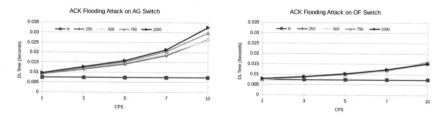

Fig. 5. Average download times seen by the client during ACK attacks on AG and OF switches. Download times without the attacks, labeled N, are plotted for reference.

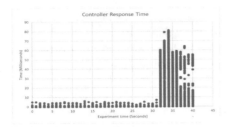
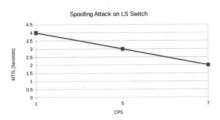

Fig. 6. Controller response times during an ACK attack on the AG switch. The graph shows the first 40 s of the experiment. CPS is 5, and ACK volume is 500.

Fig. 7. The MTTL of client-server connectivity under a TCP SYN spoofing attack to exploit the blacklisting feature of the LS switch.

- Let A be an ACK received and does not have a suitable FT entry to process it. Let $s = a - 1$, where a is the value in the ACK field of A. (If the received ACK is a legitimate response to a SYN-ACK the switch sent, then a is the SYN cookie incremented by 1 as per the TCP specification.)
- If s is found in ACK cache, the ACK, A, is considered to be a duplicate and is dropped.
- Otherwise, A is verified to ensure that it is a response to a SYN-ACK the switch generated recently. The switch recomputes the SYN cookie with the current timestamp and the received ACK information and checks if s is within a specified range from the recomputed SYN cookie.
- If the verification is successful, s is recorded in the ACK cache and a packet_in message is sent to the controller for processing as per the AG logic.

To evaluate the effectiveness of the proposed technique, we implemented an ACK cache with $k = 1024$ entries, each entry capable of storing a 32-bit value. Let $s = (s_{31}s_{30}\ldots s_0)_2$ and $s_{(10)} = s_{23}\ldots s_{14}$ be a 10-bit sequence from s. First, $s_{(10)}$ is used to index into the cache and the 32-bit value stored in that location retrieved. If that value matches s, then the ACK A is dropped. If it does not match s, then a SYN cookie, say t, based on the information from A is computed and is verified that the difference $s - t$ is within an acceptable range of time difference. (Even if A is a legitimate ACK, t may not match s exactly since the latter was computed using the timestamp at that time of the corresponding SYN-ACK generation.) If the verification is successful, the ACK cache entry at $s_{(10)}$ is updated to record s and the ACK is processed by AG's connection migration. If the verification is not successful, A is not considered to be part of AG's connection migration and is handled using the default OF logic for such packets. (This verification step prevents SYN cookie spoofing, which can lead to populating ACK cache with bogus entries and victimization attacks.)

Figure 8 presents the MTTL of client-server connectivity and average client download times during the attack. With the proposed ACK mitigation technique,

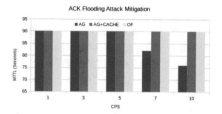 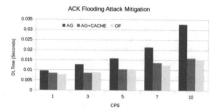

Fig. 8. MTTL of client-server connectivity and average client download times during ACK attacks on the AG switch with and without the proposed ACK attack mitigation. The performance of the OF switch, which is not susceptible to these attacks, is shown for comparison.

the AG switch's performance is similar to that of the OF switch, with SYN flood attack mitigation capability undiminished.

We have not seen any prior discussion of ACK flooding attacks on the use of SYN proxy technique by SDN switches. Consequently, there are no prior mitigation techniques for ACK flooding attacks to compare with the proposed ACK cache technique.

ACK Cache Vulnerabilities. The proposed ACK cache has a potential vulnerability (VL7) since an ACK value not found in the cache triggers a packet_in message to the controller. We discuss a possible attack that exploits this vulnerability and make the cache ineffective. We assume that SYN cookies are generated using cryptographically strong hash functions and that the hash values generated by them are practically random values.

Suppose the attacker received two different SYN-ACKs from a switch with cookies that have the same 10-bit sequence used to index into the ACK cache. Then, by sending ACKs to these two SYN-ACKs alternatively to overwrite the ACK cache entry, the attacker can create conflict misses and render the ACK cache checks useless. (Note that if one of these SYN cookies is sent to a legitimate client, then the conflict miss occurs at most twice since the legitimate client sends only one ACK.)

Since SYN cookies are assumed to be random bit sequences, two SYN cookies will have the same 10-bit sequence in positions 23 through 14 with probability 2^{-10}. Therefore, on the average, 1024 SYN cookies from one switch need to be received by the attacker within a short time to launch this attack.

With a 1024-entry ACK cache, the 32-bit space of SYN cookie is mapped to a 10-bit space similar to the direct-mapping used for processor cache memories in computer systems [32]. Overwriting an ACK cache location with different SYN cookies is equivalent to conflict misses in a processor cache memory.

We can make the conflict misses even more unlikely by using the set-associative cache design principles. For example, if s is the value found in the ACK field of a received ACK, we could use a 9-bit sequence, $s_{(9)} = (s_{23} \ldots s_{15})$ to index into cache but check two locations $2 * s_{(9)}$ and $2 * s_{(9)} + 1$ for the matching cookie. This would require the attacker to have 3 SYN-ACKs from the same switch that have SYN cookies with the same 9-bit sequence to render the ACK cache checks useless. The probability of three SYN cookies having the same 9-bit sequence embedded is $2^{-9} \times 2^{-9} = 2^{-18}$; so it takes an average of 2^{18} SYN cookies to create a conflict miss. We could use an 8-bit sequence to index into the ACK cache and check 4 consecutive locations to find the matching cookie, with probability 2^{-32} to create a conflict miss.

The proposed ACK attack mitigation avoids the vulnerability (VL5) —using simplistic indicators—since a received ACK is verified before dropping it or modifying the ACK cache. This, in turn, makes the blacklist (the ACK cache) robust and difficult exploit.

*ACK Cache Advantages.*It takes only a small and constant amount of time to verify if the ACK was already received. The ACK cache requires very little maintenance. Just like a processor cache, the ACK cache can use FIFO replacement policy if set associativity is 2 or higher.

A small amount of memory is sufficient since SYN cookies have a short lifetime and the entries in it are populated by received ACKs that are verified to be replies to the SYN-ACKs the switch sent. This ensures that the old ACK cache entries are purged at the rate at which distinct and true ACKs are received, which is self-pacing, thereby increasing its effectiveness in a switch congested with SYN or ACK floods. If at most k distinct ACKs are expected to processed during the lifetime of a SYN cookie, then an ACK cache of size k or slightly larger is sufficient.

The probability of a false positive is considered to be extremely low since the probability of generating two SYN cookies with the same value is extremely low owing to the cryptographic hash function used for SYN cookie generation. The cost of a false positive is low if the client sends data first; if the server is expected to initiate data exchange, then the cost of a false positive is a hung client. The probability of a false negative is low as discussed above in constructing an attack on the ACK cache. Furthermore, the cost of a false negative is very low: an extra packet_in message is sent to the controller.

5   Related Work

Software-defined networking and OpenFlow design were introduced in a landmark paper by McKeown et al. [1]. The OpenFlow protocol and OF switch design were discussed in [20]. OpenFlow vSwitch code [33,34] spurred the adaption of SDNs for campus and data center networks.

TCP SYN flooding attacks are the most frequently investigated DDoS attacks for legacy networks [5,7,35]. The DDoS problem in the context of SDNs is investigated by many researchers and several solutions were proposed [9,12,13,15–19,36–42]. Some solutions suggest architectural or protocol modifications to mitigate DDoS attacks; some propose machine learning and deep learning techniques to train attack detection methods using large amounts of network data with attack and normal traffic; some propose SDN security framework to cope with a wide variety of DDoS and other attacks at network, transport, and application level; and some address internal attacks, while others external attacks.

The works that are most relevant to our work are SYN flood mitigation techniques using architectural modifications and protocol tweaks such as the Avant-Guard technique by Shin et al. [13] and the LineSwitch by Ambrosin et al. [15] and other works such as SDN Shield [16] which use these or similar techniques. We have described these results and the vulnerabilities they introduce in Sect. 2.

Many other results on DDoS defense for SDNs involve elaborate frameworks and complex techniques including machine learning for DDoS detection. Some of the recent works on DDoS defense frameworks for SDNs can be found in [17, 18, 36, 38–41]. Anomaly detection based on machine learning, statistical analysis, Shannon's entropy, generalized (Renyi's) entropy, and information distance were presented in [36, 43–45].

6   Conclusions

Distributed denial of service attacks, in particular, TCP SYN flooding attacks are particularly problematic in SDNs owing to the low cost of sending TCP SYNs and the high cost of responding to them by the targeted SDN and server. The detection and mitigation techniques proposed to counter SYN floods often introduce additional and different vulnerabilities. In this paper, we have investigated the vulnerabilities introduced by a few DDoS mitigation techniques for software-defined networks.

Our primary contribution is the identification of the sources of vulnerabilities and using them to investigate new vulnerabilities introduced by mitigation techniques. For this purpose, we identified nine sources of vulnerabilities commonly introduced by mitigation techniques and could be exploited by DDoS attacks. As an example of this approach, using this list, we analyzed the new vulnerabilities of Avant-Guard, a well-known technique proposed for TCP SYN flood mitigation. We demonstrated the impact of the new vulnerabilities using experiments on an SDN testbed. We proposed a solution to overcome Avant-Guard's vulnerabilities and analyzed the vulnerabilities of our solution. Our solution solves the vulnerability it aimed to solve while any vulnerabilities it introduces are hard to exploit.

In future, we would like to use the vulnerability list to develop new solutions to TCP SYN flood detection and mitigation that are resistant to exploitation by new attacks. We are interested in evaluating the performance issues and vulnerabilities caused by the SYN proxy mechanism implemented by network vendors in their products. We are also interested in investigating the vulnerabilities or limitations introduced by machine learning, entropy, and statistical techniques for anomaly detection and DDoS detection.

Acknowledgment. This research is partially supported by grant H98230-18-1-0335 from the National Security Agency, U.S.A. The claims and opinions expressed in this document are solely those of the authors and do not represent those by the U.S. government.

References

1. McKeown, N., Anderson, T., Balakrishnan, H., Parulkar, G., Peterson, L., Rexford, J., Shenker, S., Turner, J.: OpenFlow: enabling innovation in campus networks. SIGCOMM Comput. Commun. Rev. **38**(2), 69–74 (2008). http://doi.acm.org/10.1145/1355734.1355746
2. Kreutz, D., Ramos, F.M., Verissimo, P., Rothenberg, C.E., Azodolmolky, S., Uhlig, S.: Software-defined networking: a comprehensive survey. Proc. IEEE **103**(1), 14–76 (2015)
3. Garber, L.: Denial-of-service attacks rip the internet. Computer **33**(4), 12–17 (2000)
4. Kolias, C., Kambourakis, G., Stavrou, A., Voas, J.: DDoS in the IoT: mirai and other botnets. Computer **50**(7), 80–84 (2017)
5. Zargar, S.T., Joshi, J., Tipper, D.: A survey of defense mechanisms against distributed denial of service (DDoS) flooding attacks. IEEE Commun. Surv. Tutor. **15**(4), 2046–2069 (2013)
6. Chang, R.K.: Defending against flooding-based distributed denial-of-service attacks: a tutorial. IEEE Commun. Mag. **40**(10), 42–51 (2002)
7. Mirkovic, J., Reiher, P.: A taxonomy of DDoS attack and DDoS defense mechanisms. ACM SIGCOMM Comput. Commun. Rev. **34**(2), 39–53 (2004)
8. Blankenship, J.: DDoS mitigation solutions, Q4 2017. In: The Forrester Wave^{TM}, December 2017
9. Akhunzada, A., Ahmed, E., Gani, A., Khan, M.K., Imran, M., Guizani, S.: Securing software defined networks: taxonomy, requirements, and open issues. IEEE Commun. Mag. **53**(4), 36–44 (2015)
10. Alsmadi, I., Xu, D.: Security of software defined networks: a survey. Comput. Secur. **53**, 79–108 (2015)
11. Zhang, P., Wang, H., Hu, C., Lin, C.: On denial of service attacks in software defined networks. IEEE Netw. **30**(6), 28–33 (2016)
12. Yan, Q., Yu, F.R., Gong, Q., Li, J.: Software-defined networking (SDN) and distributed denial of service (DDoS) attacks in cloud computing environments: a survey, some research issues, and challenges. IEEE Commun. Surv. Tutor. **18**(1), 602–622 (2016)
13. Shin, S., Yegneswaran, V., Porras, P., Gu, G.: AVANT-GUARD: scalable and vigilant switch flow management in software-defined networks. In: Proceedings of the ACM Conference on Computer & Communications Security, pp. 413–424 (2013)
14. Postel, J.: Transmission control protocol. STD 7, RFC Editor, September 1981. http://www.rfc-editor.org/rfc/rfc793.txt
15. Ambrosin, M., Conti, M., Gaspari, F.D., Poovendran, R.: Lineswitch: tackling control plane saturation attacks in software-defined networking. IEEE/ACM Trans. Netw. **25**(2), 1206–1219 (2017)
16. Chen, K., Junuthula, A.R., Siddhrau, I.K., Xu, Y., Chao, H.J.: SDNShield: towards more comprehensive defense against DDoS attacks on SDN control plane. In: IEEE Conference on Communications and Network Security (CNS), pp. 28–36. IEEE (2016)

17. Mohammadi, R., Javidan, R., Conti, M.: Slicots: an SDN-based lightweight countermeasure for TCP SYN flooding attacks. IEEE Trans. Netw. Serv. Manag. **14**(2), 487–497 (2017)
18. Fichera, S., Galluccio, L., Grancagnolo, S.C., Morabito, G., Palazzo, S.: OPERETTA: an openflow-based remedy to mitigate TCP synflood attacks against web servers. Comput. Netw. **92**, 89–100 (2015)
19. Sahay, R., Blanc, G., Zhang, Z., Debar, H.: ArOMA: an SDN based autonomic DDoS mitigation framework. Comput. Secur. **70**, 482–499 (2017)
20. Specification, O.S.: Version 1.0. 0 (wire protocol 0x01). Open Networking Foundation (2009). https://www.opennetworking.org/wp-content/uploads/2013/04/openflow-spec-v1.0.0.pdf
21. Schabel, L.: Working with synproxy. https://github.com/firehol/firehol/wiki/Working-with-SYNPROXY
22. Bernstein, D.J.: Syn cookies. https://cr.yp.to/syncookies.html
23. Cisco, Inc.: Configuring Firewall TCP SYN Cookie, chap. Cisco ASR 1000 Series Aggregation Services Routers. Cisco, Inc., August 2018
24. Cisco Systems, Inc.: Implementing Open Flow Agent, chap. Cisco ASR 9000 Series Aggregation Services Router System Management Configuration Guide, Release 5.1.x. Cisco Systems, Inc., October 2017
25. Juniper Inc.: OpenFlow Support on Juniper Networks Devices, chap. Juniper, Inc., Junos OS (2018)
26. Juniper Networks, Inc.: Junos Space Network Management Platform Hardening, chap. Junos Space Service Automation, Release 16.1. Juniper Networks, Inc., August 2017
27. CVE details, May 2019. https://www.cvedetails.com/vulnerability-list/vendor_id-13628/product_id-36893/Opendaylight-Openflow.html
28. NIST: Cve-2018-1000155 detail. https://nvd.nist.gov/vuln/detail/CVE-2018-1000155. Accessed May 2019
29. Bifulco, R., Cui, H., Karame, G.O., Klaedtke, F.: Fingerprinting software-defined networks. In: 2015 IEEE 23rd International Conference on Network Protocols (ICNP), pp. 453–459. IEEE (2015)
30. cURL client. https://github.com/php-http/curl-client
31. BoNeSi: the DDoS botnet simulator. https://github.com/Markus-Go/bonesi
32. Hennesy, J.L., Patterson, D.A.: Computer Architecture: A Quantitative Approach, 6th edn. Morgan Kaufmann Publishers, Burlington (2017)
33. Openflow. https://github.com/mininet/openflow
34. Pfaff, B., Pettit, J., Koponen, T., Jackson, E., Zhou, A., Rajahalme, J., Gross, J., Wang, A., Stringer, J., Shelar, P., et al.: The design and implementation of Open vSwitch. In: USENIX Symposium on Networked Systems Design and Implementation (NSDI), pp. 117–130 (2015)
35. Wang, H., Zhang, D., Shin, K.G.: Detecting SYN flooding attacks. In: Twenty-First Annual Joint Conference of the IEEE Computer and Communications Societies, vol. 3, pp. 1530–1539. IEEE (2002)
36. Wang, B., Zheng, Y., Lou, W., Hou, Y.T.: DDoS attack protection in the era of cloud computing and software-defined networking. Comput. Netw. **81**, 308–319 (2015)
37. Shang, G., Zhe, P., Bin, X., Aiqun, H., Kui, R.: FloodDefender: protecting data and control plane resources under SDN-aimed DoS attacks. In: IEEE INFOCOM 2017-IEEE Conference on Computer Communications, pp. 1–9. IEEE (2017)

38. Sonchack, J., Smith, J.M., Aviv, A.J., Keller, E.: Enabling practical software-defined networking security applications with OFX. In: NDSS, vol. 16, pp. 1–15 (2016)
39. Porras, P., Shin, S., Yegneswaran, V., Fong, M., Tyson, M., Gu, G.: A security enforcement kernel for OpenFlow networks. In: Proceedings of the First Workshop on Hot Topics in Software Defined Networks, pp. 121–126. ACM (2012)
40. Hu, H., Han, W., Ahn, G.J., Zhao, Z.: FLOWGUARD: building robust firewalls for software-defined networks. In: Proceedings of the Third Workshop on Hot Topics in Software Defined Networking, pp. 97–102. ACM (2014)
41. Wang, H., Xu, L., Gu, G.: Floodguard: a DoS attack prevention extension in software-defined networks. In: Proceedings of the 2015 45th Annual IEEE/IFIP International Conference on Dependable Systems and Networks, DSN 2015, pp. 239–250. IEEE Computer Society, Washington (2015). http://dx.doi.org/10.1109/DSN.2015.27
42. Shin, S.W., Porras, P., Yegneswara, V., Fong, M., Gu, G., Tyson, M.: Fresco: modular composable security services for software-defined networks. In: 20th Annual Network & Distributed System Security Symposium, NDSS (2013)
43. Mousavi, S.M., St-Hilaire, M.: Early detection of DDoS attacks against SDN controllers. In: 2015 International Conference on Computing, Networking and Communications (ICNC), pp. 77–81. IEEE (2015)
44. Bhuyan, M.H., Bhattacharyya, D., Kalita, J.K.: An empirical evaluation of information metrics for low-rate and high-rate DDoS attack detection. Pattern Recognit. Lett. **51**, 1–7 (2015)
45. Sahoo, K.S., Puthal, D., Tiwary, M., Rodrigues, J.J., Sahoo, B., Dash, R.: An early detection of low rate DDoS attack to SDN based data center networks using information distance metrics. Futur. Gener. Comput. Syst. **89**, 685–697 (2018)

Investigating Crowdsourcing to Generate Distractors for Multiple-Choice Assessments

Travis Scheponik[1]([✉]), Enis Golaszewski[1], Geoffrey Herman[2],
Spencer Offenberger[2], Linda Oliva[3], Peter A. H. Peterson[4],
and Alan T. Sherman[1]

[1] Cyber Defense Lab, Department of Computer Science and Electrical
Engineering, University of Maryland, Baltimore County (UMBC),
Baltimore, USA
{tschepl,golaszewski,sherman}@umbc.edu
[2] Department of Computer Science, University of Illinois at Urbana-Champaign,
Urbana, USA
{glherman,sol0}@illinois.edu
[3] Department of Education, UMBC, Baltimore, USA
oliva@umbc.edu
[4] Department of Computer Science, University of Minnesota Duluth,
Duluth, USA
pahp@d.umn.edu

Abstract. We present and analyze results from a pilot study that explores how crowdsourcing can be used in the process of generating distractors (incorrect answer choices) in multiple-choice concept inventories (conceptual tests of understanding). To our knowledge, we are the first to propose and study this approach. Using Amazon Mechanical Turk, we collected approximately 180 open-ended responses to several question stems from the Cybersecurity Concept Inventory of the Cybersecurity Assessment Tools Project and from the Digital Logic Concept Inventory. We generated preliminary distractors by filtering responses, grouping similar responses, selecting the four most frequent groups, and refining a representative distractor for each of these groups.

We analyzed our data in two ways. First, we compared the responses and resulting distractors with those from the aforementioned inventories. Second, we obtained feedback from Amazon Mechanical Turk on the resulting new draft test items (including distractors) from additional subjects. Challenges in using crowdsourcing include controlling the selection of subjects and filtering out responses that do not reflect genuine effort. Despite these challenges, our results suggest that crowdsourcing can be a very useful tool in generating effective distractors (attractive to subjects who do not understand the targeted concept). Our results also suggest that this method is faster, easier, and cheaper than is the traditional method of having one or more experts draft distractors, building on talk-aloud interviews with subjects to uncover their misconceptions. Our results are significant because generating effective distractors is one of the most difficult steps in creating multiple-choice assessments.

Keywords: Concept inventories · Crowdsourcing ·
Cybersecurity assessment tools (CATS) project · Cybersecurity education ·

© Springer Nature Switzerland AG 2020
K.-K. R. Choo et al. (Eds.): NCS 2019, AISC 1055, pp. 185–201, 2020.
https://doi.org/10.1007/978-3-030-31239-8_15

Cybersecurity concept inventory (CCI) ·
Digital logic concept inventory (DLCI) · Distractors ·
Amazon mechanical turk · Multiple-choice questions

1 Introduction

Generating effective multiple-choice questions (MCQs) is a difficult, time-consuming, and expensive process. For example, in the Cybersecurity Assessment Tools (CATS) Project [She17a], after identifying core concepts and student misconceptions [Par16, Tho18], a group of approximately three to five experts spent several hours per assessment item devising scenarios, question stems, and alternatives (answer choices). A significant part of that effort went into creating the distractors (incorrect alternatives). An effective distractor is attractive to subjects who do not thoroughly understand the targeted concept. We present and analyze results from a pilot study that investigates how crowdsourcing can be used to facilitate the process of generating distractors for concept inventories (conceptual tests of understanding).

Crowdsourcing, for example through *Amazon Mechanical Turk (AMT)* [Kit08], offers several attractive capabilities. One can obtain results from a significant number of diverse respondents inexpensively and with fast turnaround. Stems can be easily pilot tested, and responses provide data about the difficulty of the item and patterns of misconceptions. One could collect such data through in-person interviews, but crowdsourcing facilitates collecting the data more easily, quickly, and cheaply.

We find it especially helpful to ask respondents to enter a free response to a specific stem, without providing any alternative—a technique that can also be used in in-person interviews. If the crowdsourcing responses come from the desired subject population, the responses inherently reflect genuine conceptions and misconceptions. When multiple subjects respond with the same or similar incorrect responses, the response is likely an effective distractor, based on a common misconception. In contrast, in the CATS Project, we identified student misconceptions through interviews before generating some of the stems and all of the distractors. Limited time for interviews prevented us from asking subjects all stems, and we generated or modified some of the stems after analyzing the results of the interviews.

It is necessary to obtain feedback on the development of any assessment item. Generating effective distractors benefits from an iterative process. Crowdsourcing greatly facilitates this iterative process.

One way to use crowdsourcing is to collect and analyze a set of responses to a specific stem. First, a team of experts can filter the collected responses to exclude ones that appear to be unresponsive or lacking genuine thought. Second, the team can group similar responses. Third, the team can select disjoint groups of incorrect responses that appeared frequently. Fourth, the team can refine a representative alternative for each of the selected groups. This process can be useful if it yields a feasible number of quality distractors.

As we explain in Sect. 2, MCQs in concept inventories should have the following specific properties [McD18]: each MCQ should target one important concept; there should be exactly one best answer; the alternatives should be homogeneous and disjoint; subjects who understand the concept should find it easy to select the best alternative; and subjects who do not understand the concept should find the distractors attractive [Hes92].

To our knowledge, we are the first to propose using crowdsourcing to facilitate the creation of distractors and the first to conduct a study of this technique. Our hypothesis is that crowdsourcing can be a useful tool for efficiently generating high-quality distractors for multiple choice assessments.

In our experiments, we asked subjects to provide free responses to stems drawn from the *Digital Logic Concept Inventory (DLCI)* [Her10] and the *Cybersecurity Concept Inventory (CCI)* of the CATS Project. Using two different concept inventories helps explore whether crowdsourcing can be useful across diverse domains. In addition, the DLCI provides a well-validated reference against which newly generated distractors can be compared. Validation studies of the CCI are currently underway [Off19].

We analyze subject responses and the resulting distractors in two ways. First, we compare the responses and resulting distractors to those from the DLCI and CCI. Second, also using AMT, we obtain feedback on the resulting new draft test items (including distractors) from additional subjects. Our experiments show that AMT responses from a global set of diverse subjects can be collected and analyzed to generate effective distractors for multiple-choice concept inventories quickly, cheaply, and easily.

In the rest of this paper, we review background and previous work, explain the purpose and methods of our study, present and analyze our results, and discuss issues raised by our study. Our contributions include proposing and investigating the strategy of using crowdsourcing to generate distractors for MCQs in concept inventories.

2 Background

We review background material on multiple-choice questions (MCQs), concept inventories, crowdsourcing, the Digital Logic Concept Inventory (DLCI) and the Cybersecurity Concept Inventory (CCI), and the traditional process we used to generate test items in the CCI.

Multiple-Choice Questions. Each multiple-choice assessment item comprises a stem (question) and some number (e.g., five) alternatives (answer choices). There is one best alternative; the others are distractors (incorrect choices). Optionally, one or more stems may share a common scenario that provides detailed context, possibly including an artifact such as a figure or table.

Concept Inventories. A concept inventory is a multiple-choice criterion-referenced test that assesses student understanding of core concepts and reveals misconceptions (e.g., The Force Concept Inventory [Her10]). They can be used to measure the effectiveness of various teaching approaches.

Crowdsourcing. In crowdsourcing, a potentially large number of workers are hired to perform tasks [How06]. Typically, these workers are unaffiliated with the organization requesting the work and they might be anonymous.

"Cheating" in Crowdsourcing. A well-known limitation of crowdsourcing is that some subjects do not make a genuine effort to complete the task in good faith. For example, some subjects—including humans and non-humans (bots)—try to earn money by completing surveys as quickly as possible. Subjects who cheat might respond by copying the question, inserting arbitrary text, or leaving the response blank. For example, one of our subjects responded with news about the fashion industry to a question about computer science. The anonymous nature of crowdsourcing motivates and enables some subjects to act unencumbered by social norms [Joi99].

DLCI. The Digital Logic Concept Inventory (DLCI) is a concept inventory that has been carefully validated [Her14].

CCI and CATS Project. The Cybersecurity Assessment Tools (CATS) Project is developing rigorous instruments that can measure student learning and identify best practices. The first tool is the Cybersecurity Concept Inventory (CCI), which assesses how well students in any first course in cybersecurity understand core concepts.

Generating MCQs for the CCI. The CATS team developed assessment items for the CCI using a four-step process. First, we identified core concepts through a Delphi process of cybersecurity experts [Par16]. Second, we conducted talk-aloud interviews to expose student misconceptions [Tho18]. Third, we generated stems and alternatives through in-person and on-line discussions. Fourth, we refined draft questions based on cognitive interviews, expert review, and pilot testing.

3 Previous and Related Work

To our knowledge, we are the first to propose using crowdsourcing to generate distractors. For reading comprehension, Araki [Ara16] discusses how natural language processing techniques can be used to generate distractors automatically by extracting events from reading passages. Welbl et al. [Wel17] explore how crowdsourcing can be used to gauge the plausibility of automatically-generated questions. Guo et al. [Guo16] study how automatically-generated questions can enable motivated students to explore additional topics and to reinforce their learning objectives.

Bucholz et al. [Buc11] devise methods for dealing with "cheaters" in their study in which they use AMT to convert text to speech. They develop two metrics for excluding cheaters based on whether the workers were qualified for a task or whether they put forth an honest effort to complete the assigned task. Alonso [Alo08] shows that, by quickly filtering responses for relevance, he can use crowdsourcing to yield high-quality results at low cost.

4   Experiment: Purpose and Methods

We explain our experiment, including its purpose, subject population, question stems, participant steps, how we generated distractors using the AMT platform, and how we analyzed the resulting distractors. UMBC's Institutional Review Board approved the protocol.

4.1   Purpose

The purpose of our experiment is to investigate the feasibility and desirability of using crowdsourcing to generate distractors for MCQ concept inventories, including assessing the quality of the resulting distractors and the difficulty and cost of generating them.

4.2   Subject Population

We invited any 200 AMT workers to complete our survey by posting a "Human Intelligence Task (HIT)" on the AMT site, hoping to obtain at least 50 genuine responses. Subjects were able to find our task from a web user interface that has search capabilities. Seeking greater geographic diversity, we advertised the task in three separate batches, each made available during a separate time period. We have no demographic information about the responding subjects, except that each has an Internet connection and an AMT account. We paid $0.25 to each subject who completed the task, setting a twenty-minute time limit to complete it.

4.3   Question Stems

We asked each subject to answer eight question stems (see Appendix A), four from the DLCI (Questions DLCI 1–DLCI 4) and four from the CCI (Questions CCI 1–CCI 4). Figure 1 gives two representative stems, one from each concept inventory. Using stems

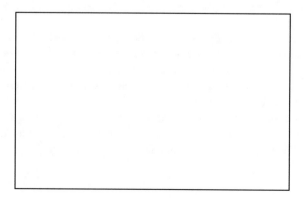

Fig. 1. Two representative stems, one from the Digital Logic Concept Inventory (DLCI), and one from the Cybersecurity Concept Inventory (CCI).

from these two concept inventories helps establish whether crowdsourcing can be used across diverse disciplines. Also, the DLCI provides a useful reference in that it has been carefully validated [Her14].

DLCI 1. A sequential circuit T that has 0 inputs, 3 flip-flops, and 2 outputs. What is the maximum number of distinct states T can potentially be in over time?

CCI 3. Alice runs a top-secret government facility where she has hidden a USB stick, with critical information, under a floor tile in her workspace. The facility is secured by guards, 24/7 surveillance, fences, electronically locked doors, sensors, alarms, and windows that cannot be opened. To gain entrance to the facility, all employees must present a cryptographically hardened ID card to guards at a security checkpoint. All of the computer networks in the facility use state-of-the-art computer security practices and are actively monitored. Alice hires Mark (an independent penetration tester) to exfiltrate the data on the USB stick hidden in her workspace.

Choose the strategy that best avoids detection while effectively exfiltrating the data:
Definitions:
24/7: Twenty-four hours a day, seven days a week.

4.4 Methods: Participant Steps

Each subject carried out the following four steps:

1. The subject logged into AMT.
2. The subject entered search criteria to find the survey (hosted via SurveyMonkey).
3. Upon finding the survey in AMT, if the subject accepted the terms of informed consent, they were given access to the remainder of the survey.
4. The participant answered eight free-response questions: four from the DLCI, and four from the CCI.

4.5 Methods: Processing Responses

We generated candidate distractors by filtering responses, grouping similar responses, selecting the four largest groups, and refining a representative distractor for each of these four groups.

We filtered out responses that appeared not to reflect a genuine effort to answer the question. Specifically, we filtered out responses that were blank, appeared to answer a different question, responded with a question (perhaps reflecting lack of understanding of what the questions was asking), or copied a question from the survey. We made this determination separately for each response.

To group the remaining AMT responses after filtering each question, we first identified all responses that appeared similar to some original alternative from the DLCI or CCI. Then, we arranged the remaining new alternatives in groups of similar responses.

4.6   Methods: Analysis of Distractors

We analyzed the resulting distractors in two ways. First, we compared the free responses from AMT to existing data showing alternatives selected by students in pilot studies of the DLCI and CCI, reporting the results in a series of bar graphs. In this comparison, we report the number of filtered responses, number of responses that overlap existing alternatives from the DLCI and CCI, and number of new alternatives generated from AMT.

Second, we collected data from additional AMT workers who answered the eight questions from the DLCI and CCI, presented in multiple-choice format. We collected these data in two separate surveys: one offered the original alternatives from the DLCI and CCI; the second offered the correct answer and four possibly new distractors from our AMT experiment. We report our results in bar graphs.

As part of the second analysis step, we compared the performance of university students on the DLCI and CCI to the performance of AMT workers on these instruments. We report these results in a stacked bar graph.

5   Results

We present results from three surveys that we conducted on AMT in April 2019: Survey 1 collected open-ended responses to eight stems from the DLCI and CCI; Survey 2 administered eight existing test items from the DLCI and CCI to AMT workers; and Survey 3 administered eight modified test items from these concept inventories using distractors generated from Survey 1. After presenting results from these surveys, we compare the performance of AMT workers with that of college and university students. We also give examples of AMT responses to one of the CCI stems.

5.1   Survey 1: Generating Distractors from Open-Ended Questions

In Survey 1, without providing any alternatives, we asked AMT workers to answer four open-ended stems from the DLCI and four open-ended stems from the CCI. 539 workers completed the survey, of which 113 (21%) answered in good faith. We discarded 426 (79%) of the responses because they were blank or otherwise nonresponsive.

Figure 2 shows how we categorized the 539 responses. For each of the 113 genuine responses, we classified it as either correct, (incorrect and) overlapping an existing distractor, or (incorrect and) a new distractor.

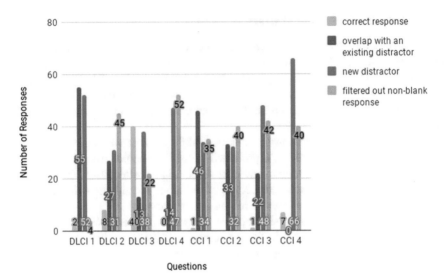

Fig. 2. Categorization of 536 AMT responses collected from open-ended stems from the Digital Logic Concept Inventory (DLCI) and Cybersecurity Concept Inventory (CCI). We filtered out 426 responses as blank or otherwise nonresponsive, leaving 113 genuine responses.

DLCI Distractor Generation

Questions DLCI 1–4 can be answered with numerals or short sentences.

Filtering. DLCI Question 1 had the smallest percentage of filtered responses, perhaps because it asked for a numeric answer.

Grouping. Each set of genuine responses resulted in at least four new potential distractors.

Overlap. Of the genuine responses, over 50% represent potential new distractors.

Example. Question DLCI 2 asked: "Define the word state when used to describe a sequential circuit?" Filtered responses include: "2," "Andhra Pradesh," "tech," "New York," and "Sequential logic is a form of binary circuit design that employs one or more inputs and one or more outputs." Responses that overlap various existing distractors or the correct answer include: "either input output or flip-flop," "State of the circuit depends on the inputs it gets," "memory of element," and "output." The similar responses "If the circuit is on or off," "off-on," "on or off," "whether it is on or off," and others were combined and polished, resulting in the new distractor "State is whether the circuit is 'on' or off'."

CCI Distractor Generation

Questions CCI 1–4 seek short sentences as answers.

Filtering. For each CCI question, we filtered out approximately 40% of the responses.

Grouping. Each set of genuine responses resulted in potential new distractors. For each question, the most prevalent two potential new distractors accounted for over 50% of the genuine responses.

Overlap. Of the genuine responses, the percent that overlapped existing distractors varied from 0% to 45%.

Example. See Sect. 5.4

5.2   Surveys 2–3: AMT Workers Answer Original and Modified Questions from the DLCI and CCI

Figures 3, 4 show the performance of AMT workers answering eight original (Fig. 3) and modified (Fig. 4) test items from the DLCI and CCI. The modified questions use the four most popular distractors generated from Survey 1.

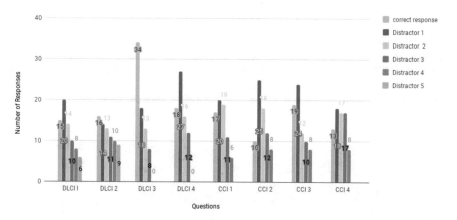

Fig. 3. Distribution of 131 responses by AMT workers to eight of the original expert-created multiple-choice questions from the Digital Logic Concept Inventory (DLCI) and Cybersecurity Concept Inventory (CCI). The correct alternative is listed first, followed by the other distractors ordered by their observed frequency. For these test items, AMT workers answered DLCI 3 correctly more frequently than they did for the other questions.

5.3   Comparison of Students and AMT Workers on Original Questions from the DLCI and CCI

Using eight original multiple-choice test items from the DLCI and CCI, Fig. 5 compares the performance of AMT workers with that of college and university students who participated in validation studies of these concept inventories. These students were taking, or had taken, introductory classes in digital logic or cybersecurity, which studied topics on which the concept inventories are based. On every question, the students outperformed the AMT workers.

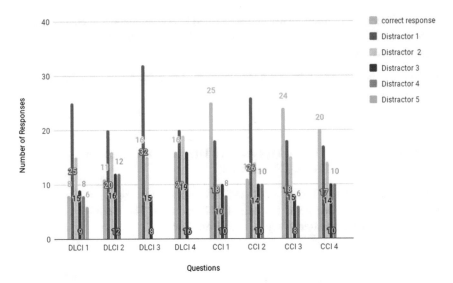

Fig. 4. Distribution of 131 responses by AMT workers to eight modified multiple-choice questions from the Digital Logic Concept Inventory (DLCI) and Cybersecurity Concept Inventory (CCI), using distractors generated from Survey 1. For DLCI 3, AMT workers performed worse on the modified question than on the original question (Fig. 3), probably because many of them found the AMT-generated distractor very attractive.

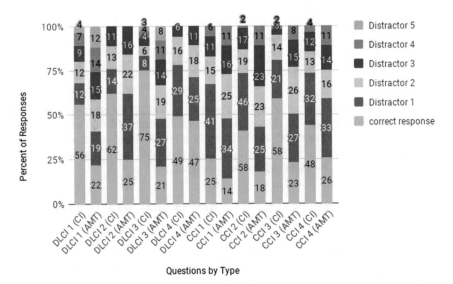

Fig. 5. Comparison of student (CI) and AMT worker (AMT) responses to eight original multiple-choice questions from the DLCI and CCI. For every question, students answered correctly more frequently than did AMT workers.

5.4   Example of Alternatives Generated from AMT Workers

To illustrate the process and results of using crowdsourcing to generate alternatives, we summarize AMT responses from Question CCI 3 (see Fig. 1).

After a scenario describing an elaborate security system and penetration-testing task, CCI 3 asked, "How can Mark best avoid detection while exfiltrating the data?" First, we filtered AMT responses that did not reflect genuine effort, excluding, for example, "90," a lengthy quote from a Michelin guide interview with a famous chef, "NO", "yes," "c," "24/8" and many others. Second, we categorized the remaining responses as correct, overlapping an original distractor, or new. The original alternatives are:

A. Convince an authorized employee to remove the USB stick and give it to Mark. (correct)
B. Compromise the facility's network and add Mark as an authorized guest.
C. Unlock electronically-locked doors using malware.
D. Climb over the perimeter fence at night and sneak into Alice's workspace.
E. Fabricate a fake ID to fool the guards at the security checkpoint.

Examples of responses that overlap various existing alternatives include: "bribe another employee to acquire the item for him" (the correct answer), "Create a copy of an ID card" (overlapping E), and "Get in as guest with current approved employee" (overlapping B). Third, we grouped the new distractors. For example, we grouped the similar responses, "say no," "to not do it," "avoid for safety," and others. Fourth, we refined each of the four most popular groups to yield the following refined AMT alternatives:

A. Convince an authorized employee to remove the USB stick and give it to Mark. (correct)
B. Steal an ID card and use this ID to gain access to the facility. (new)
C. Refuse to take the job as security is too strong. (new)
D. Write malicious code to read the data from the USB. (new)
E. Get hired as an employee of the facility. (new)

Filtering and categorizing AMT responses requires human judgment and subject expertise. For example, some AMT responses were somewhat similar to existing distractors but differed in some important details. For example, we feel that stealing an ID is different from fabricating an ID.

As described above, some AMT responses overlapped existing distractors. The AMT responses with new distractors, however, appeared more frequently, so we used them in the test with AMT-derived alternatives.

For this question, AMT workers produced the correct response, responses that overlapped with the original distractors, and new distractors.

6   Discussion

We now discuss several issues raised by our pilot study in which, using AMT, we generated distractors for four questions from the DLCI and four questions from the CCI. Specifically, we discuss the quality of distractors generated using AMT, the financial cost and difficulty of generating these distractors, challenges with using crowdsourcing, lessons learned using AMT, limitations of our study, suggestions for improving AMT, and open problems.

6.1   Quality of Distractors Generated Using Mechanical Turk

Our pilot study generated many quality distractors, including ones similar to distractors on the DLCI and CCI and new distractors. Subsequent testing of the resulting complete test items on AMT with new subjects confirmed the attractiveness of the distractors to AMT workers. It remains to be determined whether other target audiences will also find these distractors appealing.

6.2   Cost and Difficulty of Generating Distractors Using Mechanical Turk

Using AMT, we can easily collect many (e.g., 200) responses overnight, paying $0.25 a worker per task (set of responses). The resulting data can be processed and analyzed in a small number of hours.

In this study, for the CCI, a we spent a total of four hours and twenty-five minutes to analyze, filter the responses, and generate new distractors. For the DLCI, we spent a total of one hour and ten minutes. This difference in time results primarily because the DLCI answers are largely numeric, whereas the CCI answers are mostly prose.

In comparison, as explained in Sect. 2, the traditional process we used to develop the CCI involved many hours of work by a team of experts, carried out over a period of weeks. The process of interviewing subjects and analyzing the resulting transcripts is difficult.

6.3   Challenges of Crowdsourcing

It is difficult to control the selection of subjects and to ensure that subjects are human. Motivated in part by financial gain, some may try to "cheat" by plagiarizing responses or by not making a genuine attempt to answer the stem. The potentially large set of responses must be processed. Unlike in-person interviews, crowdsourcing does not permit the interviewer to deviate from the script interactively.

For some questions, AMT workers produced responses that were "too good" to include, given that there is supposed to be one best (not necessarily ideal) alternative. In some of these cases, while developing the CCI, we had previously considered and rejected these responses. Test developers sometimes find it convenient to direct a subject's reasoning about a chosen concept by intentionally excluding certain powerful alternatives.

6.4   Lessons Learned

The choice of the worker reward and deadline can influence the likelihood of obtaining the desired data and possibly the quality of results. We offered $0.25 a task with a twenty-minute deadline and found that those choices typically produced at least one hundred responses by the deadline. Currently, a reward of $7 is needed to appear on the first page of advertised tasks, but being on this first page does not seem to be very important.

Responses suggest that many workers were candid, for example stating that they did not understand a question. We speculate that the anonymity of the process encouraged many workers to write statements that some subjects in in-person interviews might feel embarrassed to utter.

We found it useful to direct workers to an on-line survey (using SurveyMonkey) to facilitate collecting the desired information. We did not restrict subjects by answers to demographic questions because we did not have confidence that they would answer truthfully. Given that AMT is available throughout the world, in an attempt to increase the diversity of the respondents across time zones, we performed multiple batches of work at different times of the day. For example, for each task, we performed three batches each separated by approximately six to eight hours. In different batches, the team tended to receive responses from users in different time zones, for example, students from the University of Hawaii and Eastern Kentucky University responded in different batches. We, however, have not scientifically analyzed the effects of this strategy.

6.5   Study Limitations

Limitations of our study using AMT include very limited control of the subject population and the impossibility of verifying whether subjects had the credentials we sought and the absence of traditional validation studies of the resulting test items. Other limitations include sample size and that the process of grouping the distractors inherently requires some subjective judgement.

6.6   Suggestions for Improving Mechanical Turk

It would greatly improve the value of the services offered by AMT if it were possible to control more reliably the characteristics of the subjects. For example, we would have liked to restrict subjects to people who have recently taken a first course in cybersecurity. AMT does offer "premium qualifications," but we would like more choices and greater assurance that subjects have the characteristics that they purport to have. For example, AMT offers qualifications based on income, but this qualification has no bearing on our study.

Relatedly, it would be helpful to be obtain more information about each worker who completes any task. For example, it would be useful to know the reputation of the worker based on their history and whether the worker is human. Such information could be used to filter responses.

6.7   Open Problems

It would be interesting to explore in greater detail how crowdsourcing can be used to support the entire process of creating a concept inventory, from identifying core concepts through validating draft inventories. It would be helpful to learn more about different methods for generating distractors, the quality of the distractors produced, and the effects of reward pricing. The most robust way to evaluate the quality of distractors is to engage in traditional validation studies of resulting test items, including expert review and large-scale psychometric testing [Her10].

Early in our work we had considered the possibility of evaluating distractors by experts, where the experts would rate distractors not knowing their source. One could thereby compare distractors generated by different methods, including using crowdsourcing and traditional methods. Ultimately, we chose not to carry out such an evaluation because experts are known to be weak arbiters of the attractiveness of distractors [Nat11].

7   Conclusion

Our pilot study shows that crowdsourcing can be a tremendously efficient method for generating distractors for multiple-choice concept inventories, saving developers significant amounts of time and money when compared to the more common and much more costly approaches of discussions and in-person interviews.

Using AMT, we generated distractors for the DLCI and CCI, including new distractors and ones similar to those on these concept inventories. The process also produced useful feedback from subjects on stems that they found unclearly worded. Crowdsourcing is very well suited for repeating work tasks with modified material.

Challenges of crowdsourcing include limited control of subjects, the need to filter out responses that do not reflect genuine effort, and the effort required to process the responses and to refine the resulting distractors. We found that filtering and processing the responses was fairly easy to do, and all distractors must be refined regardless of how they were generated. Even though a high percent of our responses were of low quality, the process was still very useful because it generated a sufficient number (at least four) of quality distractors.

Although we focus on concept inventories, we conjecture that crowdsourcing is also useful to develop other types of MCQs. Our promising initial results inspire us to continue to use, study, and refine crowdsourcing to generate distractors. We encourage others to do likewise.

Acknowledgments. This work was supported in part by the U.S. Department of Defense under CAE-R grants H98230-15-1-0294, H98230-15-1-0273, H98230-17-1-0349, and H98230-17-1-0347; and by the National Science Foundation under SFS grants 1241576, 1753681, and 1819521, and DGE grant 1820531.

Appendix A: DLCI and CCI Question Stems

A.1 Selected Stems from the Digital Logic Concept Inventory

DLCI 1. Asequential circuit T that has 0 inputs, 3 flip-flops, and 2 outputs. What is the maximum number of distinct states T can potentially be in over time?

DLCI 2. Which statement best defines the word state when used to describe a sequential circuit?

DLCI 3. A combinational circuit is specified by the truth table below. For three input combinations, the output of the circuit does not matter ("don't-care"). The specification is implemented as a circuit using the following Boolean expression: $f = ac + b$. What will the circuit output when it receives the input combination $<a,b,c> = <1,1,0>$?

```
0 0 0 | 1
0 0 1 | 0
0 1 0 | 1
0 1 1 | 1
1 0 0 | 0
1 0 1 | X (don-t care)
1 1 0 | X (don-t care)
```

DLCI 4. Why do computers use two's complement number representation?

A.2 Selected Stems from the Cybersecurity Concept Inventory

CCI 1. Bob's manager Alice is traveling abroad to give a sales presentation about an important new product. Bob receives an email with the following message: "Bob, I just arrived and the airline lost my luggage. Would you please send me the technical specifications? Thanks, Alice."

Upon receiving Alice's message, choose what action Bob should take?

CCI 2. A company delivers packages to customers using drones. The company's command center controls the drones by exchanging messages with them. The company's command center authenticates each message with a keyed message authentication code (MAC), using a key that is known by the command center and installed in each drone at initialization. The command center stores this key encrypted in a database.

Choose the most promising action for a malicious adversary to masquerade as the command center:

Definitions:
to masquerade: To pretend to be someone else.

CCI 3. Alice runs a top-secret government facility where she has hidden a USB stick, with critical information, under a floor tile in her workspace. The facility is secured by guards, 24/7 surveillance, fences, electronically locked doors, sensors, alarms, and windows that cannot be opened. To gain entrance to the facility, all employees must present a cryptographically hardened ID card to guards at a security checkpoint. All of the computer networks in the facility use state-of-the-art computer security practices and are actively monitored. Alice hires Mark (an independent penetration tester) to exfiltrate the data on the USB stick hidden in her workspace.

Choose the strategy that best avoids detection while effectively exfiltrating the data:

Definitions:
24/7: Twenty-four hours a day, seven days a week.

CCI 4. When a user Mike O'Brien registered a new account for an online shopping site, he was required to provide his username, address, first and last name, and a password. Immediately after Mike submitted his request, you – as the security engineer – observe a database input error message in the logs.

For this error message, choose the potential vulnerability that warrants the most concern:

References

[Alo08]   Alonso, O., Rose, D., Stewart, B.: Crowdsourcing for relevance evaluation. ACM SIGIR Forum **42**(2), 9 (2008). https://doi.org/10.1145/1480506.1480508

[Ara16]   Araki, J., Rajagopal, D., Sankaranarayanan, S., Holm, S., Yamakawa, Y., Mitamura, T.: Generating questions and multiple-choice answers using semantic analysis of texts. In: Proceedings of COLING 2016, The 26th International Conference on Computational Linguistics: Technical Papers, Osaka, Japan, pp. 1125–1136 (2016). http://aclweb.org/anthology/C16-1107

[Buc11]   Bucholz, S., Latorre, J.: Crowdsourcing preference tests, and how to detect cheating. In: Conference INTERSPEECH (2011)

[Guo16]   Guo, Q., Kulkarni, C., Kittur, A., Bigham, J.P., Brunskill, E.: Questimator: generating knowledge assessments for arbitrary topics. In: IJCAI, pp. 3726–3732 (2016)

[Her10]   Herman, G.L., Zilles, C., Loui, M.C.: Creating the digital logic concept inventory. In: Proceedings of the Forty-First ACM Technical Symposium on Computer Science Education, Milwaukee, WI, 10–13 March, pp. 102–106 (2010)

[Her14]   Herman, G.L., Zilles, C., Loui, M.C.: Psychometric evaluation of the digital logic concept inventory. J. Comput. Sci. Educ. **24**(4), 277–303 (2014)

[Hes92]   Hestenes, D., Wells, M., Swackhamer, G.: Force concept inventory. Phys. Teach. **30**, 141–166 (1992)

[How06]   Howe, J.: A definition. Crowdsourcing, 2 June 2006. crowdsourcing.typepad.com/cs/2006/06/crowdsourcing_a.html

[Joi99]   Joinson, A.: Social desirability, anonymity, and Internet-based questionnaires. https://link.springer.com/content/pdf/10.3758/BF03200723.pdf

[Kit08]   Kittur, A., Chi, E., Suh, B.: Crowdsourcing user studies With Mechanical Turk. In: Proceedings of CHI 2008, pp. 453–456 (2008). https://doi.org/10.1145/1357054.1357127

[McD18]   McDaniel, R.: Writing good multiple-choice test questions. Vanderbilt University, 7 May 2018, cft.vanderbilt.edu/guides-sub-pages/writing-good-multiple-choice-test-questions/

[Nat11]   Nathan, M.J., Koedinger, K.R., Alibali, M.W.: Expert blind spot: when content knowledge eclipses pedagogical content knowledge (2011)

[Off19]   Offenberger, S., Herman, G., Sherman, A.T., Oliva, L., Peterson, P., Scheponik, T., Golaszewski, E.: Initial validation of the cybersecurity concept inventory: pilot testing and expert review. IEEE Front. Educ. (2019, submitted)

[Par16]   Parekh, G., DeLatte, D., Herman, G.L., Oliva, L., Phatak, D., Scheponik, T., Sherman, A.T.: Identifying core concepts of cybersecurity: results of two Delphi Processes. IEEE Trans. Educ. **61**, 1–10 (2017)

[She17a] Sherman, A.T., Oliva, L., Delatte, D., Golaszewski, E., Neary, M., Patsourakos, K., Phatak, D., Scheponik, T., Herman, G.L., Thompson, J.: Creating a cybersecurity concept inventory: a status report on the CATS Project, 2017 National Cybersecurity Summit (2017). http://arxiv.org/abs/1706.05092

[Tho18]   Thompson, J.D., Herman, G.L., Scheponik, T., Oliva, L., Sherman, A., Golaszewski, E.: Student misconceptions about cybersecurity concepts: analysis of think-aloud interviews. J. Cybersecur. Educ. Res. Pract. **2018**(1), (2018). https://digitalcommons.kennesaw.edu/jcerp/vol2018/iss1/5

[Wel17]   Welbl, J., Liu, N.F., Gardner, M.: Crowdsourcing multiple choice science questions. In: Proceedings of the 3rd Workshop on NoisyUser-generated Text, NUT@EMNLP, pp. 94–106. ACL

A Sequential Investigation Model for Solving Time Critical Digital Forensic Cases Involving a Single Investigator

Jatinder N. D. Gupta[1(✉)], Ezhil Kalaimannan[2], and Seong-Moo Yoo[3]

[1] College of Business, University of Alabama in Huntsville,
Huntsville, AL 35899, USA
guptaj@uah.edu
[2] Department of Computer Science, University of West Florida,
Pensacola, FL 32514, USA
ekalaimannan@uwf.edu
[3] Department of Electrical and Computer Engineering,
University of Alabama in Huntsville, Huntsville, AL 35899, USA
yoos@uah.edu

Abstract. The number of evidences found in a digital crime scene has burgeoned significantly over the past few years. In addition, the demand for delivering accurate results within a given time deadline has increased. The major challenges coinciding with these aforementioned objectives are to investigate the right set of evidences and to allocate appropriate times for their investigation. In this paper, we present a mixed integer linear programming (MILP) model to analyze the problem of allocating optimal investigation times for evidences involving a single investigator. The objective is to maximize the overall effectiveness of a forensic investigation procedure. We particularly focus on the time critical digital forensic cases, in which results have to be finalized in a court of law within a specified time deadline. While the general problem is NP-hard, two special cases are illustrated to be optimally solvable in polynomially computational effort. Two heuristic algorithms are proposed to solve the general problem. Results of extensive computational experiments to empirically evaluate their effectiveness in finding an optimal or near-optimal solution are reported. Finally, this paper concludes with a summary of findings and some fruitful directions for future research.

Keywords: Digital forensics · Np-hardness · MILP · Heuristic solution · Computational results

1 Introduction

Digital Forensics can be defined as "the use of scientifically derived and proven methods toward the preservation, collection, validation, identification, analysis, interpretation, documentation and presentation of digital evidence derived from

K.-K. R. Choo et al. (Eds.): NCS 2019, AISC 1055, pp. 202–219, 2020.
https://doi.org/10.1007/978-3-030-31239-8_16

digital sources for the purpose of facilitating or furthering the reconstruction of events found to be criminal" [1]. The need for digital forensic investigation procedures has become apparent with the proliferation of cyber crime or fraud (see, for example, the recent review conducted by James and Gladyshev [2]). According to a recent survey conducted by the Crime Scene Investigation (CSI) Computer Crime group, about 46% of the respondents were affected by at least one major form of digital crime [3]. Correspondingly, there is a significant growth in the number of investigations which in turn demands accurate results within shorter time limits [4].

According to Pollitt [5], the path taken by any digital evidence comprises of media (Physical context), data (Logical context) and information (Legal context). The first basic crime scene investigation model proposed by Pollitt [5] includes three primary steps namely acquisition, identification, and evaluation of computer forensic investigation before admission of the evidence in court. The major theme of subsequent research in digital forensic investigations has been related to the identification of the number of stages and phases of a digital crime scene investigation [6]. For example, the Abstract Digital Forensics model (ADFM) proposed by Reith et al. [7] provides a generalized methodology that judicial members can use to relate technology to non-technical observers.

Contrary to earlier models, the Computer Forensic Field Triage Process Model (CFFTPM) developed by Rogers et al. [8] does not require the use of the lab for an in-depth examination of the systems/media or acquisition of a complete forensic image. As a result, this model completes digital forensic investigation in a shorter time frame. Freiling and Schwittay [9] proposed a model with a two-fold purpose: Incident Response to focus on restoration of normal service and Computer Forensics to focus on the presentation of evidence in a court of law. Walls et al. [10] strongly argued and illustrated that an effective digital forensic investigation process must be investigator centric. Hitchcock et al. [11] demonstrate a new tiered response hierarchical model involving digital evidence specialists and nonspecialists can better deal with the increasing number of investigations involving digital evidence. Extending the work of Overill et al. [12] and Rogers et al. [8], Bashir and Khan [13] proposed a comprehensive digital forensic investigation model and describe a forensic analysis procedure covering several phases from identification of the *victim* machine to report generation.

Sun et al. [14] proposed a forensics investigation model that applies probability theory and graph modeling approach to validate the reasoning process behind forensic investigation of digital evidences. Wang and Daniels, [15] applied graph modeling approaches to network forensic analysis to facilitate evidence presentation, automated reasoning, and interactive hypothesis testing in an attempt to identify the attacker's non-explicit attack activities from secondary evidence. A model for handling incident analysis and digital forensic investigations in SCADA (Supervisory Control and Data Acquisition) and industrial control systems was proposed by Spyridopoulos et al. [16].

The above discussion coupled with the recent review of existing research in digital forensics investigation by Agarwal and Kothari [6], Alharbi et al. [17],

James and Gladyshev [2], and Raghavan [18] show that the existing research has primarily focused on developing either a conceptual or a procedural framework to model digital investigations. In order to overcome this weakness, Gupta et al. [19] used the mathematical programming formulation and heuristic approach to solve the digital forensics investigation problem involving multiple investigators. In this paper, we build on and extend the mathematical programming approach of Gupta et al. [19] to model and solve the sequential digital forensic investigation problem involving a single investigator. The objective of this proposed model is to maximize the overall effectiveness in identifying the perpetrator of the crime in a court of law. The solution of our proposed model provides a list of the selected evidences to be investigated along with their respective investigation times that maximizes the overall effectiveness of the digital forensic investigation.

The rest of the paper is organized as follows: In light of the absence of any existing analytical model, Sect. 2 formulates an optimization model for the digital forensic investigation involving a single investigator and proves that the problem is NP-hard at least in the ordinary sense. Since the general problem is NP-hard, Sect. 3 develops polynomially bounded algorithms to optimally solve two special cases of the crime scene investigation problem. For the general digital forensic investigation problem, this section develops two polynomially bounded heuristic algorithms to generate optimal or near optimal solutions. The computational results about the effectiveness of the proposed heuristic algorithms in finding an optimal or near-optimal solution are provided in Sect. 4. Finally, Sect. 5 concludes the paper with some fruitful directions for future research.

2   Problem Formulation and Complexity

In a sequential scenario, the primary step in a digital forensic investigation starts with the identification, acquisition and examination of evidences related to the nature of crime, collectively called the *evidence collection phase* at the crime scene. An example scenario of evidence collection in digital forensic investigation cases was discussed in Divakaran et al. [20], which focuses on the problem of evidence gathering by detecting fundamental patterns in network traffic related to suspicious activities. Following the collection of evidences from the digital crime scene, the acquired evidences are analyzed, results are documented and presented for case resolution termed as the *evidence investigation phase* in the computer forensic laboratory.

In the process of analyzing these evidences, the challenges encountered by an investigator would be to make a decision whether to include or discard an evidence from the investigation and to allocate appropriate processing times for those evidences included in the investigation process.

2.1   Problem Assumptions and Definition

Forensic cases which involve digital evidences are usually complex in nature due to the massive amount of data to be analyzed from the storage devices like the acquisition and examination of a hard disk memory secured from a digital device. Three potential areas of concentration while acquiring evidences from a crime scene are the local hard drive, local user shares and network traffic. The preference is usually given to the host data over network data. To assist the forensics analyst in the evidence search process, Herrerias and Gomez [21] developed the log correlation model to collect, filter and correlate events coming from diverse log files in a computer[1]. Other areas include registry settings, log information, internet history, host-based security tools, anti-virus logs, directory information on the file system, operating system updates and patches[2]. The complexity of analyzing digital evidences is also due to the presence of latent data and files which are protected using stronger encryption algorithms. One case specific example for this scenario is the Identity (ID) theft which is being noted as a top complaint category in consumer fraud [22].

Since the specific parameters that define an investigation can only be determined by knowing its specifics, we define evidences in their generic form, separated only by their effectiveness in solving the crime scene case. The effectiveness value of investigating an evidence for a specified duration is a numerical measure of the value for apprehending the suspect and establishing a crime in a court of law. This numerical measure depends strongly on the time spent in investigating that particular evidence. The more time the investigator spends in examining a specific evidence, the better the chances, that results of the investigation will be able to stand alone in a court of law. Another important factor to be considered is the "admissibility of evidence" through which an evidence could basically stand alone in a court of law [23]. Further, if a valuable evidence is not investigated either due to its "lower effectiveness rate" or due to "poorly being investigated", it could be used in a court of law to stand against the investigator rendering his work to be practically useless. In this paper, we assume that the effectiveness rate (i.e., effectiveness per unit time spent on investigating an evidence) of an evidence is constant and independent of other evidences. Thus, the total effectiveness of a CSI is assumed to be the sum of the effectiveness of investigating the evidences.

As stated earlier, the practical digital crime scene investigation problem is quite complex and hence require consideration of several issues before a final decision can be made as to which evidences should be investigated and the time to be spent on each investigated evidence. Therefore, to define the problem considered in this paper, we assume that all evidences have been acquired successfully. Thus, our focus is on selecting the evidences to be investigated in the lab and the time to be spent on investigating each selected evidence. While such an effort divorces the problem formulation and solution from the practical scenarios being considered, such modelling efforts are a good first attempt at using mathematical models in this rapidly emerging area of digital forensic science. In

---

[1] Log files are specific files that are generated to keep a history of actions occurred on the system.
[2] Private communication with Anuj Soni from Booz Allen Hamilton Inc.

light of this, to keep the model formulation and presentation simple, we make the following additional assumptions:

- Admissibility of each evidence has already been determined.
- All evidences to be investigated are available.
- No additional evidences need be collected.
- Chain of custody for each evidence is always maintained throughout the investigation.
- No interdependencies exist between the evidences.
- No known priorities exist between the evidences.
- Effectiveness per unit time spent on investigating an evidence is constant and independent of other evidences.
- Total effectiveness generated upon investigating an evidence is a linear function of the time spent on its investigation.
- Budgetary and/or legal restrictions limit the total time available to investigate a set of evidences.

While the above assumptions describe only an ideal digital crime scene investigation scenario and may divorce this idealistic problem from a real-life digital forensic investigation problem, it does make the understanding of the model and algorithm development processes easier. However, our basic model can be extended to ensure the relevancy of the proposed mathematical programming approach to handle the practical digital forensic cases [19].

To define the digital forensic investigation problem with these assumptions, consider the following scenario: in a given digital crime scene, a set $N = \{1, 2, \ldots, n\}$ of n evidences are collected and sent to the laboratory for investigation by a single investigator. Associated with each evidence $i \in N$, there is an effectiveness rate β_i that depicts the effectiveness of investigating an evidence i for one time unit. However, if evidence i is selected for investigation, a minimum amount of time a_i must be spent on it. Further, the effectiveness of investigating an evidence i reaches its maximum when $b_i \geq a_i$ time units are spent on its investigation. Thus, there is no need to spend more than b_i time units in investigating evidence i. Because of budgetary and/or legal restrictions, the total time available to complete the investigation of this digital crime scene investigation is limited to a total of T units.

With above notations, definitions, and assumptions, the digital forensic crime scene investigation problem involving a single investigator consists of finding the amount of time t_i to be spent on each evidence i such that $\sum_{i=1}^{n} t_i \leq T$ and the total effectiveness of investigating all evidences, $\sum_{i=1}^{n} \beta_i t_i$ is maximized where for each evidence $i \in N$, either $t_i = 0$ or $a_i \leq t_i \leq b_i$.

2.2   Problem Formulation

To mathematically formulate the above digital forensic crime scene investigation problem, for each evidence $i \in N$, let X_i be a binary variable that takes on a value of 1 if evidence i is selected for investigation and 0 otherwise. Then, the digital forensic crime scene investigation problem (called **Problem P**) can be formulated as the following mixed integer linear programming problem (MILP).

$$\max \quad E = \sum_{i=1}^{n} \beta_i t_i \tag{1}$$

subject to:

$$\sum_{i=1}^{n} t_i \leq T, \tag{2}$$

$$t_i - X_i b_i \leq 0; \qquad\qquad \text{for all} \quad i \leq n, \tag{3}$$

$$t_i - X_i a_i \geq 0; \qquad\qquad \text{for all} \quad i \leq n, \tag{4}$$

$$X_i = 0 \quad \text{or} \quad 1; \qquad\qquad \text{for all} \quad i \leq n. \tag{5}$$

The objective function (1) seeks to maximize total effectiveness of the digital fraud investigation. The first constraint (2) ensures that total time spent on all evidences does not exceed the maximum allowable time T. Constraint set (3) states that the time spent on each evidence i is no more than its upper bound b_i. Constraint set (4) ensures that the time spent on each selected evidence i is no less than its lower bound a_i. The final integer (binary) constraints (5) determine whether the evidence is included or discarded from the investigation.

For the formulation mentioned above, the total number of binary variables is n while the total number of constraints is $2n + 1$. Thus, the overall size of the MILP seems quite manageable. An interesting feature of Problem P is the fact that the binary integer variables X_i are needed only to assure the feasibility of the solution (i.e., to resolve the disjunctive constraints on the selection of any given evidence). As we will show in the next section, if these disjunctive constraints are not needed, the problem is solvable in a polynomially bounded computational effort. Further, because of this compact formulation, the existing MILP software like AIMMS[3] can solve moderately sized problem instances in a fraction of a second.

2.3   Problem Complexity

We now show that the sequential digital forensic investigation problem considered in this paper is NP-hard, at least in the ordinary sense. To do so, we first show that the Problem P is NP-hard at least in the ordinary sense.

Lemma 1. *Problem P is NP-hard at least in the ordinary sense.*

Proof. For special case of Problem P with $a_i = b_i$ for all $i \in N$, let $W = T$, $p_i = \beta_i a_i$, $w_i = a_i$, and $t_i = X_i a_i$. Then, special case of Problem P reduces to the following standard knapsack, **Problem K**:

---

[3] Advanced Integrated Multidimensional Modeling Software for building decision support and optimization applications.

$$\max \quad E = \sum_{i=1}^{n} p_i X_i \tag{6}$$

subject to:

$$\sum_{i=1}^{n} w_i X_i \leq W. \tag{7}$$

$$X_i = 0 \quad \text{or} \quad 1; \qquad\qquad \text{for all} \quad i \leq n. \tag{8}$$

Since the knapsack Problem K, which is special case of Problem P is known to be NP-hard in the ordinary sense Garey and Johnson [24], Martello and Toth [25], it follows that Problem P is NP-hard in the ordinary sense.   ∎

As a consequence of Lemma 1, we have the following result.

Theorem 1. *Sequential digital forensic investigation problem is NP-hard at least in the ordinary sense.*

Proof. If the digital forensic investigation problem is polynomially solvable, then Problem P will also be polynomially solvable. However, Lemma 1 shows that problem P is NP-hard in the ordinary sense. Therefore, it follows that the digital forensic investigation problem is also NP-hard in the ordinary sense.   ∎

While the Theorem 1 shows that the digital forensic investigation problem considered in this paper is NP-hard in the ordinary sense, its status as to its NP-hardness in the strong sense still remains open.

3   Proposed Solution Algorithms

We now turn to the development of the proposed solution algorithms that can provide optimal or near-optimal solutions to the problem considered in this paper. For this purpose, without loss of generality, we assume that evidences are arranged such that $\beta_1 \geq \beta_2 \geq \ldots \geq \beta_n$. If not, this can be done in $O(nlogn)$ computational effort by rearranging and renumbering the evidences.

3.1   Polynomially Solvable Cases

In view of the proof that the problem formulated in the previous section is NP-hard, we now describe the formulation and polynomially bounded solution procedures to optimally solve two special cases.

Case 1: Zero Minimum Investigation Time: We first consider the case where $a_i = 0$ for each evidence i. In this case, there are no disjunctive constraints and hence the integer variables X_i are not needed in the problem formulation.

Therefore, for this case 1, the optimization Problem P reduces to the following **Problem P1**:

$$\max \quad E = \sum_{i=1}^{n} \beta_i t_i \tag{9}$$

subject to:

$$\sum_{i=1}^{n} t_i \leq T, \tag{10}$$

$$t_i - b_i \leq 0; \qquad\qquad \text{for all} \quad i \leq n. \tag{11}$$

We solve Problem P1 by allocating b_i time units for all $(s-1)$ evidences until $\sum_{i=1}^{s} b_i$ exceeds the total investigation deadline T and the remaining time, $T - \sum_{i=1}^{s-1} b_i$ for the sth evidence. We do so by using the following Algorithm P1:

Algorithm P1

Inputs. A set $N = \{1, 2, \ldots, n\}$ of n evidences. For each evidence $i \in N$, effectiveness rate is β_i and upper bound on the investigating time is b_i. Total available investigation time is T.

Assumption. Effectiveness rates are in a decreasing order i. e., $\beta_1 \geq \beta_2 \geq \cdots \geq \beta_n$. If not, evidences are renumbered so that this assumption is valid.

Step 1. Find s such that $\sum_{i=1}^{s-1} b_i \leq T < \sum_{i=1}^{s} b_i$.

Step 2. The optimal solution is as follows: The optimal investigation time t_i^* of each evidence $i \in N$, is: $t_i^* = b_i$ for $i \leq (s-1)$, $t_s^* = T - \sum_{i=1}^{s-1} b_i$ and $t_i^* = 0$ for all $i > s$.

Step 3. Calculate $E^* = \sum_{j=1}^{n} \beta_j t_j^*$ and STOP.

Outputs. Time spent on each evidence $j \in N$ is the best solution found by Algorithm P is t_j^* and total effectiveness as E^*.

Theorem 2. *Algorithm P1 optimally solves the Problem P1 in $O(n)$ computational effort.*

Proof. We note that Problem P1 is a *continuous bounded knapsack problem*. Developments by Dantzig [26] show that this continuous bounded knapsack problem can be solved in $O(n)$ computational effort. Algorithm P1 is simply the formalization of Dantzig's developments and hence can be executed in $O(n)$ computational effort. ∎

Case 2: Must Investigate All Evidences: In the second case, all evidences must be investigated. This implies that $\sum_{i=1}^{n} a_i \leq T$, i. e., enough time is available to investigate all evidences for their minimum possible times. In formulating this special case, the binary variable, X_i is not needed as all n evidences are selected. Therefore, problem P reduces to the following polynomially solvable linear programming **Problem P2**:

$$\max \quad E = \sum_{i=1}^{n} \beta_i t_i \tag{12}$$

subject to:

$$\sum_{i=1}^{n} t_i \leq T, \tag{13}$$

$$t_i - b_i \leq 0; \qquad \qquad \text{for all} \qquad i \leq n, \tag{14}$$

$$t_i - a_i \geq 0; \qquad \qquad \text{for all} \qquad i \leq n. \tag{15}$$

We propose the following algorithm to optimally solve Problem P2.

Algorithm P2

Inputs. A set $N = \{1, 2, \ldots, n\}$ of n evidences. For each evidence $i \in N$, effectiveness rate is β_i and lower and upper bounds on investigating times are a_i and b_i respectively. Total available investigation time T.

Assumption. Effectiveness rates are in a decreasing order i. e., $\beta_1 \geq \beta_2 \geq \cdots \geq \beta_n$. If not, evidences are renumbered so that this assumption is valid. Further, $T \geq \sum_{i=1}^{n} a_i$ which implies selected evidences must at least be investigated until their minimum time limits are reached.

Step 1. For each evidence $i \in N$, $t_i' = t_i - a_i$, $b_i' = b_i - a_i$, and $T' = T - \sum_{i=1}^{n} a_i$.

Step 2. Find s such that $\sum_{i=1}^{s-1} b_i' \leq T' < \sum_{i=1}^{s} b_i'$.

Step 3. The optimal investigation time t_i^* of each evidence $i \in N$, is as follows: $t_i^* = b_i$ for $i \leq (s-1)$, $t_s^* = a_s + [T' - \sum_{i=1}^{s-1} b_i']$ and $t_i^* = a_i$ for all $i > s$.

Step 4. Calculate $E^* = \sum_{j=1}^{n} \beta_j t_j^*$ and STOP.

Outputs. Time spent on each evidence $j \in N$ is the best solution found by Algorithm P is t_j^* and total effectiveness as E^*.

Theorem 3. *Algorithm P2 optimally solves the Problem P2 in $O(n)$ computational effort.*

Proof. It is easy to see that problem P2 can be transformed to an equivalent Problem P1 in $O(n)$ computational effort. To do so, the decision variables and parameters of Problem P1 are defined as follows. For each evidence $i \in N$, $t_i' = t_i - a_i$, $b_i' = b_i - a_i$, and $T' = T - \sum_{i=1}^{n} a_i$. Therefore, the proof of Theorem 3 follows from Theorem 2. ∎

3.2   Heuristic Algorithms for Problem P

While the compact MILP formulation can optimally solve moderately large sized problem instances in a fraction of a second, the Problem P is NP-hard. This implies that it is likely that several problem instances cannot be optimized using MILP formulation in a finite and acceptable amount of computational effort to be of practical use. Therefore, in view of the NP-hard nature of the general Problem P, we develop appropriate heuristic algorithms to find an optimal or a near-optimal solution.

Our first heuristic algorithm is adapted from the truncation of the solution obtained in Algorithm P1 proposed in Sect. 3.1. Since all a_i values are not zero in the general case, it is possible that the solution obtained by algorithm P1 is not feasible for problem P. Therefore, if needed for feasibility, the solution is truncated by ignoring the remaining time available after evidence $s - 1$. The steps of this truncated algorithm are as follows:

Algorithm T

Inputs. A set $N = \{1, 2, \ldots, n\}$ of n evidences. For each evidence $i \in N$, effectiveness rate β_i and lower and upper bounds on investigating times are a_i and b_i respectively. Total available investigation time T.

Assumption. Effectiveness rates are in a decreasing order i. e., $\beta_1 \geq \beta_2 \geq \cdots \geq \beta_n$. If not, evidences are renumbered so that this assumption is valid.

Step 1. Find s such that $\sum_{i=1}^{s-1} b_i \leq T < \sum_{i=1}^{s} b_i$.

Step 2. Set $t_j^* = b_j$ for $j \leq (s - 1)$ and $t_j^* = 0$ for all $j > s$. If $T - \sum_{i=1}^{s-1} b_i \geq a_s$, set $t_s^* = T - \sum_{i=1}^{s-1} b_i$; otherwise set $t_s^* = 0$.

Step 3. Calculate $E^* = \sum_{j=1}^{n} \beta_j t_j^*$ and STOP.

Outputs. Time spent on each evidence $j \in N$ is t_j^* and total effectiveness as E^* is the best solution found by Algorithm T.

The effectiveness of the naive truncated algorithm T above can be improved by finding an evidence which can use the remaining time and hence can increase the total effectiveness of the investigation. For this purpose, the following developments help us in proposing an effective and polynomially bounded heuristic algorithm for the general Problem P.

Theorem 4. *An optimal solution exists where at most one evidence falls strictly in between its minimum and maximum time limits for investigation.*

Proof. Follows from contradiction. Assume that two evidences j and k are in an optimal solution such that, $\beta_j > \beta_k$, $a_j < t_j^* < b_j$ and $a_k < t_k^* < b_k$. The best value of effectiveness contributed by these two evidences is $E_{jk}^* = t_j^* \beta_j + t_k^* \beta_k$. Since $\beta_j > \beta_k$, we note that a simultaneous decrease in t_k^* and increase in t_j^* will increase the value of E_{jk}^*. Therefore, we decrease the value of t_k^* and increase the value of t_j^* by the same amount such that either $t_k^* = a_k$ and/or $t_j^* = a_j$. In either case, there is at most one evidence left which is strictly between its minimum and maximum value. Further, the value of total effectiveness for this new solution is higher than that of the original solution implying that the original solution cannot be optimal.   ∎

Theorem 4 provides an effective way to search for an optimal solution. To do so, we note that problem P is a *disjunctive continuous doubly bounded knapsack problem*. The major difficulty to seek an optimal solution to this problem is caused by the disjunctive nature of this problem as it gives rise to constraint sets (4) and (5) involving binary integer variables. As our first attempt, we propose to use Algorithm P1 presented earlier. If the solution so obtained is feasible, then using Theorem 2, it must be optimal as the lower limits on the investigation times do not affect the feasible solution. However, if the solution obtained by Algorithm P1 is not feasible, then we attempt to find a feasible solution with best total effectiveness. To do so, we attempt to decrease the investigation time assigned to evidence $s-1$ and attempt to use the remaining time such that either the time assigned to investigation $s-1$ equals a_{s-1} or an evidence is found that can be assigned its minimum investigation time. Let $T' = T - \sum_{i=1}^{s-1} b_i$ where $\sum_{i=1}^{s-1} b_i < T < \sum_{i=1}^{s} b_i$ and that the evidences are arranged in the decreasing order of their effectiveness rates. Further, consider an evidence candidate $i \geq s$ for inclusion in the evidences to be investigated. Then, we have the following result.

Lemma 2. *If* $E_i = \max[a_i, \min(b_i, T')]\beta_i - \max[a_i - T', 0]\beta_{s-1} > 0$, *then the inclusion of evidence* i *in those selected for investigation will increase the total effectiveness.*

Proof. We note that $\max[a_i, \min(b_i, T')]$ represents the maximum time that can be allocated to evidence i by giving up some time from evidence $s-1$. Similarly, $\max[a_i - T', 0]$ is the corresponding time to be given up from evidence $s-1$. Therefore, total change in the effectiveness of the changed solution is $E_i = \max[a_i, \min(b_i, T')]\beta_i - \max[a_i - T', 0]\beta_{s-1}$ which if greater than zero will show that the inclusion of evidence i in those selected for investigation increases the total effectiveness of the solution. ∎

As mentioned earlier, for the general Problem P, use of Algorithm P1 will provide an optimal solution if after step 1 of algorithm, $T - \sum_{i=1}^{s-1} b_i \geq a_s$. If that is not the case, we infer that the solution for evidences $\{1, 2, \ldots, s-2\}$ given by Algorithm P1 can be retained. For the remaining evidences, we generate two schedules A and B to determine the best solution. In Schedule A, we attempt to reduce the time spent on evidence $s-1$ with a view to improve the solution by possibly including evidences $\{s, s+1, \ldots, n\}$ in the solution. In order to do so, we use the results from Lemma 2 to assure that the effectiveness gained by including evidence $j \geq s$ must be more than the loss incurred by reducing the time spent on evidence $s-1$. In Schedule B, we try to discard evidence $s-1$ with a view to improve the solution by possibly including evidences $\{s, s+1, \ldots, n\}$ in the solution. In order to do so, we note that the effectiveness gained by including evidence $j \geq s$ must be more than the loss incurred by excluding the evidence $s-1$. Following these computations, the best solution is obtained by comparing net effectiveness generated by schedules A and B. Using this logic, our proposed heuristic algorithm to solve Problem P is as follows.

Algorithm P

Inputs. A set $N = \{1, 2, \ldots, n\}$ of n evidences. For each evidence $i \in N$, effectiveness rate β_i and lower and upper bounds on investigating times are a_i and b_i respectively. Total available investigation time T.

Assumption. Effectiveness rates are in a decreasing order i. e., $\beta_1 \geq \beta_2 \geq \cdots \geq \beta_n$. If not, evidences are renumbered so that this assumption is valid.

Step 1. Find s such that $\sum_{i=1}^{s-1} b_i \leq T < \sum_{i=1}^{s} b_i$. For each $i \in N$, set $E_i = t_{iA}^* = 0, t_{iB}^* = 0$. If $T' = T - \sum_{i=1}^{s-1} b_i = 0$, set $t_i^* = b_i$ for $i \leq (s-1)$, and $t_i^* = 0$ for all $i \geq s$, $E^* = \sum_{i=1}^{n} \beta_i * t_i^*$, and STOP; otherwise enter Step 2.

Step 2. If $T' \geq a_s$, set $t_i^* = b_i$ for $i \leq (s-1)$, $t_s^* = T'$ and $t_i^* = 0$ for all $i > s$, $E^* = \sum_{i=1}^{n} \beta_i * t_i^*$, and STOP; otherwise set $t_{iA}^* = b_i$ for $i \leq (s-2)$, $t_{iA}^* = 0$ for $i > (s-2)$, $T'' = T - \sum_{i=1}^{s-2} b_i$ and enter Step 3.

Step 3. Let δ be a subset of N evidences such that for each $i \in \delta$, $i \geq s$ and $a_i - T' \leq b_{s-1} - a_{s-1}$. If $\delta = \emptyset$, set $t_{s-1,A}^* = b_{s-1}$ and proceed to Step 5; Otherwise enter Step 4.

Step 4. For each $i \in \delta$, calculate $E_i = \max[a_i, \min(b_i, T')]\beta_i - \max[a_i - T', 0]\beta_{s-1}$. Find k such that $E_k = \max_{i \in \delta} E_i$. If $E_k > 0$, set $t_{kA}^* = \max[a_i, \min(b_i, T')]$, $t_{s-1,A}^* = b_{s-1} - \max(a_i - T', 0)$; Otherwise set $t_{s-1,A}^* = b_{s-1}$.

Step 5. Calculate $E_A^* = \sum_{i=1}^{n} \beta_i * t_{iA}^*$ and enter Step 6.

Step 6. Set $T' = T''$ and $t_{s-1,B}^* = 0$. Let σ be a subset of N evidences such that for each $i \in \sigma$, $i \geq s$ and $a_i - T' \leq b_{s-2} - a_{s-2}$. If $\sigma = \emptyset$, set $t_{s-2,B}^* = b_{s-2}$ and proceed to Step 8; Otherwise enter Step 7.

Step 7. For each $i \in \sigma$, calculate $E_i = \max[a_i, \min(b_i, T')]\beta_i - \max[a_i - T', 0]\beta_{s-2}$. Find k such that $E_k = \max_{i \in \sigma} E_i$. If $E_k > 0$, set $t_{k,B}^* = \max[a_i, \min(b_i, T')]$, $t_{s-2,B}^* = b_{s-2} - \max(a_i - T', 0)$; Otherwise set $t_{s-2,B}^* = b_{s-2}$.

Step 8. Calculate $E_B^* = \sum_{i=1}^{n} \beta_i * t_{iB}^*$. If $E_A^* > E_B^*$, set $t_i^* = t_{iA}^*$, $E^* = E_A^*$ and STOP. Otherwise set $t_i^* = t_{iB}^*$, $E^* = E_B^*$ and STOP.

Outputs. E^* is the best total effectiveness; t_i^* is the time spent on each evidence $i \in N$.

Algorithm P requires $O(n)$ computational effort since steps 4 and 7 are executed only once in $O(n)$ computational time.

4   Computational Experiments and Results

We now describe the design and development of experimental procedures to determine the effectiveness and efficiency of the proposed heuristic algorithms in finding an optimal or near-optimal solution to the considered problem. To do so, it would be ideal to test the effectiveness of the proposed heuristic algorithm using real-time data but unfortunately, our search for obtaining them was unsuccessful as the commercial organizations we approached denied us access to obtain such data. In view of this, our computational experiments were generated randomly using various uniform distributions of data. Use of these randomly generated problem instances is justified, since it has been shown that these problem instances are relatively harder to solve than real-life combinatorial and scheduling problem instances [27]. For empirically evaluating the effectiveness and efficiency, the proposed algorithms were coded in Visual C++ and the computational experiments were executed on a PC with Intel Core i3 CPU running at 1.80 GHz. The optimal solution for each problem instance was obtained using AIMMS optimization software version 3.13[4] [28]. The design and operations of these computational experiments and results for testing the effectiveness and efficiency of the proposed algorithm are discussed below.

---

[4] The solver used for optimization is CPLEX optimizer version 12.5.

4.1   Parameters for Problem Instances

The experimental parameters were the number of evidences n, the minimum and maximum times, a_i and b_i respectively, effectiveness rate, β_i for each evidence $i \in N$, and the total available investigation time, T. The number of evidences in the test problem instances varied from 10 through 100 in increments of 10. Problem hardness is likely to depend on the values and ranges of the parameters of the problem instance: the rate of the evidence effectiveness β_i, the minimum and maximum investigation times a_i and b_i respectively, and the total time available for investigation T. Therefore, we generated the following six classes of problem instances by varying the ranges of problem parameters β_i, a_i and b_i as follows.

- **Class 1:** β_i, a_i, and b_i were generated from a uniform distribution $U(1, 99)$;
- **Class 2:** β_i was generated from a uniform distribution $U(1, 49)$ and a_i, b_i were generated from a uniform distribution $U(1, 99)$;
- **Class 3:** β_i was generated from a uniform distribution $U(1, 99)$ and a_i, b_i were generated from a uniform distribution $U(50, 99)$;
- **Class 4:** β_i was generated from a uniform distribution $U(1, 49)$ and a_i, b_i were generated from a uniform distribution $U(50, 99)$;
- **Class 5:** β_i was generated from a uniform distribution $U(1, 99)$, a_i was generated from a uniform distribution $U(1, 99)$ and b_i was generated from a uniform distribution $U(50, 99)$; and
- **Class 6:** β_i was generated from a uniform distribution $U(1, 49)$, a_i was generated from a uniform distribution $U(1, 99)$ and b_i was generated from a uniform distribution $U(50, 99)$.

In addition, to test the impact of the total available time in finding an optimal solution, we set $T = \alpha \sum_{i=1}^{n} b_i$ where $\alpha = 0.3$, 0.6, and 0.9. For each combination of a_i, b_i, β_i and n values, 25 problem instances were generated. Thus, our computational experiment consisted of solving a total of 4,500 (=10*25*6*3) problem instances.

4.2   Algorithm Effectiveness

To assess the effectiveness of the algorithm P in relation to algorithm T, we solved each problem instance using the proposed heuristic algorithms P and T. We also found the optimal solution to each problem instance by solving its corresponding MILP problem P using the AIMMS optimization software.

To find the effectiveness of the algorithm P and T, we calculated the *average percent relative deviation*, APRD defined as follows:

$$APRD_H = \frac{1}{25} \sum_{i=1}^{25} 100 * \frac{E_A - E_H}{E_A}, \qquad (16)$$

where E_H and E_A are the total effectiveness values found by using heuristic Algorithm H and AIMMS software respectively. The smaller the value of $APRD_H$, the better the performance of the algorithm H.

Tables 1 and 2 show the $APRD_H$ values for each heuristic algorithm summarized by the problem parameters and the number of evidences respectively. The computational performance measures used in the above tables are defined as below:

- OPT = Total Number of Optimal Solutions.
- % Deviation = [Optimal solution - Heuristic solution] / [Optimal solution] * 100.00.
- Average % Deviation = $\sum_{i=1}^{25}$ [Values of % Deviation].
- Maximum % Deviation = MAX [Values of % Deviations]
- Minimum % Deviation = MIN [Values of % Deviations]

Table 1. Overall Computational results by parameter values

a_i	b_i	β_i	Heuristic algorithm P				Truncated algorithm T			
			OPT	% Deviation			OPT	% Deviation		
				Min	Avg	Max		Min	Avg	Max
$U(1,99)$	$U(1,99)$	$U(1,99)$	701	0.00	0.00	3.33	390	0.00	1.32	64.26
$U(1,99)$	$U(1,99)$	$U(1,49)$	707	0.00	0.01	1.45	405	0.00	0.72	14.96
$U(50,99)$	$U(50,99)$	$U(1,99)$	408	0.00	0.34	9.21	221	0.00	1.84	45.50
$U(50,99)$	$U(50,99)$	$U(1,49)$	399	0.00	0.44	23.03	216	0.00	1.74	42.16
$U(1,99)$	$U(50,99)$	$U(1,99)$	673	0.00	0.02	3.57	318	0.00	1.30	35.78
$U(1,99)$	$U(50,99)$	$U(1,99)$	674	0.00	0.02	2.84	317	0.00	1.31	35.96
Overall			3562	0.00	0.14	23.03	1867	0.00	1.37	64.26

Table 2. Overall computational results by number of evidencesl

n	α	Heuristic algorithm P				Truncated algorithm T			
		OPT	% Deviation			OPT	% Deviation		
			Min	Avg	Max		Min	Avg	Max
10	0.6	394	0.00	0.39	23.03	231	0.00	5.40	64.26
20	0.6	371	0.00	0.25	9.01	168	0.00	2.63	51.92
30	0.6	344	0.00	0.19	5.28	180	0.00	1.33	12.11
40	0.6	352	0.00	0.11	3.46	161	0.00	1.19	8.71
50	0.6	346	0.00	0.10	3.45	178	0.00	0.81	7.40
60	0.6	340	0.00	0.10	2.92	173	0.00	0.64	7.86
70	0.6	353	0.00	0.08	2.46	201	0.00	0.57	4.73
80	0.6	346	0.00	0.06	1.67	187	0.00	0.44	4.57
90	0.6	364	0.00	0.06	2.11	201	0.00	0.40	3.26
100	0.6	352	0.00	0.06	1.45	187	0.00	0.32	2.97
Overall	-	3562	0.00	0.14	23.03	1867	0.00	1.37	64.26

The above summary Tables 1 and 2 coupled with the detailed results not included in this paper[5] show that the proposed Algorithm P performed better than Algorithm T for all six categories when measured in terms of the number of optimal solutions found or the minimum, average, and maximum APRD value. In fact, in majority of the cases, the total number of optimal solutions found by Algorithm P exceeds those found by Algorithm T by 91%. Further, for each size and category of problem instances, the average APRD value found by Algorithm P is well below 1% and is always less than that found by Algorithm T. The overall APRD value of Algorithm P is 0.14 as compared to 1.37 for Algorithm T. The overall maximum percent deviation of Algorithm P is 23.03 as compared to 64.26 for Algorithm T.

The effectiveness of the proposed heuristic algorithm P does not significantly change with the change in the range of the effectiveness rate, β_i and or the tightness of the total time available for investigations represented by α. However, these problem parameters do significantly affect the performance of Algorithm T. A reduction in the range of the minimum and maximum time required (a_i and b_i) to investigate an evidence does affect the performance of both algorithms. Further, as the total number of evidence increases, the average and the maximum APRD decreases for each algorithm. Thus, the really hard instances to optimally solve Problem P appear to be those for which the minimum and maximum investigation times are drawn from a relatively small range of values. Further, for problem instances involving a really large number of evidences which cannot be optimized using MILP formulation, proposed Algorithm P will provide solutions that are quite close to optimal.

The computational results in Tables 1 through 2 show that the proposed Algorithm P is relatively more effective in solving the digital evidence selection problem considered in this paper than Algorithm T, which could be considered an adaptation of the fractional knapsack problem. Therefore, these results show that the proposed algorithm P is quite suitable for finding an optimal or a near-optimal solution to the problem considered in this paper.

5   Conclusions and Future Work

Cyberspace is constantly evolving, which is constantly opening doors for newer threats to information systems. As a result of this, cyber crime and security incidents have become more frequent, which is creating an adverse effect on associated costs for managing and mitigating breaches to increase substantially. One of the main reasons for the rapid proliferation of cyber crime is an increasing human dependence on the Internet. The existing security models are minimally effective in protecting against cyber attacks and crimes. Therefore, in this paper, we considered the problem of investigating multiple evidences from a crime scene to maximize the effectiveness of apprehending the suspect for committing a crime. We proposed a mixed integer linear programming model for the digital forensic

---

[5] Please contact the corresponding author for a numerical illustration and the detailed computational results.

investigation process involving time critical cases and a single investigator. Since the general digital crime scene investigation problem involving a single investigator is NP-hard at least in the ordinary sense, we showed that two special cases can be polynomially solved in $O(n)$ computational effort. For the general problem, we developed heuristic algorithms that requires $O(n)$ computational effort. Results of the empirical experiments show that the proposed heuristic algorithm P is quite effective in finding an optimal or approximately optimal solution to the general problem.

While we limited the discussion in this paper to the digital forensic investigation of computer related crimes, our proposed methodology is applicable to several other crime scene investigations as well. For example, in the case of a physical theft in a house, evidences collected at the crime scene are taken to the laboratory. Our proposed algorithms are equally applicable to solve such problems provided the assumptions made in this paper are valid.

Several issues are worthy of future investigations as the proposed mathematical and analytical approach suggested in this paper is only a first step in developing an useful methodology for digital forensic investigations. First, the proposed heuristic algorithm P can be improved by considering more than one evidence to be included after Steps 1 and 2 fail to find a feasible solution. Second, finding worse-case performance bounds of the proposed heuristic algorithms will be useful. Third, from a practical viewpoint, it will be interesting and useful to develop appropriate measures and methodologies to determine the effectiveness rates of various evidences (which are quite dependent on the specific crime scene that is being investigated). Fourth, research that relaxes the assumptions of our mathematical model as discussed earlier in this paper, like consideration of non-linear effectiveness of evidence investigation times will increase the practical application of optimization techniques to the field of digital forensic investigations. Finally, extension of our results to more complex digital crime scene investigation problems, like those involving multiple investigators and/or simultaneous evidence collection and investigation and considering interdependencies among evidences being investigated is important to solve problems found in practice. Thus, the approach suggested in this paper identifies digital forensic crime scene investigation as an additional avenue for research and application in the area of Operations Research.

References

1. Palmer, G.: A road map for digital forensic research (2001). Technical Report DTR-T001-0
2. James, J.I., Gladyshev, P.: A survey of digital forensic investigator decision processes and measurement of decisions based on enhanced preview. Digit. Investig. **10**(2), 148–157 (2013)
3. Richardson R.: CSI Computer crime and Security survey (2011). http://www.ncxgroup.com/wp-content/uploads/2012/02/CSIsurvey2010.pdf. Accessed 29 Nov 2013

4. Casey, E., Katz, G., Lewthwaite, J.: Honing digital forensic processes. Digit. Investig. **10**(2), 138–146 (2013)
5. Pollitt, M.: Computer forensics: an approach to evidence in cyberspace. In: Proceedings of the National Information Systems Security Conference, pp. 487–491 (1995)
6. Agarwal, R., Kothari, S.: Review of digital forensic investigations frameworks. In: Kim, K.J. (ed.) Information Systems and Applications, pp. 561–571. Springer, Heidelberg (2015)
7. Reith, M., Carr, C., Gunsch, G.: An examination of digital forensic models. Int. J. Digit. Evid. **1**(3), 1–12 (2002)
8. Rogers, M.K., Goldman, J., Mislan, R., Wedge, T., Debrota, S.: Computer forensics field triage process model. J. Digit. Forensics Secur. Law **1**(2), 27–40 (2006)
9. Freiling, F.C., Schwittay, B.: A common process model for incident response and computer forensics. In: Proceedings of Conference on IT Incident Management and IT Forensics, pp.19–40 (2007)
10. Walls, R.J., Levine, B.N., Liberatore, M., Shields, C.: Effective digital forensics research is investigator-centric. In: Proceedings of the 6th USENIX Conference on Hot Topics in Security, pp. 1–11 (2011)
11. Hitchcock, B., Le-Khac, N., Scanlon, M.: Tiered forensic methodology model for digital field triage by non-digital evidence specialists. Digit. Invest. **16**(S), S75–S85 (2016)
12. Overill, R., Silomon, J., Roscoe, K.: Triage template pipelines in digital forensic investigations. Digit. Invest. **10**(2), 168–174 (2013)
13. Bashir, M.S., Khan, M.N.: A triage framework for digital forensics. Comput. Fraud Secur. **3**(1), 8–18 (2015)
14. Sun, G.Z., Dong, Y., Liu, J.P., Shen T.: The validity of trusted forensics based on probability. In: Proceedings of the International Conference on Information Engineering and Computer Science, pp. 1–4 (2009)
15. Wang, W., Daniels, T.A.: Graph based approach toward network forensics analysis. ACM Trans. Inf. Syst. Secur. (TISSEC) **12**(2), 4–33 (2008)
16. Spyridopoulos, T., Tryfonas, T., May, J.: Incident analysis & digital forensics in SCADA and industrial control systems. In: Proceedings of the 8th IET International System Safety Conference incorporating the Cyber Security Conference, pp. 1–6 (2013)
17. Alharbi, S., Weber-Jahnke, J., Traore, I.: The proactive and reactive digital forensics inveitgation process: a systematic literature review. Int. J. Secur. Appl. **5**(4), 59–71 (2011)
18. Raghavan, S.: Digital forensic research: current state of the art. CSIT **1**(1), 91–114 (2013)
19. Gupta, J.N.D., Kalaimannan, E., Yoo, S.M.: A heuristic for maximizing investigation effectiveness of digital forensic cases involving multiple investigators. Comput. Oper. Res. **69**(1), 1–9 (2016)
20. Divakaran, D.M., Fok, K.W., Nevat, I., Thing, V.L.L.: Evidence gathering for network security and forensics. Digit. Invest. **20**(S), S56–S65 (2017)
21. Herrerias, J., Gomez, R.: A log correlation model to support the evidence search process in a forensic investigation. In: Second International Workshop on Systematic Approaches to Digital Forensic Engineering, pp. 31–42 (2007)
22. US Department of Justice: Forensic Examination of Digital Evidence: A Guide for Law Enforcement. National Institute of Justice Report No. NCJ 187736 (2004)

23. Williams, J.: ACPO good practice guide for digital evidence (2011). http://www.acpo.police.uk/documents/crime/2011/201110-cba-digital-evidence-v5.pdf. Accessed 25 Oct 2014
24. Garey, M.R., Johnson, D.S.: Computers and Intractability: A Guide to the Theory of NP-Completeness. Freeman, Dallas (1979)
25. Martello, S., Toth, P.: Knapsack Problems: Algorithms and Computational Implementation. Wiley, Hoboken (1990)
26. Dantzig, G.: Discrete varaible extremum problems. Oper. Res. **5**(1), 266–277 (1957)
27. Amar, D.A., Gupta, J.N.D.: Simulated versus real life data in testing the efficiency of scheduling algorithms. IIE Trans. **18**(1), 16–25 (1986)
28. AIMMS 3.13: Paragon Decision Technology B.V., Netherlands. http://www.AIMMS.com

A Cloud Based Entitlement Granting Engine

David Krovich$^{(\boxtimes)}$, Austin Cottrill, and Daniel J. Mancini

West Virginia University, Morgantown, WV 26505, USA
dmkrovich@mail.wvu.edu

Abstract. Cyber Sandbox Software Portal (CSSP) is a web application built on the Ruby on Rails framework which provides an entitlement granting engine to assign AWS Instances to different users on demand. The motivation behind this effort is to create an easy way to give users access to virtual machines while also monitoring their work. CSSP would be useful in a classroom setting that requires students to have their own machines to work on. It would also be useful in training for cybersecurity competitions such as the Mid-Atlantic Collegiate Cyber Defense Competition. By utilizing the cloud and cloud programming API's, virtual instances can be allocated on demand which can not only save a huge amount of time, but also greatly reduce costs of setup, deployment, delegation, and usage of resources.

Keywords: Cybersecurity · Amazon web services · Ruby on Rails

1 Introduction

The need for trained cybersecurity professionals is growing immensely as threats evolve. Reports have shown that unemployment in the industry is effectively zero percent with an estimated shortfall of 1.5 million workers needed [13]. There simply are not enough qualified professionals to meet the demand.

Challenge Based Learning (CBL) [7] methodology can be effective in preparing industry-ready workers to meet the growing demand for qualified cybersecurity professionals. In CBL, participants are presented with problems or challenges where they need to work together in a team environment to come up with a solution. The Mid-Atlantic Collegiate Cyber Defense Competition (MACCDC) is one example of the CBL concept [11]. In this competition, student teams inherit real world environments and attempt to defend them from attack. Teams are also given real world tasks known as "injects" that they must attempt to complete during the competition.

Hands on experience in environments that mimic real world infrastructure is an effective way to train students for competitions like MACCDC. Setting up and configuring these infrastructures have all the same problems as setting up infrastructures in the real world. Hardware costs, network concerns, electrical power, and HVAC are a few of the barriers to setting up real world training environments for cybersecurity education.

Cloud computing platforms such as Amazon Web Services (AWS) offer an option for training that minimizes some of the infrastructure costs. AWS allows for a pay on demand model with no upfront costs. Infrastructures can be setup and torn down on

© Springer Nature Switzerland AG 2020
K.-K. R. Choo et al. (Eds.): NCS 2019, AISC 1055, pp. 220–231, 2020.
https://doi.org/10.1007/978-3-030-31239-8_17

demand by using cloud providers such as AWS. Software can be developed to auto-mate the setup and termination of cloud-based infrastructures by using a cloud com-puting API.

The Cyber Sandbox Software Portal (CSSP) is a software prototype that provides an interface to effectively delegate cloud resources for training or teaching purposes. CSSP is a Ruby on Rails web application that utilizes the AWS Ruby API to manipulate resources in the cloud. CSSP effectively allows an administrator to create AWS instances and delegate them to any user they desire via an email address. This paper will go into more about the motivations, implementation, and future work related to CSSP.

2 Motivation and Related Works

The motivation behind this project began due to the difficulties in training for cyber-security competitions and in teaching classes that are based on an engagement with a machine and its operating system. This section will go into detail about the motivations behind the development of CSSP, some of the related works to CSSP, and how CSSP differentiates itself from existing technologies.

2.1 Motivation

Like many Universities around the nation WVU (West Virginia University) has been competing in The Collegiate Cyber Defense Competition (CCDC) [5]. CCDC is a national competition with participants from all 50 states across the country. The country is split into regions and WVU competes in the Mid Atlantic Collegiate Cyber Defense competition which is also known as MACCDC.

Training and preparing for MACCDC is a difficult undertaking. Students must learn a variety of system administration skills across Windows and Linux operating systems. Students are tasked with inheriting an existing infrastructure and are asked to defend it against a red cell intent on breaking into their systems. This is all done while com-pleting tasks known as injects that simulate real world business type requests.

The difficulty in preparing for MACCDC every year is giving students exposure to the systems and services they will be asked to inherit and defend. The challenge is to provide a similar training experience to what the students will experience in the competition. To do that requires a lot of system administration knowledge. Server hardware needs to be acquired, configured, and maintained. This requires connection to a network which means network switches, routers, and cabling all become concerns. There are also many software costs beyond the hardware costs. Some software must be purchased, and licenses must be managed. Other software packages such as Apache [9] are free to download and install but still require knowledge and expertise to properly configure.

While MACCDC was the initial reason for the creation of CSSP, it became apparent that there were uses for CSSP outside of just training for cybersecurity competitions. Many classes based in Computer Science, Cybersecurity, Computer Engineering, and the likes, require hands on training with specific software configurations. CSSP can also be used to deliver training or testing platforms for courses or activities in those disciplines.

Currently, there are four main ways instructors often use for resource delegation: the use of Open Virtual Appliances (OVA), through a public cloud (such as AWS), through a custom private cloud, and locally in a configured lab of machines. The first way of handling this is by doing the assignment directly on a local machine. A lab of computers would have each machine configured in respect to the assignment. Once every student completes the assignment on their own machine, the administrator then goes to every machine, logs on, and grades their work. This method requires each individual machine to be configured properly. This would also require the instructor to grade each machine individually. With public labs, or labs that will be used for more than one class or session, the machines that are used would need to persist in the state that the student left them for grading. Issues arise with public labs as a machine could be changed by the time the instructor is able to grade it. Along with this, the machine state would need reconfigured after use to allow consistent experience between sessions.

Another way to handle assignments would be through distributing an Open Virtual Appliance (OVA). An OVA is an archive format that allows for easy distribution of virtual machines. For an assignment or training session, a student typically sets up a virtual machine from a custom OVA created by the administrator. The student will complete the assignment, create a new OVA from the final state of their virtual machine, and upload the new OVA to the administrator. The administrator then downloads each student's OVA and evaluates them to see if the assignment or training session was completed successfully. This process can be bandwidth intensive and time consuming. OVA files are often large and can take a long time to transfer. In addition to transfer times there is time to import each OVA and start each virtual machine.

Setting up a private cloud for the classroom that provides a more flexible training environment is another approach. To setup and configure a private cloud requires elaborate networking and physical hardware configurations that are very cost and time intensive. The resources required for a private cloud would also include administrators proficient in networking and the system administration to maintain and provide upkeep for the system.

Finally, utilizing the public cloud is another way to accomplish this. Administrators can manually make an Amazon Machine Image (AMI) and create an individual instance for each student to connect to and do their work on. The administrator can stop and start the instances at any time and connect to them one by one to grade the students work. This cuts time and money and only requires the administrator to pay on demand for the instances and has minimal startup costs. However, delegation of resources via AWS brings some overhead itself. It would still take a lot of manual configuration to use AWS itself to facilitate teaching and training environments.

If cybersecurity training environments and virtual instances could be created, deployed, delegated, and terminated all within the cloud with an easy to use web interface, many costs and a lot of wasted time would be eliminated. Utilizing public cloud resources would minimize costs of maintaining a private cloud or individual machines. The pay on demand model of a public cloud will allow the administrators to only pay for what is necessary, while the service provides their own maintenance and support. With all of this, there would be no need to manage an OVA or maintain elaborate physical machines and networks, allowing for ease of use in training environments. CSSP is an attempt to implement this type of model in software that uses a public cloud and an on demand pay as you go model to facilitate teaching and training.

2.2 Related Works

With such a glaring need for cybersecurity training, there is a lot of research in this area. There are efforts to create a universally adopted Cyber Operations Simulation Standard [8] to bring a common framework to teaching cybersecurity. The SANS Institute is leading the development of a framework. They are defining components as objects in the system and then defining interactions between objects via a Scenario Definition Language (SDL). They then use the model to illustrate a CBL approach to cybersecurity education and give an example of how to architect an environment in the cloud.

There are several services that offer cloud-based cybersecurity training. A complete cloud-based lab solution offered by the National Cyberwatch Center [6] is one example of this. This solution allows students to train in a cloud-based simulation environment and comes preloaded with many labs but is not free and costs are per student. Symantec is another example of a cloud-based cybersecurity training application [12]. They offer a product called the "Cyber Security: Simulation Platform" which is a web hosted exercise geared towards offensive or red team activities. This service does not appear to be free and requires a subscription to be purchased to use.

In comparison to existing service offerings from Symantec and the National Cyberwatch Center, CSSP varies itself by providing a way to easily create, associate, and terminate instances which ultimately gives the administrator more freedom and the ability to use this product to fulfill their exact needs. Additionally, rather than becoming a hosted service the ultimate plan for CSSP will be to release the source code under an open source software license [10]. Anyone that wants to use CSSP will be able to download the software and run it using their own cloud provider. Ideally an active community will develop around CSSP and a vibrant product will be produced. When it comes to providing an easy way for teachers to distribute and monitor custom virtual machines, there are no products that are similar or compete with CSSP.

3 Cyber Sandbox Software Portal

AWS was used for all cloud resources in CSSP. AWS is a leading provider of cloud resources and has a wealth of features and functionality to aid in the construction of a cybersecurity environment. AWS provides several APIs to allow manipulation of cloud resources through many popular programming languages. CSSP is developed in Ruby on Rails so the AWS Ruby API was used [1]. The Virtual Private Cloud for Blue Team instances is created using a setup script included, while the application is run in a separate VPC. User experiences of CSSP were positive. Participants reported no issues with the functionality of CSSP when being surveyed on the effectiveness of the application.

3.1   CSSP Architecture

CSSP is deployed as a Docker container running an Ubuntu 16.04 base environment. Using Docker containers allows for ease of deployments as well as development environments. CSSP can be run from an instance in the cloud, a private server, or locally from any computer supporting Docker containers. CSSP uses AWS access keys to make calls to the AWS Ruby API to manipulate resources in the cloud. All CSSP local data is stored in a SQLite database.

CSSP is implemented using the Ruby on Rails framework. This framework has clear segmentation between presentation, logic, and data layers. The presentation layer is referred to as views, the logic layer is referred to as controllers, and the data store layer is implemented in an abstraction called models. In Rails typically for each controller there is an associated view and model that are used together to segment out into a class like structure. The controllers for CSSP are: accounts_controller, blue_team_controller, instances_controller, user2_instance_maps_controller, ami_controller, omniauth_callbacks_controller, and a welcome_controller.

The welcome_controller handles the main landing page for the CSSP application. This is where users land after they have authenticated to CSSP. The welcome controller provides links to other functionality of the CSSP application.

The omniauth_callbacks_controller is used for interfacing with Google's oAuth2 authentication. This controller handles the callback from Google and google sends attributes of the authenticated user. The controller than takes those attributes and loads them into a local user database managed by CSSP.

The ami_controller is used to create, manage, and delete AMI machine images managed by CSSP.

The user2_instance_maps_controller handles the association of an instance to a user. When a new association is made a new entry is made in the local database to store that mapping. This mapping is used to determine which instances a user can access.

The instances_controller handles all tasks related to instances in AWS. It can create instances from an existing AMI. It can start an existing instance. It can stop an existing instance. And lastly it can be used to terminate an instance and delete it from existence.

The blue_team_controller is used for user level access in CSSP. This is where individual users can see which instances are assigned to them. They get a list of all instances. Users can click on an individual instance and then get connection information such as IP address and username and password to connect to the instance.

The accounts controller is used to modify local database information on users. Typically, user account information is imported directly from Google and imported into the local database. This controller does let you view local account information as well as delete local users from the database.

When first setting up CSSP, a Virtual Private Cloud (VPC) must be created to host CSSP instances. CSSP comes with a Ruby script that uses the AWS Ruby API to automatically configure the VPC for use by CSSP. The VPC is tagged in such a way that allows the CSSP application to find and use it as the home for instances that are created and managed. This VPC keeps all CSSP instances separate from other resources that may be running in the cloud. The CSSP application itself runs outside of this VPC but manipulates resources within the created VPC (Fig. 1).

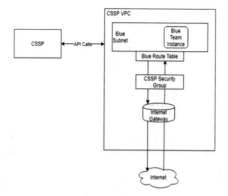

Fig. 1. CSSP VPC diagram

3.2 CSSP Roles

The CSSP web application was built with 3 main user roles: Unauthenticated User, Blue Team User, and Administrative User. The Unauthenticated User role has no access to the system, the Blue Team role is meant for students, and the Administrative User role which deals with administrative functions of CSSP.

Unauthenticated Users have the least amount of functionality and merely serves as a limbo for users who have not logged in yet. An Unauthenticated User in the system is presented with a login screen where they will enter their credentials and become either a Blue Team User or an Administrative User.

A Blue Team User can list instances that has been associated with them and gain information about an instance by clicking on it (See Fig. 2). A remote IP address and credentials to log in to the instance are some of the information that can be retrieved by clicking on an instance (See Fig. 3). It should be noted that Blue Team Users only have power to access the instances that are associated with them by an Administrative User. In the current version of the prototype, Blue Team Users can only view instances. They have no ability to start and stop them on their own.

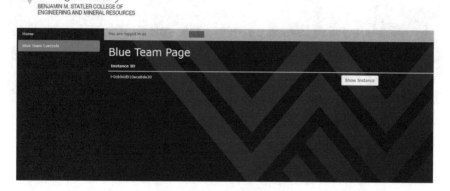

Fig. 2. Blue team controls page

Fig. 3. Blue team show instance page

An Administrative User handles all administrative functions in the CSSP application. This includes listing all users, creating an instance, starting or stopping an instance, and associating or disassociating an instance to or from a Blue Team User. An Administrative User can create either Windows or Linux instances in the cloud (See Fig. 4) and associate it to the desired Blue Team User to grant that person access (Fig. 5).

Fig. 4. Admin instance creation and listing screen

Fig. 5. Admin instance management screen

3.3 Identity and Authentication Management

Managing user identification and user authentication can be a big challenge for any system. CSSP relies on the OAuth2 protocol and uses Google as a backend to authenticate users. This allows users to log into the CSSP web application using their Gmail credentials. Once authenticated through Google, a user entry is made in CSSP to track the user through the application. It is during this process that an Unauthenticated User will become either a Blue Team User or an Administrative User.

Using Google OAuth2 for user identification and authentication, offloads those issues to Google. We no longer had to concern ourselves with password resets or implementing two factor authentication. Using Google as part of the back-end makes associating instances to users easy. The admin user will use the Gmail address of the user to determine which user to associate a specific instance with. This can be done before the user logs into the system for the first time so it would be possible to preload the system with associations.

3.4 Image Creation

When an instance is launched by CSSP, it is based upon an existing Amazon Machine Image (AMI). The AMI is an image of a machine saved in a specific state. Many new instances can be launched from the same AMI and they will have the same exact configurations. Within CSSP there are two ways to create AMIs for use by CSSP.

The first process is a manual process done outside of CSSP using the AWS web interface. An administrator can use the AWS interface to create an AMI from an existing running virtual machine instance. The administrator then tags the created AMI with a tag that the CSSP can find and use the AMI. Once tagged, CSSP can be used to launch new instances from the AMI.

CSSP also can create AMIs from within the CSSP application itself. The administrator can select any instance currently managed by CSSP and use that as the basis to create an AMI from (See Fig. 6). After the AMI is created, it will then become available to launch new instances from within CSSP.

Fig. 6. AMI creation page

3.5   Instance Access Control List

CSSP needs to ensure that only desired users can connect to the instances where they have been granted an entitlement. Additionally, the instances that are created by CSSP cannot be made available on the general internet and mechanisms need to be put in place to restrict access. CSSP handles this by use of security groups in AWS.

When a Blue Team User connects to the CSSP web application, the IP address that they are connecting from is noted and an AWS security group is created containing that address. Any instance that is associated to the Blue Team User is updated with the security group that has just been created. This allows the user to connect to the instance without anyone else being able to. If the Blue Team User moves to an alternate location, they simply need to log out and log back into CSSP to make note of the new IP address at the new location. CSSP only stores one remote IP per user at a time to further protect unauthorized users from connecting to an instance.

3.6   Application Deployment

CSSP is meant to be run from an AMI. A virtual machine is configured with all the CSSP software and then an AMI image is created from the virtual machine. When a new CSSP instance is desired it should be launched from the AMI. After the virtual machine starts up, the administrator needs to connect to the machine and provide keys for Google API for oAuth2 authentication and AWS keys for manipulating AWS resources. The administrator must also set which AWS region the CSSP VPC should live in. After the keys are put in place the VPC setup script is run to create the initial VPC and tag it. After that the CSSP web application is started and can be used to start managing cloud resources.

4   Live Simulation

CSSP was used for the final exam of sophomore level cybersecurity course at West Virginia University. The exam required students to inherit the system that was given to them, setup a web server, and hunt for back doors in the virtual instance that they were working.

For the final exam, the administrator created a custom AMI to be used as the basis for each of the student's instances. The AMI was then tagged in AWS so that it could discovered in CSSP. CSSP was then used to create an instance and associate it to each student in the class with just a few clicks in the web portal.

During the exam students logged into the CSSP web portal with their email address to discover the connection information for the instance that was assigned to them. The students then used Windows Putty to SSH into the instance and completed the assignment. When they were done, all they had to do was logout of their computers and head home.

At the end of the time limit for the final exam, the administrator was able to stop all instances from running at once and effectively kicked off any students that were trying to work past the timer. To grade them, the administrator could start up the instances and see the work that the students had done. No need to transfer large files around, every file was left in the cloud in the exact state students left them.

This usage of CSSP showcased just how effectively the application can be used in a teaching setting to save time and money. It provided an easy to use platform to provide a real-world testing environment at a fraction of the cost and time of other methods. It shows promise for future work.

4.1   Acceptable Use Policy of Cloud Resources

Using AWS as the backend brings with it some concerns when performing cybersecurity related actions. Amazon provides an acceptable use policy [2] that defines acceptable and unacceptable use of AWS. Many actions that would normally be taken by a red team performing Penetration Testing would be in direct violation of the acceptable use policy. Some of the activities that are prohibited include: dns walking via Route53 hosted zones, denial of service or distributed denial of service, port flooding, protocol flooding, and request flooding against either logins or the AWS API itself.

In order to allow for AWS resources to be used for cybersecurity practice, Amazon allows customers to request authorization [4]. Details provided in the request are time and dates of the event, accounts involved, assets involved, contact information, and a detailed description of planned events. Amazon reviews each request on a case by case basis. After receiving authorization from Amazon, the actions in question can go forward.

CSSP is now geared more towards being a passive sandbox for teaching and training rather than providing a platform for penetration testing. In most use cases CSSP should not be at risk in violating the AWS terms of service. When it comes to using CSSP in a classroom, the only concern would be if a class specifically incorporates prohibited actions in an assignment or training session. The delegation of instances to students should pose little threat to breaking these terms if the sandbox is used correctly as a passive environment.

5 Conclusions

Exploring the feasibility of building a cloud enabled application to facilitate distribution and management of virtual instances as well as controlling cybersecurity activities was the goal of this project. A Ruby on Rails application named CSSP was built using AWS to fulfill these ambitions. CSSP handled authentication using OAuth2 and Google which greatly simplified identity and authentication management.

Using CSSP addressed a lot of the concerns with handling assignments that required students to have their own machine as well as providing a training test ground for students to utilize. Giving access to a Windows server or a Linux server becomes a very easy process. Log in as an administrator, click a button, and then associate the newly created instance with a user's Gmail address. The admin can stop or terminate the instance once the user is finished with it. The process can easily scale from a few students to hundreds of students if desired.

6 Future Work

CSSP shows promise and production will progress. The features and functionality discussed in this section will be implemented in future iterations.

Automatic shutdown of resources is one feature that needs to be added into CSSP. Currently the administrator role can manually start and stop instances but if they forget to shut down instances when they are finished, this could incur unwanted costs. Functionality needs to be built into CSSP to address this problem. Preferably, this would allow administrators to set up a timer within CSSP that will shutdown instances at specific times.

Group management needs to be added to CSSP to allow multiple Blue Team Users to access the same instance as part of a group. This would make it easier for group projects in a class to be accomplished on one virtual machine. This would also allow multiple users to practice cybersecurity activities on one virtual machine so that users can practice or learn as a team.

Allowing administrators to view and edit security groups within CSSP could be added as well. This would also allow manual changes within CSSP or AWS to be reflected within the other. Viewing and editing these security groups may be a necessary feature for some instance association tasks.

The long-term goal of this project is to release an open source version of CSSP that anyone could download and use to grant entitlement with ease and run their own cybersecurity activities. Work must be done to easily package and distribute the application to accomplish this goal. During demonstrations, this has been achieved by packaging an AMI on AWS beforehand, and using it to quickly stand up CSSP in well under half an hour. Using CSSP organizations that do not have a lot of computing resources would still be able to delegate instances on demand and run effective cybersecurity activities. The hope is that a tool like CSSP can help solve the problem of a severe shortage of cybersecurity professionals by providing more opportunities for people to train.

Acknowledgements. The author would like to thank Dr. Brian Woerner, Dr. Roy Nutter, Dr. Hany Ammar, and Mr. Dale Dzielski for their guidance and support as CSSP was developed. Thanks to Brian Sweeney, Cameron Morris, Brandon Phillips, and Michael Petik for their contributions to CSSP. Thanks also to Mrs. Cindy Tanner for her assistance in proofreading and comments. Thanks to Dr. Saiph Savage for ideas and motivational support. Thanks as well to Mr. Terry Ferrett for his thoughts and suggestions as the paper was being developed. Additionally, thanks to Kevin Knopf and Brian Shafer for their help in the initial development stages of CSSP. Last but not least the author would like to thank all the students past and present of CyberWVU, WVU's Student Organization that focuses on cybersecurity.

References

1. AWS: AWS SDK for Ruby. https://docs.aws.amazon.com/sdk-for-ruby/v3/api/
2. AWS: AWS Acceptable Use Policy (2017). https://aws.amazon.com/aup/
3. AWS: Elastic IP Addresses (2017). http://docs.aws.amazon.com/AWSEC2/latest/UserGuide/elastic-ip-addresses-eip.html
4. AWS: Penetration Testing (2017). https://aws.amazon.com/security/penetration-testing/
5. CCDC: National Collegiate Cyber Defense Competition (2017). http://nationalccdc.org
6. National Cyberwatch Center: Complete Cloud-Based Solution (2017). https://www.nationalcyberwatch.org/programs-resources/complete-cloud-based-lab-solution/
7. Cheung, R.S., Cohen, J.P., Lo, H.Z., Elia, F.: Challenge based learning in cybersecurity education (2011)
8. Fite, B.K., Filkins, B.L.: Simulating cyber operations: a cyber security training framework (2014)
9. The Apache Software Foundation: Welcome to The Apache Foundation (2017). http://apache.org
10. Open Source Initiative: Open Source Licenses by Category (2017). https://opensource.org/licenses/category
11. MACCDC: Mid Atlantic Collegiate Cyber Defense Competition (2017). http://maccdc.org
12. Symantec: Cyber Security: Simulation Platform (2017). https://www.symantec.com/content/en/us/about/downloads/b-cyber-security-simulation-platform-service-description-en.pdf
13. Wisniewski, M.: Wanted: hackers. Am. Bank. Mag. **126**(7), 30–33 (2016)

Machine Learning Cyberattack Strategies with Petri Nets with Players, Strategies, and Costs

John A. Bland[1]([:envelope:]) [iD], C. Daniel Colvett[1] [iD], Walter Alan Cantrell[2] [iD],
Katia P. Mayfield[3] [iD], Mikel D. Petty[1] [iD], and Tymaine S. Whitaker[1] [iD]

[1] University of Alabama in Huntsville, Huntsville, AL 35899, USA
bjohn316@gmail.com
[2] Lipscomb University, Nashville, TN 37204, USA
[3] Athens State University, Athens, AL 35611, USA

Abstract. A model of a structured query language injection attack was designed using the Petri nets with players, strategies, and costs formalism. The formalism models the attacker and defender as competing players that can observe a specific subset of the net and act by changing the transition firing rates that the respective player can control. This model of the attack was based on the Common Attack Pattern and Enumeration Classification database and was validated by a panel of subject matter experts to be representative of a structured query language injection attack. The model was simulated with a reinforcement learning algorithm using an ε-greedy selection method. The algorithm learned within each iteration an optimal solution by varying player-controlled transitions rates. This paper describes the validation of the model, the design of the algorithm, and the results from 4 different ε-values.

Keywords: Cybersecurity · Modeling · Petri nets · SQL injection · Reinforcement learning

1 Introduction

Cybersecurity incidents occur on a daily basis. Modern society has put an ever increasing amount of trust into computer systems. The designers and defenders of these systems have a responsibility to protect the data and those that may be impacted from misuse of the systems. Motivated by the growing impact and importance of cybersecurity issues, modeling cyberattacks is an active area of research [5,10,14].

With the amount of attack surfaces increasing as new networks are deployed, managing defense resources has become a costly burden. Applying machine learning techniques to models of cyberattacks allows for resourcing decisions to be made to maximize the impact a defender can have. This paper shows how a machine learning variant known as reinforcement learning can be applied to a cyberattack model. Two players, an attacker and defender, compete against one

© Springer Nature Switzerland AG 2020
K.-K. R. Choo et al. (Eds.): NCS 2019, AISC 1055, pp. 232–247, 2020.
https://doi.org/10.1007/978-3-030-31239-8_18

another to achieve their individual goal. For an attacker, the goal would be to compromise the system. For a defender, the goal is to stop the attacker as early as possible.

In this paper we detailed a machine learning algorithm that is able to learn optimal decision making strategies for an attacker or defender. Following this introduction, Sect. 2 of this paper briefly describes several related works. Section 3 provides explanatory background of the reinforcement learning algorithm. Section 4 describes how a web application interface was created to run the algorithm. Section 5 describes a SQL injection attack as modeled by the Petri net formalism. Section 6 details the results of the experiments. Finally, Sect. 7 reports the conclusions from this research.

2   Related Works

Over the years, many models have been developed that represent cyberattacks and their effects on systems. These models vary widely in their intended audience and functionality. Petri nets are one type of modeling technique. Petri nets have been extended to add functionality and computing power. This section specifically looks at Petri nets as they have been used in cybersecurity.

The authors in [7] developed stochastic game nets (SGNs) to model a computer network. SGNs are based on stochastic Petri nets (SPNs). SGNs are Petri nets that are extended with average transition rates for exponentially distributed transition-firing times. Based on stochastic game theory, the goal of the SGN is to find a stationary strategy that maximizes rewards for a given player. In order to determine if a given strategy is the best, the Nash equilibrium is computed. The Nash equilibrium is a prediction of the stationary state for the network. Zakrzewska and Ferragut, developers of the PNPS formalism that was the starting point for this work, assert that while the SGNs are similar to their formalism, PNPS nets differ from SGNs in two ways. In PNPS nets, incomplete information is modeled by limiting the players' observability of places and utility is measured as a function of Petri net markings [14]. An additional shortcoming of SGNs is that they allow the players to have full knowledge of all available actions, which is potentially unrealistic.

Henry et. al. [6] presents an application of Petri nets to Supervisory Control and Data Acquisition (SCADA) systems. In this work, the authors present a new notion of risk with regards to cybersecurity. According to the authors, previous notions of risk were resolved to be estimates of mean time to compromise, or ease of access from an attacker's viewpoint. These measures are difficult to evaluate in all but the most basic of examples.

Instead, Henry et. al. attempt to define a risk measure that is based on reachable attack states, given initial access conditions. These measures enabled the authors to account for all high-level consequence attack states without regards to likelihood. This also allowed for a more flexible notion of risk that can be defined as a computable measure in the attack space [6].

In [2,8,9], the Petri net with players, strategies, and costs (PNPSC) formalism is detailed. A player as defined by the formalism competes against an opponent in order to achieve a predefined goal. For cybersecurity modeling, the goal of the attacker is to compromise they system under attack. Alternatively, the defender's goal is to stop the attacker and prevent the compromise.

PNPSC models of cyberattacks were validated in [3,4]. SQL injection was performed by hand to prove the attack was possible as modeled. Additionally, subject matter experts performed a face validation of the model for SQL injection and determined that it was sufficient for research.

3   Reinforcement Learning Algorithm

Machine learning techniques can be categorized into two broad classes, supervised and unsupervised. A supervised learning algorithm uses a training set to determine correct actions to take when a given situation occurs. A reinforcement learning algorithm does not require a training set. Reinforcement learning uses experience gained during performance to make future decisions.

The reinforcement algorithm framework is composed of five elements. The first is the agent. The agent uses knowledge from previous actions to make decisions to improve performance. The agent attempts to achieve his or her goal. The agent is able to observe and understand the current observable state of the system. The agent can perform an action based on the current state of the system in order to influence the environment, which is the complete state of the system. This action causes the environment to change. As part of the learning process, the environment produces a reward when the environment changes. The reward measures the benefit to the agent that resulted from the action that was taken. This reward can take any integer value, either positive or negative. This process repeats until a problem-specific termination condition is achieved. This research uses the reinforcement framework as shown in Fig. 1 and described in detail by [13].

The reinforcement learning algorithm is designed to either exploit the action that gives the highest reward, or explore another action to attempt to improve the overall average reward. The percent chance to explore is known as the ε-value. The reward is generated at the end of each episode, which is one complete execution of the simulation. The reward is averaged with previous rewards for that action to determine the expected value of choosing the selected action.

4   Web Application Interface

A web application was developed to extend Bland's [2] research into a more user-friendly format. The web application does not change any of Bland's original work, but instead provides a create, read, update, and delete (CRUD) graphical user interface (GUI) that will allow for wider user adoption. The web application was developed using full-stack development best practices.

The Python Flask library serves as the backend component of the web application. Flask acts as a web server responsible for responding and collecting user interactions via HTTP requests. The SQLAlchemy library is an object

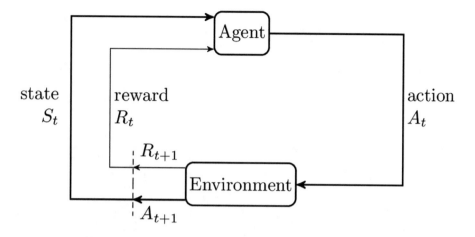

Fig. 1. Reinforcement learning framework [13]

relational mapper (ORM) that allows for CRUD actions to be performed in a database agnostic approach allowing for backend mobility. SQLite is the relational database management system (RDBMS). The RDBMS is responsible for data storage as the user provides inputs to the application. An overview of the web application's architecture can be seen in the Fig. 2 below.

5   SQL Injection Modeled

Structured query language (SQL) injection is a common attack technique. It consistently ranks high in the Open Web Application Security Project (OWASP) Top Ten security vulnerabilities [12].

To model an attack, a common standard representation must be used. The Common Attack Pattern Enumeration Classification (CAPEC) database hosted by MITRE was selected to act as a baseline for describing an attack [11]. The entry for SQL injection was categorized into four phases corresponding to the flow of the attack described. The phases are the explore phase, experiment phase, exploit phase, and goal phase. For SQL injection, Figs. 3, 4, 5 and 6 shows the phases for explore, experiment, exploit, and goals respectively [2]. Tables 1 and 2 describe the places and transitions for the explore phase given in Fig. 3. Similarly, Tables 3 and 4 correspond to the experiment phase shown in Fig. 4. Tables 5 and 6 likewise detail the places and transitions for the exploit phase shown in Fig. 5. Finally, Tables 7 and 8 correspond to Fig. 6, the goals phase. The complete details of the model are given in [2].

Each phase was modeled using the Petri net with players, strategies, and costs (PNPSC) formalism. PNPSC was chosen because of its ability to model multiple players (attackers and defenders) and the choices they could make. The details of the PNPSC formalism are given in [9]. Using the PNPSC, places and transitions are constructed from the CAPEC entry. When marked, places represent truth

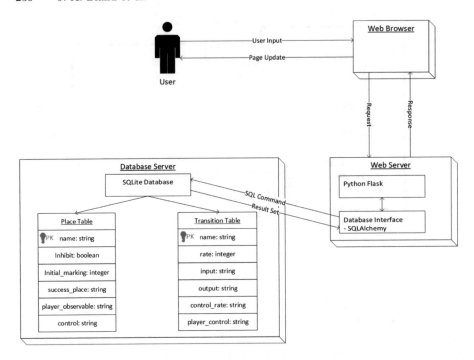

Fig. 2. Web application diagram

Table 1. SQL injection Petri net explore phase places

Place	Description
aP1	SQL queries used by the application to store, retrieve, or modify data
aP2	User Controllable Input that is not Properly Validated by the Application as part of SQL Queries
aP3	Explore Phase of the Attack Occuring
aP4	Spidering Occuring
aP5	Spidering Successful
aP6	Spidering Attempt Complete
aP7	Sniffing Occuring
aP8	Sniffing Successful
aP9	Sniffing Attempt Complete
aP10	Defender Detected Sniffing
aP11	Survey Application Complete
aP12	Defender Detected Spidering

conditions of the model, such as "spidering occurring" or "defender detected sniffing." The transitions represent actions that can be taken to change the state or condition of the system.

Table 2. SQL injection Petri net explore phase transitions

Transition	Description	Rates
aT1	Attacker surveys functionality exposed by the application	3
aT2	Attacker attempts to spider website	9
aT3	Attacker succeeds in spidering	6
aT4	Attacker fails in spidering	9
aT5	Attacker completes Explore phase via spidering	6
aT6	Attacker attempts to sniff network communications	8
aT7	Attacker fails in sniffing	8
aT8	Attacker succeeds in sniffing	5
aT9	Defender Detects Sniffing Occurring	4
aT10	Attacker completes Explore phase via sniffing	2
aT11	Defender Detects Spidering Occurring	8

Table 3. SQL injection Petri net experiment phase places

Place	Description
bP1	Experiment phase of the attack occurring
bP2	Use of web browser attack occurring
bP3	Use of web browser attack successful
bP4	Experiment phase of the attack successful
bP5	Use of web browser attack flag
bP6	Use of web application debugging tool occurring
bP7	Use of web application debugging tool successful
bP8	Web application debugging tool flag
bP9	Use of network-level packet injection tool occurring
bP10	Use of network-level packet injection tool successful
bP11	Use of network-level packet injection tool flag
bP12	Use of modified client attack occurring
bP13	Use of modified client successful
bP14	Use of modified client flag
bP15	Experiment phase failed
bP16	Defender Detected Web Browser Attack
bP17	Defender Detected Web Application Debugging Tool Use
bP18	Defender Detected Network Level Packet Injection Tool Use
bP19	Defender Detected Modified Web Client Use

The rate is interpreted as the number of times the transition will fire, on average, per time unit, or more generally, as the likelihood that the action or

Table 4. SQL injection Petri net experiment phase transitions

Transition	Description	Rates
bT1	Attacker moves to experiment phase	1
bT2	Use web browser to inject input through text fields	10
bT3	Attacker succeeds in web browser attack	4
bT4	Attacker fails to use web browser	6
bT5	Use a web application debugging tool	5
bT6	Attacker succeeds in debugging tool attack	10
bT7	Attacker fails in using debugging tool	10
bT8	Use network-level packet injection tool	8
bT9	Attacker succeeds in using network-level packet injection tool	10
bT10	Attacker fails in using network-level packet injection tool	4
bT11	Use modified client	9
bT12	Attacker succeeds in using modified client	7
bT13	Attacker fails in using modified client	3
bT14	Attacker gives up in experiment phase	2
bT15	Defender Detects Web Browser Attack Occurring	8
bT16	Defender Detects Web Application Debugging Tool Use Occurring	1
bT17	Defender Detects Network Level Packet Injection Tool Use Occurring	1
bT18	Defender Detects Modified Web Client Use Occurring	6

Table 5. SQL injection Petri net exploit phase places

Place	Description
cP1	Exploit phase of the attack occurring
cP2	Exploit phase aborted
cP3	SQL Injection Cheat Sheet attack occurring
cP4	SQL Injection Cheat Sheet attack successful
cP5	SQL Injection Cheat Sheet attack flag
cP6	Exploit phase successful
cP7	Add logic to query attack occurring
cP8	Add logic to query attack successful
cP9	Add logic to query flag
cP10	Blind SQL Injection attack occurring
cP11	Blind SQL Injection attack successful
cP12	Blind SQL Injection attack flag
cP13	Stacking queries: (DOS) attack occurring
cP14	Stacking queries: (DOS) attack successful
cP15	Stacking queries: (DOS) attack flag
cP16	Defender Detected Use of SQL Injection Cheat Sheet
cP17	Defender Detected Added Logic to Query
cP18	Defender Detected Blind SQL Injection Techniques
cP19	Defender Detected Stacked Queries (Denial of Service Attempt)

Table 6. SQL injection Petri net exploit phase transitions

Transition	Description	Rates
cT1	Attacker moves to exploit phase	10
cT2	Attacker unable to exploit	2
cT3	Attacker uses SQL Injection Cheat Sheet	7
cT4	Attacker succeeds in using SQL Injection Cheat Sheet	4
cT5	Attacker fails in using SQL Injection Cheat Sheet	7
cT6	Attacker adds logic to query	3
cT7	Attacker succeeds in adding logic to query	6
cT8	Attacker fails in adding logic to query	1
cT9	Attacker uses Blind SQL Injection techniques	10
cT10	Attacker succeeds in using Blind SQL Injection techniques	7
cT11	Attacker fails in using Blind SQL Injection techniques	7
cT12	Attacker tries stacking queries: (DOS)	3
cT13	Attacker succeeds in stacking queries: (DOS)	4
cT14	Attacker fails in stacking queries: (DOS)	8
cT15	Defender Detects Use of SQL Injection Cheat Sheet Occurring	5
cT16	Defender Detects Added Logic to Query Occurring	7
cT17	Defender Detects Blind SQL Injection Techniques Occurring	2
cT18	Defender Detects Stacked Queries (Denial of Service Attempt) Occurring	10

Table 7. SQL injection Petri net goals phase places

Place	Description
dP1	Goal phase of the attack occurring
dP2	Read application data flag
dP3	Failed attack
dP4	Goal phase failed
dP5	Gain privileges flag
dP6	Goal decided
dP7	Attacker/Defender strategies
dP8	Goal successful
dP9	Defense has knowledge of system attack
dP10	Attacker's goal blocked
dP11	Modify application data flag
dP12	Execute unauthorized code flag

event the transition is modeling will occur. Higher rates result in increased likelihood of occurring. During each execution cycle, a firing time is generated for each enabled transition as an exponentially distributed random variate, using each transition's rate as the exponential distribution's rate parameter λ (see [1] for details of how to do this). Then the enabled transition with the earliest firing time is selected for firing.

The implementation of the reinforcement learning algorithm for the PNPSC model uses the rates of player-controlled transitions as the actions. Each set of transitions has a rate that can be modified to be either zero or ten, which either disallows or allows for the enabling of the transition respectively. Each set of possible rates for each phase is tried at least once out of a 500,000 episode experiment. The ε-value is varied between experiments.

A reward of 100 is given if a player reaches a success place. Likewise a reward of zero is given if the player fails to reach a success place by the end of the episode. A cost as described in [9] is incurred whenever a player-controlled transition changes its rate. This cost is subtracted from the reward to give the final reward value for that episode. It is possible for a reward of less than zero to be awarded if the player changes transition rates and fails to reach a success place. Every transition set that is chosen for the current episode is given the same reward at the end of the episode. This is averaged with the previous reward for that specific transition set.

Table 8. SQL injection Petri net goals phase transitions

Transition	Description	Rates
dT1	Attacker moves to goal phase	7
dT2	Attacker attempts to read app	6
dT3	Attacker fails to read app data	10
dT4	Attacker tries again	8
dT5	Attacker aborts goal attempt	7
dT6	Attacker attempts to gain privileges	1
dT7	Attack moves to Attacker/Defender strategies	5
dT8	Attacker fails to gain privileges	5
dT9	Attacker succeeds in reaching goal	3
dT10	Defender monitors attack	1
dT11	Defender blocks actively	4
dT12	Defender blocks passively	9
dT13	Attacker fails to modify application data	3
dT14	Attacker fails to execute unauthorized code	8
dT15	Attacker attempts to execute unauthorized code	2
dT16	Attacker attempts to modify app	5

The model that was created was validated by a panel of subject matter experts. The experts reviewed the CAPEC database entry to determine if it accurately reflects a SQL injection cyberattack. Next the experts compared the CAPEC entry with the places and transitions for the Petri net model. They determined if the Petri net model accurately reflected the content of the CAPEC entry. Finally, the experts compared the Petri net model to how a SQL injection attack is performed to determine if the Petri net accurately represents how that attack would occur. The overall determination was that the Petri nets were of sufficient quality to be useful for research [4].

6   Results

For the experiment, several ε-values were chosen. Each experiment was run for 500,000 episodes in which a terminating condition was reached. The ending conditions was either the attacker succeeded, gave up, or was stopped by the defender. Place aP3 is the top most place where an attacker can make a decision. As can be seen in Fig. 7a, the average reward for place aP3 reaches a stable state well before the 500,000th episode. Figure 7b is a zoomed in section of Fig. 7a in which the majority of the learning occurs. The positive trend shows that reinforcement learning algorithm is improving the attacker's strategy thereby maximizing the likeliness of the attacker's success.

The SQL injection attack has an average reward of 31.35 when using an ε of 0.02 and is shown in Table 9. Table 9 is a sample of strategies developed by the reinforcement learning algorithm. These strategies give the average reward for the current marking when the player-controlled transitions are set as provided. In some cases, it is better to not allow for transitions to be enabled. This minimizes the costs as described in [9].

Table 9. Sample strategy for SQL injection attacker determined by the maximum average reward

Average	Marking	Rates
31.35	(aP3)	(aT2 = 0, aT6 = 10)
28.45	(aP8, bP1)	(bT2 = 10, bT5 = 0, bT8 = 10, bT11 = 10)
20.49	(aP5, bP1)	(bT2 = 0, bT5 = 10, bT8 = 10, bT11 = 10)
44.12	(bP10, cP1)	(cT3 = 0, cT6 = 10, cT9 = 10, cT12 = 10)
43.76	(bP7, cP1)	(cT3 = 0, cT6 = 10, cT9 = 10, cT12 = 10)
23.49	(aP8, bP3, cP1)	(cT3 = 10, cT6 = 0, cT9 = 10, cT12 = 10)
33.81	(aP8, bP13, cP1)	(cT3 = 10, cT6 = 0, cT9 = 10, cT12 = 10)
47.95	(aP8, bP13, cP4, dP1)	(dT2 = 10, dT6 = 0, dT15 = 10, dT16 = 10)
44.61	(aP5, bP7, cP8, dP1)	(dT2 = 0, dT6 = 0, dT15 = 10, dT16 = 10)
40.55	(bP7, cP8, dP1)	(dT2 = 10, dT6 = 10, dT15 = 10, dT16 = 10)

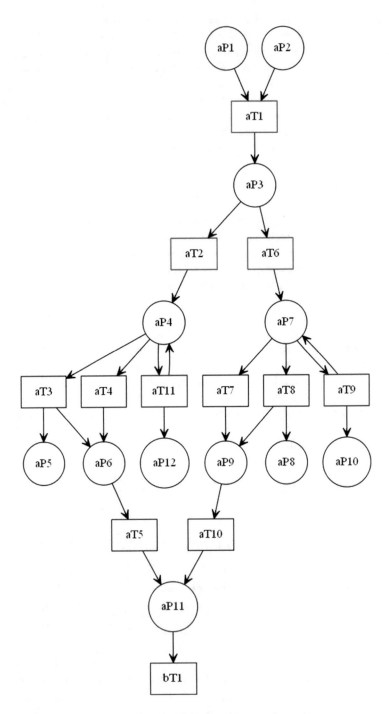

Fig. 3. CAPEC 66 - SQL injection explore phase

There is a balancing act that must occur between exploration and exploitation. Too much exploration results in higher costs from changing transitions. Too much exploitation may miss the opportunity to improve the strategy, thereby missing improvements to the likelihood of success.

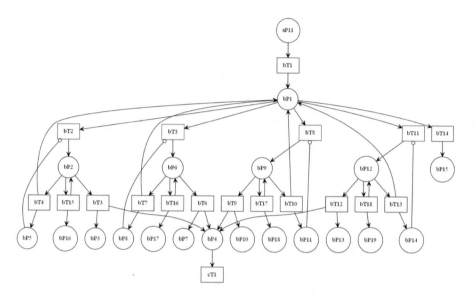

Fig. 4. CAPEC 66 - SQL injection experiment phase

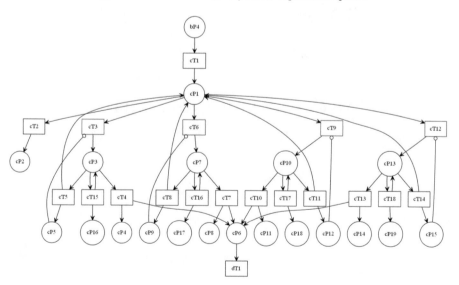

Fig. 5. CAPEC 66 - SQL injection exploit phase

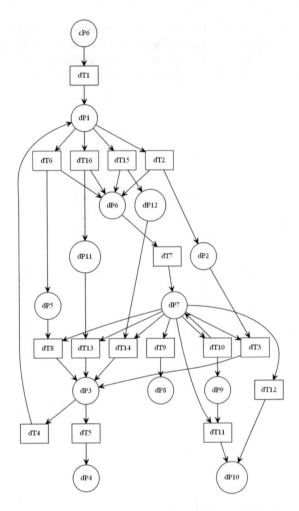

Fig. 6. CAPEC 66 - SQL injection goals phase

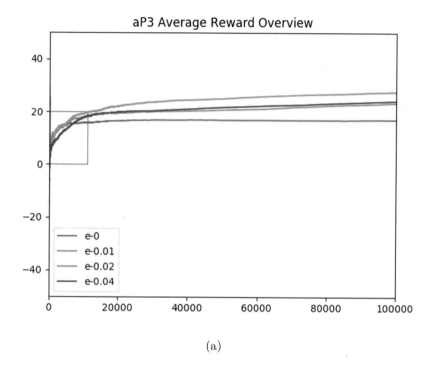

(a)

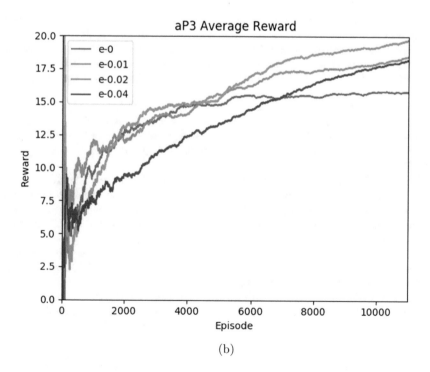

(b)

Fig. 7. Attacker average reward over 100,000 episodes for SQL injection

7 Conclusions

A model of a SQL injection cyberattack was shown to implementable into a Petri net variant known as PNPSC. This model was implemented in a reinforcement learning framework and simulated for 500,000 episodes with different ε-values. The positive trend of the average reward shows that the algorithm is improving over time. Multiple iterations showed that exploitation and exploration must be considered to optimize the reinforcement learning algorithm.

The results of the strategies that are determined by the machine learning algorithm can inform decision makers on best utilization of resources. Implementing hardware, software, and personnel all have costs that when realized may not be easily justified. Using the results of the strategies, a defender could implement intrusion detection systems at the places that have the highest average success for the attacker. This would reduce the likelihood of success for the attacker.

References

1. Banks, J., Carson, J.S., Nelson, B.L., Nicol, D.M., Banks, C.M.: Discrete-Event System Simulation, 5th edn. Prentice-Hall, Upper Saddle River (2012)
2. Bland, J.A.: Machine learning of cyberattack and strategies. Ph.D. thesis, University of Alabama in Huntsville (2018)
3. Bland, J.A., Mayfield, K.P., Petty, M.D., Whitaker, T.S.: Validating Petri Net models of common attack pattern enumeration and classification. In: Proceedings of the 2017 AlaSim International Conference and Exposition, Huntsville, AL (2017)
4. Cantrell, W.A., Mayfield, K.P., Petty, M.D., Whitaker, T.S., Bland, J.A.: Structured face validation of extended Petri Nets for modeling cyberattacks. In: Proceedings of the 2017 AlaSim International Conference and Exposition, Huntsville, AL (2018)
5. Ferragut, E.M., Brady, A.C., Brady, E.J., Ferragut, J.M., Ferragut, N.M., Wildgruber, M.C.: HackAttack: game-theoretic analysis of realistic cyber conflicts. In: Proceedings of the 11th Annual Cyber and Information Security Research Conference, no. 1, p. 8. https://doi.org/10.1145/2897795.2897801. http://arxiv.org/abs/1511.04389
6. Henry, M.H., Layer, R.M., Snow, K.Z., Zaret, D.R.: Evaluating the risk of cyber attacks on SCADA systems via Petri Net analysis with application to hazardous liquid loading operations. In: 2009 IEEE Conference on Technologies for Homeland Security, HST 2009, pp 607–614 (2009). https://doi.org/10.1109/THS.2009.5168093
7. Lin, C., Wang, Y.: A stochastic game nets based approach for network security analysis. In: 29th International Conference on Application and Theory of Petri Nets and other Models of Concurrency, pp. 21–34 (2008)
8. Mayfield, K.P., Petty, M.D., Bland, J.A., Whitaker, T.S.: Composition of cyberattack models. In: Proceedings of the 31st International Conference on Computer Applications in Industry and Engineering, New Orleans, LA (2018)

9. Mayfield, K.P., Petty, M.D., Whitaker, T.S., Bland, J.A., Cantrell, W.A.: Component-based implementation of cyberattack simulation models. In: Proceedings of the ACM SouthEast 2019 Conference. Association for Computing Machinery, Kennesaw, GA (2019). https://doi.org/10.1145/3299815.3314435. ISBN 978-1-4503-6251-1
10. Mitchell, R., Chen, I.R.: Modeling and analysis of attacks and counter defense mechanisms for cyber physical systems. IEEE Trans. Reliab. **65**(1), 350–358 (2016). https://doi.org/10.1109/TR.2015.2406860
11. MITRE: CAPEC - Common Attack Pattern Enumeration and Classification (CAPEC) (2018). https://capec.mitre.org/
12. OWASP: Open Web Application Security Project (2019). https://www.owasp.org/
13. Sutton, R., Barto, A.: Reinforcement Learning: An Introduction, 2nd edn. MIT Press, Cambridge (2017)
14. Zakrzewska, A.N., Ferragut, E.M.: Modeling cyber conflicts using an extended Petri Net formalism. In: IEEE SSCI 2011: Symposium Series on Computational Intelligence - CICS 2011: 2011 IEEE Symposium on Computational Intelligence in Cyber Security, pp. 60–67 (2011). https://doi.org/10.1109/CICYBS.2011.5949385

Safety and Consistency of Subject Attributes for Attribute-Based Pre-Authorization Systems

Mehrnoosh Shakarami[✉] and Ravi Sandhu

Department of Computer Science,
Center for Security and Privacy Enhanced Cloud Computing (C-SPECC),
Institute for Cyber Security (ICS), University of Texas at San Antonio,
San Antonio, USA
mehrnoosh.shakarami@my.utsa.edu, ravi.sandhu@utsa.edu

Abstract. Attribute-based access control (ABAC) systems typically enforce pre-authorization, whereby an access decision is made once prior to granting or denying access. This decision utilizes multiple components: subject's, object's and environment's attribute values as well as the authorization policy. Here, we assume that the policy, object and environment attribute values are known with high assurance while subject attributes are collected incrementally from multiple attribute authorities. This incremental assembly with differing validity periods for subject attribute values creates potential for inconsistency leading to incorrect access decisions. This problem was studied in context of trust negotiation systems by Lee and Winslett (LW), who define four different notions of consistency which are partially ordered in strictness. In this paper, we propose an alternate set of five consistency levels, also partially ordered in increasing strictness. Three of our levels are equivalent to counterparts in LW. The third LW level is differentiated by receive time, to which we are agnostic. Our fifth and highest level is new in that it utilizes request time which is not recognized in LW. We define the formal specification of each of our consistency levels and identify the properties guaranteed by each level. We discuss implication of these consistency levels in different practical scenarios and compare our work with related previous research.

Keywords: ABAC · Pre-Authorization · Safety · Consistency · Revocation

1 Introduction

In attribute-based access control (ABAC) access decisions are made on basis of attribute values of subjects, objects and environment with respect to a policy. For convenience we understand the term attributes to mean attribute values. Attributes and policy are susceptible to change. Ideally the decision point should know their real-time values, which is not practically feasible. Even if attributes

© Springer Nature Switzerland AG 2020
K.-K. R. Choo et al. (Eds.): NCS 2019, AISC 1055, pp. 248–263, 2020.
https://doi.org/10.1007/978-3-030-31239-8_19

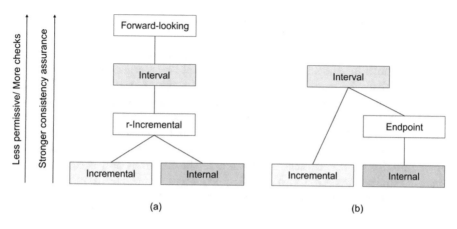

Fig. 1. (a) Our consistency levels (b) LW consistency levels. Equivalence is color coded.

and policy are queried from appropriate authorities immediately prior to every decision, there are irreducible network latencies. Realistically, some values will be cached due to performance, cost, and failures. Consequently, some access decision may be incorrect. We call this the safety and consistency problem.

This paper investigates this problem with focus on subject attributes. We assume that the policy and object/environment attributes are known in real-time at the decision point. This is reasonable since the decision and enforcement points are typically co-located with the object's custodian who maintains these values. This reduces the problem to safety and consistency of subject attributes. These attributes are obtained as credentials (a.k.a certificates) issued by an Attribute Authority (AA). Credentials may be signed or unsigned depending on how they are acquired. The closest prior work is by Lee and Winslett (LW) [8,9]. Our work is inspired by LW but takes a significantly different perspective and provides novel insights as discussed in Sect. 2.

This paper is organized as follows. In Sect. 2, we review the LW model and compare it to our work. Section 3 documents our system model and assumptions. In Sect. 4 we formalize proposed consistency levels specifications along with the properties guaranteed by each level. We review other relevant related works in Sect. 5. The implications of our consistency levels in context of different architectures and practical scenarios are discussed in Sect. 6. Section 7 gives our conclusion.

2   Lee-Winslett (LW) Model and Comparison

Figure 1 summarizes the relationship between the consistency levels defined in this paper (Fig. 1-(a)) versus the LW definitions (Fig. 1-(b)). A higher level requires additional checks, so it imposes a performance cost at the benefit of stronger consistency. Although formulated differently, three of the levels on both sides are equivalent as indicated by the same name and colors. Proofs of these equivalences are given in the appendix.

Common to both sets of definitions, a credential specifies the value of a single attribute for the subject. Each credential has a *start time* and *end time*, which establish the overall lifetime for validity of the attribute value given in the credential. The credential may be revoked by the AA during this putative lifetime. Thus the relying party (the decision point) must make one or more revocation checks with the AA to gain additional assurance of the credential's validity.[1] The consistency problem arises when multiple credentials of a subject are required to make an access decision. If the lifetimes and *revocation check times* of required credentials do not align properly it is possible to make incorrect access decisions, both in allowing access that should be disallowed and vice versa. This is discussed further and elaborated by examples in Sect. 3. The consistency levels of Fig. 1 define different criteria for aligning the lifetimes and revocation check times for multiple credentials.

The start, end and revocation check times are common to both sets of definitions.[2] The LW definitions all utilize the notion of *receive time*, which is the time when a credential is received by the relying party. Top two LW levels (endpoint and interval) utilize the notion of decision time. We consider decision time as central while receive time is irrelevant. We are agnostic about the delivery path and timing of credential receipt. The decision time is the instant where an access decision is committed and intuitively should be central to consistency consideration. Three of the LW levels are equivalently reformulated in our definition as indicated in Fig. 1. If receive time is ignored the endpoint and interval levels of LW are equivalent and hence not differentiated in Fig. 1-(a). The two new levels in our definitions, r-incremental (short for restricted incremental) and forward-looking are discussed below.

The incremental and internal levels allow the use of credentials that are known to be expired or revoked. So, we do not recommend general use of these two levels.[3] The explicit use of decision time enables us to formulate the r-incremental level that eliminates use of expired/revoked credentials. It is the lowest level we would generally recommend, and is missing in LW. The highest forward-looking level in our definitions requires the use of *request time*, which is the time when a subject makes a request for access and is missing in LW.

---

[1] Credential lifetimes can range widely from months to seconds. For very short-lived credentials revocation checks may not be useful. For simplicity we consider that for a short lived credential there is an implicit and successful revocation check at its start. Thus we can uniformly assume there is at least one revocation check by the relying party for each credential that it uses in making an access decision. For long-lived credentials there should be at least one revocation check after start time.

[2] Note that start and end times are determined by the AA, while revocation check times are determined by relying party actions.

[3] In a risk-based approach it may be acceptable to use expired/revoked credentials, but general use is not recommended.

3   Problem Statement and System Assumptions

The value of an attribute of a subject is represented by a credential which must be coupled to a specific subject, which is typically achieved by embedding the subject's identity in the credential. The identity of the subject must be authenticated before the credential is coupled with that subject. The details of these processes can be complex and susceptible to security vulnerabilities and flaws. All the same there are multiple well-known standards such as X.509 [1], SAML [11] and OAuth [15] in this arena. We assume that suitable mechanisms exist to bind credentials to subjects without requiring any specific technique for this purpose. Regardless of the way through which the attributes are presented, we assume proposed attributes to be authentic and tied to the subject.

We require every credential to have a determined lifetime interval which has been specified by its *start time* and *end time*. For short-lived credentials this interval is small, say minutes, seconds or even less, while for long-lived credentials this interval could span days, months or years. In either case we recognize the possibility that the credential may get revoked during its lifetime, although this is especially germane for long-lived credentials. We forbid use of a credential outside its lifetime. The revocation status of a credential may be checked as appropriate by the decision point, and a credential that is known to be revoked cannot be unrevoked. We also assume that attribute values do not change as a result of credential usage, so that attributes are immutable in the sense of [13].

Following LW we refer to the set of a subject's credentials used to make an access decision as the *view* of the decision point (V_{DP}). The appropriate view depends upon the policy being evaluated. In general, the view might change during evaluation. Consider a policy P which is a disjunction of two predicates A and B, i.e., $P = A \vee B$. The decision point may choose only credentials included in the A predicate as relevant to P in order to perform first step of evaluation; if A fails, B will replace it in the view and the set of relevant credentials to P will change as a result.

Definition 1. *The set of attributes included in the view of decision point related to the policy P at time t is called the set of relevant credentials and denoted by $V_{DP}^{P,t}$.*

Since required credentials for evaluating a policy would be collected incrementally and lifetimes of different credentials might not be the same, there is no guarantee that previously collected credentials are still valid while the latter ones are acquired, which might cause the safety and consistency problem. Following example illustrates the inconsistency problem.

Example 1. Alice is a portal manager in the sales department of a company. If she wants to communicate with clients through the portlet website, the decision point needs credentials attesting her sales group membership and user role. If she wants to utilize higher levels of access, for example editing and approving contracts with clients, both user and manager role credentials are required. Suppose Alice has the user role since January 1st (start time) until March 1st (end time). Her

Table 1. Table of symbols

Symbol	Meaning	Symbol	Meaning
c_i	i^{th} credential	t_{req}	Request time
$t^i_{r,k}$	Time of kth revocation check for c_i	t_d	Decision time
$t^i_{r,max}$	Last time of revocation status check for c_i	t_e	Enforcement time
$t^i_{invalid}$	First time c_i has been found to be revoked	t^i_{start}	Start time of c_i
t^i_{revoc}	Actual revocation time for c_i (if any)	t^i_{end}	End time of c_i

sales group member certificate is valid since January 25th till February 24th and she is given the manager role from Feb 10th (start time), which is valid until March 9th. Suppose the decision point acquired and validated the user role and sales group certificates most recently at January 25th and Feb 8th respectively. It also collected manager certificate at Feb 10th which was verified to be valid (via revocation check) on the same day. So, the decision point would honor the manager access to Alice if she request an access afterward. Due to a reorganization in the company, Alice may no longer be a manager after Feb 17th. Also, suppose Alice's user role certificate has been prematurely revoked at Feb 9th. But if the decision point still relies on the previous revocation checks, Alice would be able to exercise manager's rights after revocation of her relevant credentials, which results in an access violation.

Violation by relying on outdated validation information is a common problem in access control enforcement. While this example illustrates inconsistency with long-term cached credentials, similar problems can arise even if all needed credentials are accumulated over a short period of time.

Table 1 defines a set of self-explanatory symbols to refer to important time stamps used in this paper. There are some common assumptions which have been made in both LW and our work as follows.

1. Once a credential is revoked, it cannot be un-revoked. However, a new credential can be issued for the same attribute for that subject.
2. There is a single instantaneous decision time t_d. The access decision may be re-evaluated subsequently with, say, different credentials but this latter evaluation is treated as a separate and distinct decision.
3. V^{P,t_d}_{DP} is the only view of interest, in which P is the policy which should be satisfied to grant the access at decision time t_d and DP stands for the decision point.

We always use the latest revocation check results for making access decision. So, if we have max_i revocation checks for c_i, the $t^i_{r,max}$ indicates the latest revocation check ($max_i{}^{th}$) of c_i and it would be utilized in decision making.

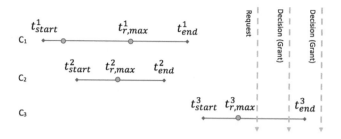

Fig. 2. Incremental consistency with unrestricted decision time

4   Consistency Levels

In this section, we develop five consistency levels relative to the view of the decision point ordered as shown in Fig. 1-(a). We say that a credential is in its validity interval or is *valid*, provided that the current time is not before the credential's start time, nor after the credential's end time and the credential is not known to be revoked at any time before the current time. A revocation check is never done after the end time since the credential has already expired. Once it is revoked a credential cannot become valid again. It follows that if validity of a particular certificate is confirmed via revocation check at time t after its start time, the credential has been valid for all times between its start time and t.

Let C be the set of all credentials in the system, and T the set of all possible time stamps. We formalize the notion of a credential's validity status at time t by defining the following 3-valued function. We call a credential *Invalid*, if following function returns *False*.

$$Valid : C \times T \rightarrow \{\,True,\,False,\,Unknown\,\}$$

$$Valid(c_i, t) = \begin{cases} True \iff & Valid(c_i, t^i_{r,k}) \wedge (t^i_{start} \leq t \leq t^i_{r,k}) \\ Unknown \iff & Valid(c_i, t^i_{r,max}) \wedge (t^i_{r,max} < t \leq t^i_{end}) \\ False \iff & (\neg Valid(c_i, t^i_{r,max}) \wedge (t \geq t^i_{r,max})) \\ & \vee (t \notin [t^i_{start}, t^i_{end}]) \end{cases} \quad (1)$$

In the rest of this section we propose five consistency levels. For each consistency level, we provide a formal specification along with the properties which are guaranteed if we apply the proposed specification.

4.1   Incremental Consistency

This level requires each relevant credential to be found valid by a revocation check before the decision time. In Example 1 suppose Alice wants to access the portal on Feb 25th, so user and sales group certificates should have been checked. Although one of her relevant credentials (sales group) has expired one day ago, the system would let her in. This access violation happens because in this level

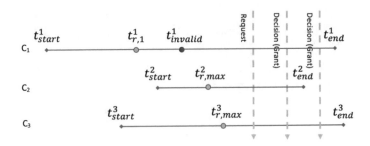

Fig. 3. Internal consistency

we may use a credential for an access decision after its t^i_{end} time. As shown in Fig. 2, two credentials C_1 and C_2 have been used after their corresponding end times.

Specification. Every credential in the view of decision point is valid at its latest revocation check which has been done before the decision time.

$$(\forall c_i \in V^{P,t_d}_{DP})\,[(t^i_{start} \leq t^i_{r,max} < t^i_{end}) \wedge (\max_{\forall c_i \in V^{P,t_d}_{DP}} t^i_{r,max} < t_d) \wedge Valid(c_i, t^i_{r,max})] \tag{2}$$

Property 1. For every relevant credential, there is at least one point in time before the decision time, at which that credential has been found (via revocation check) to be valid.

$$(\forall c_i \in V^{P,t_d}_{DP})(\exists t_i)\,[(t^i_{start} \leq t_i < t^i_{end}) \wedge (t_i < t_d) \wedge Valid(c_i, t_i)]$$

Proof. Without loss of generality, we can assume $t_i = t^i_{r,max}$. Moreover, we know that $\max_{\forall c_j \in V^{P,t_d}_{DP}} t^j_{r,max} < t_d \implies t_i < t_d$.

4.2 Internal Consistency

In order to enforce lifetime overlap for all relevant credentials, internal consistency requires every relevant credential to be started before the end point of any other relevant credential. Furthermore, if a credential is revoked, this revocation should happen after all credentials have started. As shown in Fig. 3, it is possible to deliberately utilize an already revoked credential in this level. Moreover, it is still possible to use a credential beyond its end time, as in incremental consistency. In case of Example 1, Alice would be granted access to the portlet even if we know her user role credential has been revoked at Feb 9. The formal specification is as follows.

Specification. Every credential in the view of decision point has to be started before the minimum endpoint of all credentials and has to be valid at some point before the decision time. The minimum known revocation of any relevant credential occurs after all credentials have been started.

$$(\forall c_i \in V_{DP}^{P,t_d})(\exists t_{r,k}^i)\,[(t_{start}^i \le t_{r,k}^i < t_{end}^i) \wedge Valid(c_i, t_{r,k}^i) \wedge (\max_{\forall c_i \in V_{DP}^{P,t_d}} t_{r,max}^i < t_d)$$

$$\wedge\,(\max_{\forall c_i \in V_{DP}^{P,t_d}} t_{start}^i < \min_{\forall c_i \in V_{DP}^{P,t_d}} t_{invalid}^i) \wedge (\max_{\forall c_i \in V_{DP}^{P,t_d}} t_{start}^i < \min_{\forall c_i \in V_{DP}^{P,t_d}} t_{end}^i)]$$

$$(3)$$

Property 1. There is at least one point in time at which all relevant credentials are in their $[t_{start}, t_{end})$ time intervals and are not known to be *Invalid*.

$$(\exists t')(\forall c_i \in V_{DP}^{P,t_d})\,[(t_{start}^i \le t' < t_{end}^i) \wedge (Valid(c_i, t') \ne False)]$$

Proof. The last condition in Eq. 3 provides overlapping of lifetimes of all relevant credentials. Also, there is at least one revocation check for every credential at which it has been found to be valid. So, there is at least one point, namely t', in intersection of lifetime intervals of all credentials at which every credential is either checked and found to be valid before t' (its validation status is unknown at t') or it has not been checked yet (so it is valid at t'). If there is any credential which has been found to be revoked, t' should be picked from the interval: $t' \in [\max_{\forall c_i \in V_{DP}^{P,t_d}} t_{start}^i, \min_{\forall c_i \in V_{DP}^{P,t_d}} t_{invalid}^i)$.

Property 2. There is no subset relationship between incremental and internal consistency levels.

Proof. It is possible to have an incrementally consistent view in which there is no overlap between lifetime intervals of all relevant credentials. So, it would not be internally consistent. On the other hand, there may be an internally consistent view at which we recognize a credential at its latest revocation check to be prematurely revoked, so thereby not incrementally consistent.

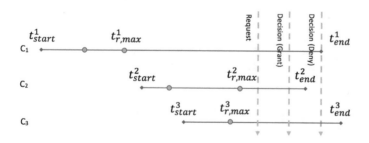

Fig. 4. Incremental consistency with restricted decision time

4.3   Incremental Consistency with Restricted Decision Time (Restricted-Incremental or r-Incremental)

In this level, we restrict the decision time to happen necessarily when all relevant credentials are in their lifetimes, say $[t_{start}^i, t_{end}^i]$ (Fig. 4). As opposed to previous levels, in this level if any of the relevant credentials has expired the access request would be denied. In case of Example 1, if Alice tries to exercise her rights at Feb 25th (after her credential expiration) the decision point would deny her access. In Fig. 4, the second decision time would result in Deny, comparing with the similar situation in Figs. 2 and 3 where both access requests resulted in Grant. Specification and guaranteed properties are given below.

Specification. Every relevant credential has to be found valid at the latest revocation check which, by assumption, happens before the decision time. Moreover, it is essential that the decision time happens before any of relevant credentials end time.

$$(\forall c_i \in V_{DP}^{P,t_d})\, [(t_{start}^i \leq t_{r,max}^i < t_d < t_{end}^i) \wedge Valid(c_i, t_{r,max}^i)] \tag{4}$$

Property 1. There is at least one point in time at which all the relevant credentials are in their $[t_{start}, t_{end})$ time intervals and are not known to be *Invalid*.

$$(\exists t')(\forall c_i \in V_{DP}^{P,t_d})\, [(t_{start}^i \leq t' < t_{end}^i) \wedge (Valid(c_i, t') \neq False)]$$

Proof. Based on Eq. 4, $(\forall c_i \in V_{DP}^{P,t_d})\, [t_{start}^i \leq t_d < t_{end}^i]$. So, $\max_{\forall c_j \in V_{DP}^{P,t_d}} t_{start}^j \leq t_d < \min_{\forall c_j \in V_{DP}^{P,t_d}} t_{end}^j$. By taking $t' = t_d$, the proof for the first part is trivial. For the second part, we know that the latest time we checked c_i's revocation status is $t_{r,max}^i$, at which we found it to valid (otherwise the access would be denied). But, we do not know about the real status of the credential after the last revocation check and the *Valid* function would return *Unknown* at these later times.

Property 2. Any incrementally consistent view with restricted decision time has the following property: $\bigcap_{\forall c_i \in V_{DP}^{P,t_d}} [t_{start}^i, t_{end}^i) \neq \emptyset$

Proof. Following previous proof, there is at least one point (t_d) that lies in the $[\max_{\forall c_j \in V_{DP}^{P,t_d}} t_{start}^j, \min_{\forall c_j \in V_{DP}^{P,t_d}} t_{end}^j)$ interval. So, this interval is not empty.

Property 3. Any r-incremental consistent view is incremental and internal consistent as well.

Proof. All three specifications have it in common that every relevant credential has to be found valid at its revocation check. The first part of the incremental consistency with restricted decision time is:

$$(\forall c_i \in V_{DP}^{P,t_d})\, [t_{start}^i \leq t_{r,max}^i < t_d < t_{end}^i \implies (t_{start}^i \leq t_{r,max}^i < t_{end}^i)$$
$$\wedge (\max_{\forall c_j \in V_{DP}^{P,t_d}} t_{r,max}^j < t_d) \wedge \exists t_d \in \bigcap_{\forall c_j \in V_{DP}^{P,t_d}} [t_{start}^i, t_{end}^i)]$$

Therefore, r-incremental is a constrained version (subset) of incremental level. Moreover, since we use the latest valid staus, we are not aware of any revocation and $t_{invalid}^i = Null$. So, all properties of internal level are also satisfied.

Property 4. It is not necessarily the case that any incrementally/internally consistent view is r-incremental as well.

Proof. In both incremental and internal levels, the decision time may be after some of the relevant credentials' endpoints, which means that we may have: $t_d > \min_{\forall c_i \in V_{DP}^{P,t_d}} t_{end}^i$, which contradicts r-incremental specification.

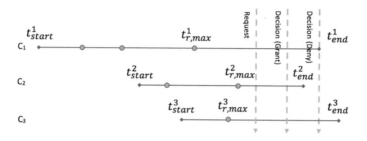

Fig. 5. Interval consistency

4.4 Interval Consistency

In Example 1, Alice's user role has been revoked at Feb 9th. If she tries to communicate at manager level with clients at that date and system still relies on the latest revocation check which happened before actual revocation she would be let in, while there is no guarantee that the credential is still valid. We know her user role has been revoked even before manager certificate starts. Interval level enforces latest revocation checks to happen in $[t^i_{start}, t^i_{end}]$ for all relevant credentials. So, it could be guaranteed that not only every credential is valid at some time, but also all credentials were simultaneously valid. The specification and properties guaranteed by this level are given below (Fig. 5).

Specification. Every relevant credential has found to be valid at the latest revocation check before the decision time. Moreover, the latest revocation check happened after all credentials have been started and before any of them ends.

$$(\forall c_i \in V_{DP}^{P,t_d}) \; [(\max_{\forall c_i \in V_{DP}^{P,t_d}} t^i_{start} \le t^i_{r,max} < t_d < \min_{\forall c_i \in V_{DP}^{P,t_d}} t^i_{end}) \wedge Valid(c_i, t^i_{r,max})] \tag{5}$$

Property 1. There is at least one point in time, after all relevant credentials have been started and before any of them ends, prior to decision time, at which all of the relevant credentials are simultaneously valid.

$$(\exists t')(\forall c_i \in V_{DP}^{P,t_d}) \; [(\max_{\forall c_i \in V_{DP}^{P,t_d}} t^i_{start} \le t^i_{r,max} < t' < \min_{\forall c_i \in V_{DP}^{P,t_d}} t^i_{end}) \wedge Valid(c_i, t')]$$

Proof. Let $t' = \min_{\forall c_i \in V_{DP}^{P,t_d}} t^i_{r,max}$. For every relevant credential to the policy, we could guarantee that it has been valid at t'. Note that if any credential has found to be revoked at t', it cannot be unrevoked at any later time. Therefore, the proof is complete.

Property 2. Every interval consistent view is r-incremental.

Proof. It is trivial that: $(t^i_{start} \le \max_{\forall c_i \in V_{DP}^{P,t_d}} t^i_{start}) \wedge (t^i_{end} \le \min_{\forall c_i \in V_{DP}^{P,t_d}} t^i_{end})$. Substituting these equation in interval specification in Eq. 5, we can deduce: $(\forall c_i \in V_{DP}^{P,t_d}) \; [t^i_{start} \le t^i_{r,max} < t_d < t^i_{end}]$. So, interval specification satisfies the specifications of r-incremental.

Property 3. Not any r-incremental consistent view is necessarily interval consistent.

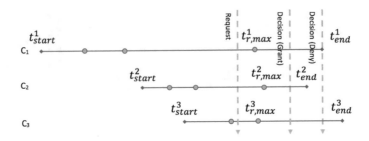

Fig. 6. Forward-looking consistency

Proof. Based on Eq. 4, latest revocation of a credential might happen before some credentials start time $(t^i_{r,max} < \max_{\forall c_j \in V^{P,t_d}_{DP}} t^j_{start})$ or a credential may be validated after some credentials expiration $(\min_{\forall c_j \in V^{P,t_d}_{DP}} t^j_{end} < t^i_{r,max})$, which contradicts with interval consistency specification.

4.5 Forward-Looking Consistency

In Example 1, suppose Alice tries to change the clients' contracts at Feb 17th (the set of relevant credentials includes sales group and manager role credentials). All relevant credentials were checked at Feb 10 at which all have been started and none of them expired yet, so the interval consistency timing constraints would be satisfied. However there is an access violation, because the decision point relied on outdated revocation status information (relying on revoked manager certificate). To solve this problem, we take the request time into account in our strongest level of consistency and impose constraints to ensure all credentials have been valid simultaneously at some point *after the request time* (Fig. 6).

Specification. Every relevant credential has to be valid at its latest revocation check time, which happens after the request time and before the decision time.

$$(\forall c_i \in V^{P,t_d}_{DP})[(\max_{\forall c_j \in V^{P,t_d}_{DP}} t^j_{start} \leq t_{req} < t^i_{r,max} < t_d < \min_{\forall c_j \in V^{P,t_d}_{DP}} t^j_{end}) \wedge Valid(c_i, t^i_{r,max})]$$

$$(6)$$

Property 1. There is at least one point in time, after the request time and before the decision time, at which all relevant credentials are valid simultaneously based upon their latest revocation checks.

$$(\exists t')(\forall c_i \in V^{P,t_d}_{DP})[(\max_{\forall c_i \in V^{P,t_d}_{DP}} t^i_{start} \leq t_{req} < t' < t_d < \min_{\forall c_i \in V^{P,t_d}_{DP}} t^i_{end}) \wedge Valid(c_i, t^i_{r,max})]$$

Proof. Suppose $t' = \min_{\forall c_i \in V^{P,t_d}_{DP}} t^i_{r,max}$. For every relevant credential, we could guarantee that it has been valid at t', because otherwise it cannot be unrevoked at any later time including its latest revocation check. So, the proof is complete.

Property 2. Every forward-looking consistent view is interval consistent as well.

Proof. The definition of forward-looking consistency is a restricted version of interval consistency, in which we restricted the latest revocation check to happen necessarily after the request time.

Property 3. An interval consistent view is not necessarily a forward-looking consistent view as well.

Proof. In case of interval consistency, it is possible to have a credential c_i with $t^i_{r,max} < t_{req}$, which contradicts with forward-looking consistency specification.

5 Related Work

Beyond the need to keep the data consistent in open and distributed systems, which has been discussed in the literature (see for example [17]), there is a crucial requirement to have access control models relying on the most recent information to grant/deny access to that data [3]. ABAC determines access based on attributes of subjects, objects and environment evaluated with respect to a policy. These attributes and the policy are exposed to change and staleness during the time which could result in inconsistency. On the other hand, delays and staleness of attributes are inherent in every distributed system owing to network latencies, caching and failures [5]. So, most practical distributed systems try to have a near-consistent [3] property, which attempts to limit the exposure of access control models to stale attributes.

The first organized work focused on consistency problem in trust negotiation proposed in [8,9]. Trust negotiation systems are specific types of distributed proof systems [6] which are appropriate when privacy is a major concern [14]. Lee and Winslett extend the proof construction in authorization systems to context-sensitive environments in [7], in which parts of the proof tree are required to remain hidden due to privacy concerns. In another research work on conversational web services [12], authors build their access control model based on user's credentials which relies on the first, most permissive level of consistency introduced in [8,9]. Authors simply assume the validity of a credential will last for the whole web service conversation duration.

Squicciarini et al. [16] present a protocol which safely performs trust negotiation during distinct negotiation sessions. Even though the authors put the probability of expired/revoked credentials during negotiation suspensions under consideration, they only mention that a synchronization algorithm would take care about updating the list of credentials without concretely describing the underlying synchronization scheme.

There is another category of research work which considers policy changes as the main concern. In [2], authors concentrated specifically on policy consistency in dynamic environments. In this paper, we consider policy inconsistency out of scope and assume that policy is known with high assurance at the decision point. Similarly, policy consistency is simply assumed as an underlying assumption in other research works [10].

Another closely related research to ours is [4], in which the authors formally specify a set of attributes in linear temporal logic in a Group-based Secure Information Sharing (g-SIS) model to express freshness of attributes. By proposing different levels of stale-safe property, authors try to limit unsafe access decisions relied on stale subjects' and objects' attributes in a distributed access control system.

Our approach differs from fail-secure access control models [18] since we assume that we can acquire fresh revocation status of credentials, but fail-secure access control applies in scenarios in which revocation states cannot be updated or accessed. Our main concern is credentials' revocation status might became obsolete since the last status update.

6   Discussion

Implementation Implications. Higher level of consistency requires additional checks. There is a tradeoff between the safety assurance provided by higher levels and cost of additional checks, as follows.

- From incremental/internal to r-incremental: the end times relate to decision time, so additional check of end time is required at the decision point in r-incremental.
- From r-incremental to interval: it potentially requires additional revocation checks, because all relevant credentials have to be checked at least once for their latest revocation status after all credentials have been started.
- From interval to forward-looking: all credentials definitely need to be checked for revocation status after the request time.

Quantitative performance evaluation is beyond the scope of this paper. It would require concrete system and workload assumptions and would be specific to the particular context.

Short-Lived Credentials. Short-lived credentials are used to obviate the need for revocation check by keeping credential lifetime very small. For our purpose we assume there is an implicit revocation check at start time, otherwise the AA would not issue the credential. No further revocation check is possible. In this case r-incremental and interval consistency will be equivalent. Forward-looking consistency could be guaranteed only if the request time has pushed prior to the start time for all credentials. The practical implication is that the decision point would need to assemble required subject credentials from appropriate AAs after the request time.

Considering Enforcement Time. After the decision point makes the access decision, it will be enforced by an enforcement point which could be the same or a different entity than the decision point. We certainly know that $t_d < t_e$. Proposed consistency levels in this paper remain unaffected by taking enforcement time into account. From another stand point, if there is a large gap between the decision and enforcement time, it is possible to utilize an access while some of the corresponding credentials have been expired; this is more probable in case of short-lived credentials. So, we can add more constraints to consistency level specifications which restricts this gap as follows: $t_e \leq \min_{\forall c_i \in V_{DP}^{P,t_d}} t_{end}^i$. Therefore, enforcement time could be considered to extend proposed levels of consistency.

7   Conclusion

Assuming an ABAC model is in place, we focused on a pre-authorization model in which our goal is to provide the decision point with the most recent status of subjects' attributes. To avoid safety and consistency problem, which is caused by relying on expired/revoked credentials, we proposed five increasingly powerful consistency specifications. At each level, we proved guaranteed properties provided by the proposed specification. We presented different implications of our proposed consistency levels in different real world scenarios and architectures. We also compared our work with the closest prior works and discussed its distinctive features and assumptions.

Acknowledgement. This work is partially supported by NSF CREST Grant HRD-1736209 and DoD ARL Grant W911NF-15-1-0518.

A   Appendix: Proof of Consistency Levels Equivalencies

We prove our claim of equivalent levels with LW model in this section. One of the distinctions is inequality of decision time with revocation check time, since we believe these two timestamps cannot be exactly the same, as the decision has to happen after revocation checks.

A.1   Incremental Levels Equivalency

As seen in Sect. 4.1, for every relevant credential in our incremental level, there is at least one point before the decision time, at which that credential has been found to be valid. The incremental level in LW model is satisfied if and only if every credential to be valid at its receive time as follows.

$$(\forall c_i \in V_e^{P,t}) \, [(s.syn = True) \wedge (revocation\text{-}check_i \neq \perp)$$
$$\wedge \, (start_i \leq receive_i \leq revocation\text{-}check_i)]$$

This could be simplified as follows: $start_i \leq receive_i \leq revocation\text{-}check_i \leq end_i$. So, there is at least one point in time (receive time) at which every relevant credential has found to be valid, which matches with our incremental level. Moreover, this revocation check at the receive time could be considered as the latest validation. Then, we need to show revocation check in LW happens before the decision time, same as its counterpart in our model. Although the decision time has not been considered explicitly in LW model, revocation checks obviously happen before the decision time, since the receive time could not occur later than decision time. So, the proof is complete.

A.2   Internal Levels Equivalency

Authors in LW define a view as internal consistent providing all relevant credentials satisfy the following conditions:

$$(\forall c_i \in V_e^{P,t}) \, [checked(credential\text{-}state) \wedge (\max_{\forall c_j \in V} start_j < \min_{\forall c_i \in V} invalidation_i)$$
$$\wedge \, (\max_{\forall c_j \in V} start_j < \max_{\forall c_i \in V} receive_i) \wedge (\min_{\forall c_j \in V} end_j > \min_{\forall c_i \in V} receive_i)]$$

Above conditions could be arranged as follows:

$$(\forall c_i \in V_e^{P,t}) \, [(start_i < revocation\text{-}check_i \leq end_i) \wedge (\max_{\forall c_j \in V} start_j < \min_{\forall c_i \in V} invalidation_i)$$
$$\wedge \, (\max_{\forall c_j \in V} start_j < \max_{\forall c_i \in V} receive_i) \wedge (\min_{\forall c_j \in V} end_j > \min_{\forall c_i \in V} receive_i)]$$

Based on our internal specification in Sect. 4.2, all conditions are the same except the last two conditions stated in LW model, which aim to provide an overlap between lifetime intervals of all relevant credentials in the view. Lifetime overlap has been provided in our model through $\max_{\forall c_i \in V_{DP}^{P,t_d}} start_i < \min_{\forall c_j \in V_{DP}^{P,t_d}} end_j$. Another distinction is the explicit consideration of decision time after all revocation checks. Even though this has not been stated in LW model, it is impossible to take revocation checks after the decision time into account while making decision, since it needs prediction of future states of credentials.

A.3   Interval Levels Equivalency

To prove equality of the properties provided by both interval levels in our work and LW model, consider their definition of interval consistency for every relevant credential in the view:

$$(\forall c_i \in V_e^{P,t})\ [checked(credential\text{-}state)$$
$$\wedge\ (start_i \le receive_i \le \max_{\forall c_i \in V} receive_i \le revocation\text{-}check_i)]$$

We can restate their interval definition as follows:

$$start_i \le receive_i \le \max_{\forall c_i \in V} receive_i \le revocation\text{-}check_i \le decision\text{-}time \le end_i$$

The following property is concluded from above definition:

$$(\forall c_i \in V_e^{P,t_d})\ [start_i \le \max_{\forall c_i \in V} receive_i \le revocation\text{-}check_i]$$
$$\implies (\forall c_i \in V_e^{P,t_d})\ [\max_{\forall c_i \in V} start_i \le \max_{\forall c_i \in V} receive_i \le revocation\text{-}check_i]$$

On the other hand, we can formally deduce the following property from interval consistency definition in LW model:

$$(\forall c_i \in V_e^{P,t_d})\ [revocation\text{-}check_i \le decision\text{-}time \le end_i]$$
$$\implies (\forall c_i \in V_e^{P,t_d})\ [revocation\text{-}check_i \le decision\text{-}time \le \min_{\forall c_i \in V} end_i]$$

Putting above concluded properties together would result in the following definition. Taking out the receive time, this definition becomes the same as our interval definition.

$$\max_{\forall c_i \in V} start_i \le \max_{\forall c_i \in V} receive_i \le revocation\text{-}check_i \le decision\text{-}time \le \min_{\forall c_i \in V} end_i$$

References

1. Housley, R., et al.: Internet X. 509 public key infrastructure certificate and CRL profile. Technical report (1998)
2. Iskander, M.K., et al.: Enforcing policy and data consistency of cloud transactions. In: ICDCSW. IEEE (2011)
3. Kortesniemi, Y., Sarela, M.: Survey of certificate usage in distributed access control. J. Comput. Secur. 44, 16–32 (2014)
4. Krishnan, R., Niu, J., Sandhu, R., Winsborough, W.H.: Stale-safe security properties for group-based secure information sharing. In: FMSE. ACM (2008)
5. Krishnan, R., Sandhu, R.: Authorization policy specification and enforcement for group-centric secure information sharing. In: ICISS. Springer (2011)
6. Lee, A.J., Minami, K., Borisov, N.: Confidentiality-preserving distributed proofs of conjunctive queries. In: ASIACCS. ACM (2009)
7. Lee, A.J., Minami, K., Winslett, M.: Lightweight consistency enforcement schemes for distributed proofs with hidden subtrees. In: SACMAT. ACM (2007)
8. Lee, A.J., Winslett, M.: Safety and consistency in policy-based authorization systems. In: CCS. ACM (2006)

9. Lee, A.J., Winslett, M.: Enforcing safety and consistency constraints in policy-based authorization systems. In: TISSEC. ACM (2008)
10. Lee, A.J., Yu, T.: Towards quantitative analysis of proofs of authorization: applications, framework, and techniques. In: CSF. IEEE (2010)
11. OASIS: Security assertion markup language (SAML) v2.0 (2005)
12. Paci, F., et al.: ACConv–an access control model for conversational web services. In: TWEB. ACM (2011)
13. Park, J., Sandhu, R.: The UCON$_{ABC}$ usage control model. In: TISSEC. ACM (2004)
14. Peisert, S., et al.: Turtles all the way down: a clean-slate, ground-up, first-principles approach to secure systems. In: New Security Paradigms Workshop (2012)
15. RFC6749: The OAuth 2.0 authorization framework (2012)
16. Squicciarini, A.C., et al.: Identity-based long running negotiations. In: DIM. ACM (2008)
17. Steen, M.V., Tanenbaum, A.S.: Distributed Systems (2017)
18. Tsankov, P., et al.: Fail-secure access control. In: CCS. ACM (2014)

Cybersecurity Framework Requirements to Quantify Vulnerabilities Based on GQM

Mohammad Shojaeshafiei[1(✉)] [iD], Letha Etzkorn[1],
and Michael Anderson[2]

[1] Department of Computer Science, University of Alabama in Huntsville,
Huntsville, USA
Ms0083@uah.edu
[2] Department of Civil and Environmental Engineering,
University of Alabama in Huntsville, Huntsville, USA

Abstract. Of particular importance for an organization in building an effective and comprehensive secure system is to addressing a mechanism to provide a standard framework that is free from vulnerabilities. Cybersecurity experts and security requirement engineers have been addressing security issues that originated from cybersecurity requirements. Many security issues can be avoided if the security requirements are configured appropriately. In this paper, we proposed a hierarchy security requirements model based on the Goal Question Metrics (GQM) and its application mapped with the security standards towards constructing vulnerability measurements at the early stage of security development of the system design.

Keywords: Cyber attack · Goal Question Metrics · Security requirements · Security standards

1 Introduction

Nowadays, almost all organizations and infrastructures such as education, health companies, and national defense depend on cybersecurity and its indispensable technology. In recent years, we are the witness of proclamations by the government to increase investments in order to bolster cybersecurity capabilities. Countless numbers of cyber attacks make us take the importance of general security and Cybersecurity as critical factors in any organization. In 2013 almost 40 million people were impacted due to the cyber breach on retail store Target [1]. Companies are doing their best to establish a reputation of trust. It is definitely important for a company to provide valuable services, for that, it is necessary to ensure that they have strong security in place. The majority of reported security attacks result from exploit against deficiencies of software code or design. Hence, ensuring the stability and integrity of software is pivotal to protecting the system from threats and vulnerabilities. Software exhibits a wide variety of vulnerabilities. Each of these vulnerabilities has a different impact on the security features of the software such as availability, accuracy, confidentiality, and integrity.

© Springer Nature Switzerland AG 2020
K.-K. R. Choo et al. (Eds.): NCS 2019, AISC 1055, pp. 264–277, 2020.
https://doi.org/10.1007/978-3-030-31239-8_20

We shall need to contemplate that the first and the most important component of any organization to build up rational and reasonable security is security requirements. Symantec conducted research [2] on security breaches on enterprises corroborates that a significant number of cyber attacks targeting web applications. It means the early stage of the development life cycle plays a pivotal role in security engineering to protect software or application against inadvertent defects. In this way, the attacker cannot easily exploit the potential or available weakness of design and development to break the system.

It is necessary to apply Goal Question metrics (GQM) approach to prepare the ground for security requirements analysis in order to achieve a quantifiable vulnerability measurement to create the system as secure as possible. Using GQM, the organization specifies the goals for itself and its projects.

This paper proposes a security requirement model for organizations based on GQM coupled with security standards such as NIST and ISO to provide a holistic consideration on various aspects of security requirements from the viewpoint of stakeholders, network security, designer and adversary. Our proof of concept will be the Department of Transportation (DOT). Considering all the crucial and sensitive components of DOT, we will design an integrated and comprehensive model based on all major aspects of the service availability, integrity, confidentiality, and accuracy.

The rest of the paper is organized as follows: Sect. 2 presents the background of the work which discusses the importance of security engineering requirements for at the initial stage of security development. Section 3 provides a description of our method to design security requirements model mapped to the security standards.

2   Background

2.1   Security Requirements

Studies show that defects of requirement engineering cost 10 to 200 times more in the correction process once the system has been deployed in the field than they would if they were detected in the requirements development [3, 4]. Therefore, requirements engineering is a crucial component in successful project development. Unfortunately, in lots of security projects requirements engineering does not receive sufficient attention. Several types of research proposed different models to support security requirement elicitation such as Multilateral Security Requirements Analysis (MSRA), misuse cases, Secure UML, Goal Base Requirements Analysis Record (GBRAM) and Security Quality Requirements Engineering(SQUARE) [27]. SQUARE has been developed at Carnegie Mellon University [5]. The purpose of that project was to define the integration of security requirements for the system.

SQUARE is a well-defined method for security requirements analysis because it has a number of advantages over the other proposed methods since it is the only method which considers Confidentiality, Availability and integrity goals explicitly while it comprises threat and risk assessment coupled with quality assurance simultaneously. Prioritization technique using a binary search tree is one of the most important factors that is provided in this method. In requirements categorization phase, it categorizes security requirements into essential and non-essential components and in requirements inspection phase, there are completeness and accuracy of the elicited requirements [28].

Moreover, it provided a formal and hierarchical attack tree "Fig. 1" to describe different types of attack that could happen. In fact, the diagram is a symbolic tree-structure representation of attacks. The root would be the attacker's goal and the leaves depicts the different methods or procedures of achieving the goal. Thus, the attack tree shows a general perspective and an outline of potential attacks on the system. The benefit of such tree is that both stakeholders and developers team have interaction with that and for each similar attack requirements engineering do not have to draw another attack tree.

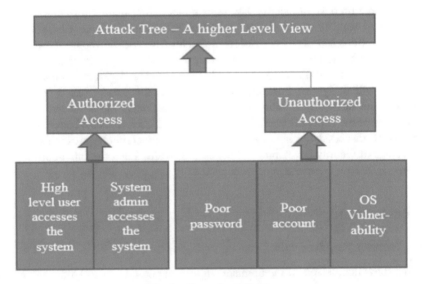

Fig. 1. General attack tree

SQUARE consists of nine steps that generate final deliverable prioritized security requirements, we came up with several major concerns as fundamental security requirements of any organization.

- Authentication
- Access Control Mechanism
- Availability
- Auditing
- Network Security
- Authorization
- Physical security
- Integrity

Hence, all of the above advantages have convinced us to choose SQUARE and apply it to our research.

2.2   Goal Question Metrics Paradigm

Basili and Weiss [6] developed Goal Question Metrics: a systematic approach to integrating goals with the models of the software product and quality perspective, and based on the specific needs of the project. In software engineering, GQM is widely accepted as the goal standard for how to create a metrics framework [7–9]. The result of GQM is the specification of a measurement system that is based on the particular set of issues and rules to interpret the measurement data. As a result, we may extract Goal, Question and Metric; a goal is defined for an object; questions try to characterize the objects with respect to security issues, and metrics provide measurement to answer the questions in a quantitative way "Fig. 2". In other words, measurement goals are defined for the specific needs of an organization and are traceable to a set of quantifiable questions and questions that lead the measurement to be based on specific set of metrics and data for collection [7].

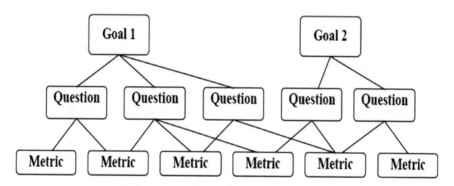

Fig. 2. Goal question metrics chart

Goals are tailored to the needs of the project and the actual trend of flow from the goals to the quantifiable questions are based on goal refinement that is traceable to the quantifiable questions [7]. Each goal will lead us to one or more questions, and the questions will be refined into either quantitative or qualitative metrics.

Security requirements identifications in SQUARE outline the security goals and prepare the ground for the next phase that we can apply GQM to provide appropriate questions to achieve metrics for the specific goals. For example, in case of security requirements for software security, we have to consider Confidentiality, Availability, Authentication, and non-repudiation of security goal. Designing questions and specifying measures to answer the questions help us to perceive whether the goal is achievable or not. Metrics for identification, authentication and authorization (goals) are shown in Table 1.

Table 1. GQM based on identification, authentication and authorization requirements

Question derived from the goal	Metrics derived from the question
How does the system identify the users?	Evaluation method by the procedure all level user identified to use the system
How does the system authenticate the users?	Common methods such as Name, password, ID, Fingerprint, etc.
How does the system lock the user account?	A certain number of failed login attempts for each user
How does the system grant authorization?	based on the predefined policy such as personal, role-based or group-based

2.3 Security Standards

ISO 27001 [10], formally known as ISO/IEC 27001:2005, was updated and revised in 2013 as ISO/IEC 27001:2013. It is a specification standard for information security management system and includes all legal, physical and technical controls and standards of organizations' information risk management processes. ISO provides checklists of controls that should be considered in security analysis. The main purposes of ISO 27001 are the preservation of confidentiality, Integrity, and availability. For example, in case of "*media handling*" in the organization in the section A.10.7 it defines the objective as "to prevent unauthorized disclosure in place for the management of removable media" and provides sub-sections as follows in Table 2:

Table 2. Mapping security standards to GQM

A.10.7.1	Management of removable media	Control: there shall be procedures in place for the management of removable media
A.10.7.2	Disposal of media	Control: media shall be disposed of securely and safely when no longer required, using formal procedures
A.10.7.3	Information handling procedures	Control: procedures for the handling and storage of information shall be established to protect this information from unauthorized disclosure or misuse
A.10.7.4	Security of system documentation	Control: system documentation shall be protected against unauthorized access

In addition, NIST SP800-53 [11] is a publication that provides security controls for federal information system and organizations. It is published by National Institute of Standards and Technology. It is a part of the Special Publication 800- series and reports Information technology Guidelines, research in information system security. It includes the procedures in the Risk Management Framework based on security controls for federal information system for security requirements.

To have a rational perspective on security requirements for each system and organization, we need to consider the NIST standards. NIST SP800-53 [11] describes system security, planning to provide an overview of security requirements of the system, and describes the controls in place for meeting those requirements. For example in the case of "*media use*" in the organization in section MP-7 it specifies each control and the related controls as follows:

- "[Selection: Restrict; Prohibit] the use of [Assignment: organization-defined types of system media] on [Assignment: organization-defined systems or system components] using [Assignment: organization-defined security safeguards]";
- "Prohibit the use of portable storage devices in organizational systems when such devices have no identifiable owner"

In order to provide reasonable security and asset protection in an organization, there are always three important questions that need to be considered;

- What assets need to be protected?
- How are those assets threatened?
- What can be done to counter those threats?

Indeed, companies need to achieve and maintain appropriate levels of factors such as confidentiality, integrity, availability, accountability, authenticity, and reliability with their functions [12]. In order to prescribe a safeguard for these factors in organizations, security controls play a vital role. Based on NIST SP800-53 Security Controls, there are three types of classes in organizations; Management, Operational and Technical classes. Consequently, there are several controls for each class in which security assessment, risk assessment for Management, awareness and information integrity for Operational, access control, audit and authentication for Technical are the most important parts of controls in each class. Moreover, risk assessment is a key component of organization-wide risk management. It is defined in NIST SP800-30 as the process of identifying, estimating and prioritizing information security risk. Contribution factors are assessed in two important ways:

- Quantitatively
- Qualitatively

The following factors need to be considered for conducting risk assessment:

- Threat source identification.
- Threat events identification that is extracted from the sources.
- Vulnerability identification that could be exploited by threats.
- Likelihood determination that the identified sources can initiate a specific threat.
- Adverse impact determination to assets.

3 Requirements Analysis Based on Security Factors and Sub-factors

3.1 Why DOT?

In "Fig. 3" Stalling and Brown [12] suggest a possible spectrum of organizational risk which shows Transportation is one of the most susceptible organizations that is prone to be more vulnerable and susceptible to risk. Based on our literature review [13–22], the DOT is very susceptible to cyber attack. To our knowledge, there has been little prior work on safety and security metrics frameworks and to our knowledge, these have never been applied to the DOT.

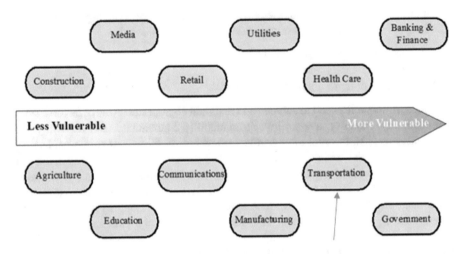

Fig. 3. Vulnerability scan mapping to the standards

Based on the security requirements of organizations in Sect. 2.1 plus several quality factors, we have designed a security requirements model comprised of all aspects and details of security requirements that are concerns of security requirements engineers, system administrators and employees of organizations (DOT in our case study). The procedure of the model design is that we mapped each major security question from with the NIST SP800, ISO 27001 2005 and ISO 27001 2013 standards to make sure the consideration and evaluation of questions are compatible with the standards.

Each question is derived from the application of GQM to quality factors and sub-factors based on the security requirements of organizations. For instance, in "Fig. 4" we depict a hierarchy chart of Network Security which is one of the major key points of security requirements and all key quality factors and sub-factors of it. Then, we mapped each question to the appropriate quality factor. (Sometimes the interaction with one quality factor has inevitable overlap with multiple questions), which is absolutely normal and proves that all major and minor security aspects are analyzed. In fact, the definition of measurable goals is the conceptual level of hierarchy in the first level and in the next level goal refinement into questions characterize the goal assessment.

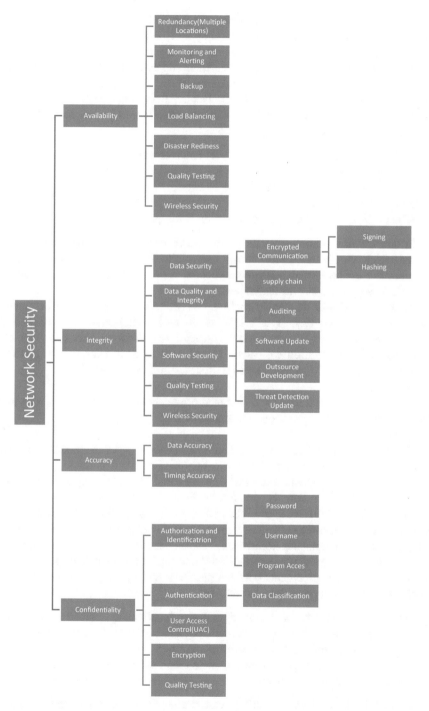

Fig. 4. Hierarchy chart of network security mapped to the security standards

Assume we want to develop a security metrics model in Network Security (one of the main goals in our GQM structure) and apply it to Integrity factor. Data Integrity contains Data Security sub-factor and finally, Supply Chain is sub-factor of Data Integrity. Organizations usually work with their supply-chain partners to correct data quality errors and associated risks. Consequently, for Mobile Devices security, Enterprise Mobile Management (EMM), Device Encryption, User Access Control (UAC), Mobile Device Management (MDM) System would be the main factors and each of them has its sub-factor(s) as follows:

EMM→ lost devices policy, patch level enforcement and encryption of SD cards and internal memory.
Device Encryption→ Virtual Private Network
UAC→ Remote data wipe
MDM→ Automatic lockout screen, jailbreaking prevention

In our security requirements hierarchy model many security questions are considered which are highly correlated with assets, attacks, and vulnerabilities. Many of them are the concerns and questions extracted from the literature, system admins, network security experts, and employees.

As we showed how the supply chain is one of the sub-factors of network security based on GQM hierarchy, it will help us to define subjective and consequently objective metrics which links us from the question to Sub-factor, Factor, and the Goal respectively. For illustration we chose two example questions for supply chain and vulnerability scan in the systems listed as follows in Tables 3 and 4:

Question 1:
Does DOT select and monitor outsourced providers in compliance with laws in the country for data storage, location, and translation?

Question 2:
How many times a year does DOT conduct network-layer vulnerability scans as prescribed by industry best practices?

The security requirements identified here are based on the security goal in Network Security and consideration is on Information Security Standard such as ISO 27001:2005, ISO 27001:2013 and NIST SP 800-53. Tables 3 and 4 shows the relationship among properties for security goals mapped with clauses in security standards and the security requirements. For instance, Table 3 shows how clauses A.6, A.9, A.10, A.13 and A.15 in ISO and CA-3, MP-5, PS-7 and SA-9 in NIST define supply chain to ensure Supply chain agreements between providers and user or customer incorporate mutually-agreed upon provisions as follows:

- The scope of services, data acquisition, usage, functionality, infrastructure network and systems components for service delivery and support, roles and responsibilities of the system or outsourced business relationships.
- Information security requirements, business relationship timeline, risk management, regulatory compliance obligations.

Table 4 illustrates the Vulnerability scan as a sub-factor of security requirements in Network Security. It shows the clauses A.12, A.14 in ISO and CM-3, CM-4, CP-10, RA-5, SA-7, SI-1, SI-2 and SI-5 in NIST that illustrates how we mapped system's

Vulnerability scan to the Security Standard clauses. Vulnerability scan defines policies and procedures to support processes and technical measures implemented for timely detection of vulnerabilities within the organization, infrastructure network, and system components to ensure the efficiency of implemented security controls. It also determines that there should be a reasonable demand for risk model prioritization for each vulnerability.

Table 3. Supply chain mapping to the sections of standards

Domain	ISO 27001 2005	ISO 27001 2013	NIST SP800-53
Supply chain management and transparency; third party agreement	A.6.2.3 A.10.2.1 A.10.8.2	A.15.1.2, A.13.2.2, A.9.4.1 A.10.1.1	CA-3 MP-5 PS-7 SA-9

Table 4. Vulnerability mapping to the sections of standards

Domain	ISO 27001 2005	ISO 27001 2013	NIST SP800-53
Threat and vulnerability management vulnerability	A.12.5.1 A.12.5.2 A.12.6.1	A.14.2.2, A.14.2.3 A.12.6.1	CM-3 CM-4 CP-10 RA-5 SA-7 SI-1 SI-2 SI-5

3.2 Other Aspects of Security Requirements

Security requirements for Network Security are only one aspect of whole security requirement consideration for the organizations. Unfortunately, here enumeration and illustration of all other factors and sub-factors are beyond the limits of this paper. We briefly mention the main parts of some aspects of security requirements.

As a matter of security, cloud security and cloud privacy would also be hotspots in technologies and industries and are important parts of system security. It has been a highly controversial issue to talk about the advantages of clouds while underestimating the confidentiality and security problems associated with clouds, based on the fact that customers are outsourcing data and computational tasks to the cloud. Therefore, there is no guarantee that these tasks can be kept confidential, e.g. a Cross VM attack is a cloud attack that could lead one to question cloud confidentiality [23]. But security requirements model can help the organizations to consider all security aspects at the very early stage to ensure the safety of their cloud storage, enhance their security implementation aligned with the goals, and improve the quality of security capability level [8].

For instance, in order to analyze the confidentiality of data in cloud storage, the goal would be the assessment and the component is confidentiality. It can be considered from the perspective of both engineer and stakeholder. Questions would be as follows:

Q1. Is there any implementation for identity management?
Q1.1 Is authorization included in implementation?
Q1.2 Is authentication included in implementation?
Q1.3 Does User Access Control work properly in the system?
Q1.4 Is encryption designated for data security?
Q1.5 Is data security architecture using industry standard?
Q1.6 If the virtual infrastructure is used, does the cloud solution include independent hardware restore and recovery capability?

Each question will be answered by experts, employees, stakeholders, and managers. Based on the answer to each question there will be a rate or metric assigned to each question for the desired goal evaluation.

Data center security is another important part of the puzzle that enterprises and organizations should consider as part of the major and vital characteristics of the security. Nowadays, the competitive environment requires organizations to think about the security of the data center as a priority when it comes to optimum network design [8, 24, 25]. Unfortunately, a hundred percent of security is never guaranteed.

All components of data center security should be considered from different perspectives. A complete security solution includes authentication, authorization, data privacy, and perimeter security. Firewalls are the main part of perimeter security for packet transmission security. Another big challenge for data center security is physical security. It has to follow some requirements such as multiple physical connections to the public power grid, adequate uninterruptible power supply, conform to standards such as bulletproof glass and reinforced walls, adequate air conditioning, and making the buildings strong enough to survive the hurricane, and flood sensors. In addition, windows should not be used in the building to prevent physical attacks [26]. "Figure 5" depicts the major security aspects of the data center for physical security.

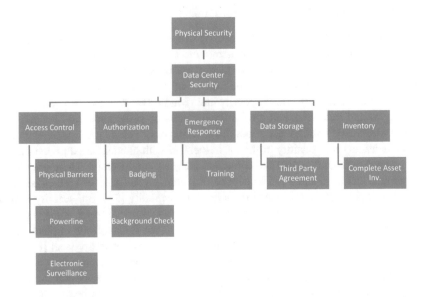

Fig. 5. Hierarchy chart of data center security mapped to the security standards

As we can apply the hierarchy GQM model to achieve the most important security requirements of a system, we can apply the same model to calculate the next stage of metric measurement. For more illustration, suppose based on the following main question we want to develop a security metrics model to be applied at the third-party stage of security in DOT.

Main Question: How are third-party software updates (Adobe, Java, etc.) installed on DOT's computers?

Answers:

- It is not important to stay current (0)
- Whenever software update prompts on the system (2)
- We use software to check for updates and prompt us to install (4)
- They are managed and updated network-wide from a console (5)
- Others:—(0–5)

Sub-Question 1: Does the software update strategy for software X correctly represent the application of third-party software update policy?

Sub-question 2: What is the ratio of the updates that do not put the software at the risk of vulnerability to the total number of updates that are predetermined by the third-party?

Consider a total set of software updates in the model as $TSU = \{so_1, ..., so_n\}$ and the successful software updates as $SSU = \{ssu_1, ..., ssu_n\}$ such that $SSU \subseteq TSU$. The metric is expressed as follows; where RSSU stands for the ratio of software updates that meet the requirements of third-party software update policy [17].

$$RSSU = \frac{SSU}{TSU}$$

4 Conclusion and Future Work

Here we have shown an initial example of how GQM can be used to create security quality factor requirements. We have shown how these quality factors map to the most recognized security standards. With the application of this method, the organizations can follow a standard framework with the presence of authentication measures throughout the system. They are required to apply UAC to handle which system elements can be active for the user. The system is required to provide such powerful protection for physical devices against destruction, natural disasters, theft, and sabotage. In the near future, we plan to tailor these as well as our quality hierarchies to the Department of Transportation (DOT). Our final goal for the current and future research is full industry adoption and integration, however, in this paper, we have just introduced the very first stage of security consideration for organizations to uphold holistic and comprehensive cybersecurity arrangements to protect the system form vulnerabilities, threats and, potential risks. The main attempt in this study was putting emphasis on consideration of security requirements based on the viewpoints of requirement engineers, stakeholders and, employees.

In the later future we plan to apply our GQM-based methodology to additional case studies to further analyze this methodology and the extent to which it is successful in different cases.

References

1. Target and Neiman Marcus hacks: The latest. CNNMoney. https://money.cnn.com/2014/01/13/news/target-neiman-marcus-hack/index.html. Accessed 11 Feb 2019
2. Symantic Inc.: Symantec Global Internet Security Threat Report Trends for 2009. Symantec Global Internet Security Threat Report, vol. XV, p. 7 (2010)
3. Boehm, B.W., Papaccio, P.N.: Understanding and controlling software costs. IEEE Trans. Softw. Eng. 14(10), 1462–1477 (1988)
4. McConnell, S.: From the editor - an ounce of prevention. IEEE Softw. 18(3), 5–7 (2001)
5. Mendonca, M.G., Basili, V.R.: Validation of an approach for improving existing measurement frameworks. IEEE Trans. Softw. Eng. 26(6), 484–499 (2000). https://doi.org/10.1109/32.852739
6. Basili, V.R., Green, S.: Software process evolution at the SEL. In: Foundations of Empirical Software Engineering, pp. 142–154 (1994)
7. Shepperd, M.: Practical software metrics for project management and process improvement. Inf. Softw. Technol. 35(11–12), 701 (1993)
8. Yahya, F., Walters, R.J., Wills, G.B.: Using goal-question-metric (GQM) approach to assess security in cloud storage. In: Enterprise Security Lecture Notes in Computer Science, pp. 223–240 (2017)
9. Abdulrazeg, A.: Security measurement based on GQM to improve application security during requirements stage. Int. J. Cyber Secur. Dig. Forensics JCSDF 1, 211–220 (2012)
10. International Organization for Standardization. Developing standards, 10 January 2019. http://www.iso.org/. Accessed 12 Feb 2019
11. National Institute of Standards and Technology. NIST, 12 February 2019. http://www.nist.gov/. Accessed 15 Feb 2019
12. Stallings, W., Brown, L.: Computer Security: Principles and Practice. Pearson, London (2018). Chp 14
13. Ernst, J.M., Michaels, A.J.: Framework for evaluating the severity of cybervulnerability of a traffic cabinet. Transp. Res. Rec.: J. Transp. Res. Board 2619(1), 55–63 (2017)
14. Ghena, B.: Green lights forever: analyzing the security of traffic infrastructure. In: Proceeding of the 8th Workshop on Offensive Technology (WOOT 2014), August 2014
15. Fok, E.: An introduction to cybersecurity issues in modern transportation systems. ITE J. (2013). https://trid.trb.org/view/1257258. Accessed 22 Oct 2018
16. Hacking US (and UK, Australia, France, etc.) Traffic Control Systems. IOActive, 15 June 2018. https://ioactive.com/hacking-us-and-uk-australia-france-etc/. Accessed 22 Oct 2018
17. Chen, Q.A., Yin, Y., Feng, Y., Mao, Z.M., Liu, H.X.: Exposing congestion attack on emerging connected vehicle based traffic signal control. In: Proceedings 2018 Network and Distributed System Security Symposium (2018)
18. Comprehensive Experimental Analyses of Automotive Attack …. http://www.autosec.org/pubs/cars-usenixsec2011.pdf. Accessed 22 Oct 2018
19. An Emerging US (and World) Threat: Cities Wide Open to …. https://ioactive.com/pdfs/IOActive_HackingCitiesPaper_CesarCerrudo.pdf. Accessed 22 Oct 2018

20. Li, Z., Jin, D., Hannon, C., Shahidehpour, M., Wang, J.: Assessing and mitigating cybersecurity risks of traffic light systems in smart cities. IET Cyber-Phys. Syst.: Theory Appl. **1**(1), 60–69 (2016)
21. Cyber Risk and Insurance for Transportation Infrastructure. https://web-oup.s3-us-gov-west-1. amazonaws.com/showc/assets/File/CIRI_Tonn_Cyber%20%Risk%20%Insurance%20%for% 20%Transportation%20%Infrastructure.pdf. Accessed 22 Oct 2018
22. Reilly, J., Martin, S., Payer, M., Bayen, A.M.: Creating complex congestion patterns via multi-objective optimal freeway traffic control with application to cyber-security. Transp. Res. Part B: Methodol. **91**, 366–382 (2016)
23. Xiao, Z., Xiao, Y.: Security and privacy in cloud computing. IEEE Commun. Surv. Tutor. **15**, 843–859 (2012)
24. Computer Security and Intrusion Detection. Intrusion Detection and Correlation Advances in Information Security, vol. 14, pp. 9–28. Springer, Boston (2005). (Chapter 2)
25. Schaen, I., Mckenney, B.: Network auditing: issues and recommendations. In: Proceedings Seventh Annual Computer Security Applications Conference. Data Centers: Best Practices for Security and Performance. http://www.echomountain.com/pdfs/CiscoBestPractices.pdf. Accessed 15 Feb 2019
26. Oivo, M., Basili, V.: Representing software engineering models: the TAME goal oriented approach. IEEE Trans. Softw. Eng. **18**(10), 886–898 (1992)
27. Ahl, V.: An experimental comparison of five prioritization methods. Master's thesis, School of Engineering, Blekinge Institute of Technology, Ronneby, Sweden (2005)
28. Fabian, B., Gurses, S., Heisel, M., Santen, T., Schmidt, H.: A comparison of security requirements engineering methods. Requirements Eng. **15**(1), 7–40 (2010)

Impact of Targeted Cyber Attacks
on Electrical Power Systems

Caroline John[1]($\boxtimes$), Bhuvana Ramachandran[2],
and Ezhil Kalaimannan[1]

[1] Department of Computer Science, University of West Florida, Pensacola,
FL 32514, USA
{cjohn, ekalaimannan}@uwf.edu
[2] Department of Electrical and Computer Engineering,
University of West Florida, Pensacola, FL 32514, USA
bramachandran@uwf.edu

Abstract. This paper formulates targeted and coordinated attacks on an electrical power system infrastructure as an optimization problem to (a) investigate the impacts of such attacks on the electric power grid and (b) to study the extent of damages to the grid depending on attacker's resources and level of protection employed in the grid. In this research, we consider the coordinated load redistribution (LR) attack, which is a variant of false data injection attack on electrical power systems. The bi-level formulation is investigated through a problem in which the goal of the hacker is to maximize load curtailment and that of the power system operator is to minimize load curtailment. The resulting nonlinear mixed-integer bilevel programming formulation is converted into an equivalent single-level mixed-integer linear program by solving the inner optimization by KKT optimality conditions. The case studies are conducted based on an IEEE 14-bus system. From the results, it is observed that the hacker could maximize the damages to the power grid, even with limited attack resources, by simultaneously targeting the physical system and deceiving the operator with a false power dispatch. This study can provide meaningful insights on the relation between hacker's available resources, existing security measures and resulting disruption in the power system. The results can be used to design methods to prevent and mitigate such coordinated attacks to improve the reliability of the electric grid.

Keywords: Bi-level programming · CPLEX solver · Cybersecurity ·
LR attacks · Mixed-Integer Linear Programming (MILP) ·
Power system security · Smart grid

1 Introduction

Over recent years, cyber-attacks on electrical power systems have increased significantly. High-voltage smart grids are particularly vulnerable to a wide variety of well-known cyber-attacks due to their wide-range accessibility and coverage [1]. For instance, if the hacker is capable of gaining entry into and compromising several generations or transmission facilities simultaneously, and target utility control centers, then the damages to the electric grid will be catastrophic. Traditional power system analysis using methods such as power flow analysis and state estimation theories could be used to assess the impact of the vulnerability of power systems to deliberate outages. However, in the recent years, there has been a two-fold purpose from the perspectives of the attacker and defender in analyzing the potential threats to a power system and to minimize the exposure to cyber-attacks [2]. The power system is made even more vulnerable to cyber-attacks due to the increasingly open architecture of electric power communication networks and the utilization of unsecured standard IP based protocols. Examples of attacks include Denial of Service (DoS) attacks, unauthorized access into the control system, database modification and misleading the operator by false data injection attack [3], etc. Moreover, as the electric grid is widely spread from generation to distribution, the power system is also vulnerable to being physically attacked by vandalism and terrorism. The hackers can easily target those exposed facilities that are unmanned and unprotected. Both cyber and physical attacks against the power grid have become more of reality [4–6]. An optimal and secure strategy must be developed to defend the power system from malicious attacks.

Even if the hacker is successful in accessing one part of the power system, since the power system is designed to withstand the N-1 redundancy criterion, no loads will be lost (curtailed). However, if the attacker launches attacks to compromise multiple parts or functions of the power grid in cyber and physical aspects simultaneously, then catastrophic power failures could be triggered. The possibility of large-scale power outages will increase and significantly weaken the reliability of power grids. In current literature, an important topic of research is to investigate the coordinated attack and to design the attack strategy. The attack strategy refers to the way to identify a few substations/transmission lines as targets. The power grid would likely experience a severe power outage if attacks are launched against these targets by either physical sabotages or cyber-attacks [7]. In earlier literature, attack strategy was mainly studied from two perspectives, substation only [8–17] and transmission line only [18–21]. Some other authors have considered load redistribution (LR) attacks [3, 22–24] which are a special type of false data injection attacks. A type of cyber-attack against state estimation through SCADA systems are the false data injection attacks. They distort the outcome of state estimation by cooperatively manipulating the measurements taken at several meters.

It is evident that some of these preliminary research studies did not consider an important aspect, the combined-LR-generation aspect, and the combined-LR-transmission line aspect when investigating the attack strategy. An attacker can target (a) loads and generators or (b) on loads and transmission lines simultaneously rather than just the transmission line only or generator only. Such an extension has a huge impact on analyzing power grid vulnerability. The coordinated attacks can significantly increase the number and level of power grids' vulnerabilities. With relevant resources, the attacker can devise a strategy that is assorted and complex.

Our main contributions in this work are (a) to study how the attacker would plan a coordinated attack on a generator and misleading data attack on a load to maximize load curtailment and damage (b) analyze the optimization problem with varying cyber-attack resource factors and limits on attack magnitude of load demand measurements to warn the operator of possible threats to the grid.

In this research, the coordination between LR attack and generation attack is investigated in depth based on a bilevel formulation of the attacks. Rest of the paper is organized as follows. Section 2 gives a brief review of the coordinated attacks in the past and the impacts of such attacks on power systems. An security constrained economic dispatch model is explained in Sect. 3. The theory of LR attack coordinated with generator attack is explained in Sect. 4. A detailed mathematical model for the application of bilevel optimization to determine the attack strategy of the attacker is provided in Sect. 5. An example case study is presented in Sect. 6 and results and discussion are given in Sect. 7. Conclusion and scope for future work are presented in Sect. 8.

2   Coordinated Attacks

The modern electric grid consists of a decentralized architecture with several renewable energy sources, energy storage elements and distributed energy resources connected. In such a cyber-physical system, the operation and control depend on a variety of measurement devices, control systems, and communication networks for monitoring, protection, and control. With multitudes of open nodes available where an attacker can inject false data or physically vandalize a substation facility, the attackers have plenty of opportunities available to induce catastrophic damage to the power system. He would be able to inflict a more severe attack if the attacker has a priori knowledge on the operation of the entire system. Enabled with this knowledge, an attacker is capable of attacking a substation that is crucial to the reliable operation of the power grid. After gaining access into a substation, a wide range of attack options are available to the attacker, that can be executed (e.g. tripping breakers, changing transformer tap settings, capacitor bank settings). He would be able to choose an attack or combination of attacks to cause maximum damage if the attacker has expertise in the controls available to him through the SCADA system. A group of attackers may be collaborating together to cause major damage to the power grid. If the attackers are knowledgeable and have well-designed plan, the reliability and security of the grid can be disrupted significantly. A coordinated cyber-attack has unique features: (1) A well-organized attack plan before the attack, and (2) Each attack step related to another step (s). As an example, in 2013, a coordinated physical attack was carried out by a few men on a PG and E transmission substation, trying to disrupt power to local customers. It was seen on a video recording from CCTV cameras at the substation that there were three or four men lowering themselves underground to cut fiber-optic cables (protected only by a heavy manhole). They fired at least 100 rounds from a high-powered rifle at equipment. They destroyed 17 transformers and damaged $16 million worth of equipment.

3   Power Flow Analysis for Security Constrained Economic Dispatch

In power system operations, it is essential for the system operator to make informed and quick decisions based on power system analysis tools such as contingency analysis, optimal power flow analysis, optimal dispatch, etc. State estimation is one such analysis that is used to make an educated estimate the state of the power systems in system monitoring. Based on the estimated state, security-constrained economic dispatch (SCED) then aims to minimize the total system operation cost through dispatching power produced by generators. The security constrained economic dispatch problem is based on dc power flow equations. Power flow equations are used by the utility as a mathematical model to plan, operate, and analyze the power grid. The results indicate how generation and demand balance, and how active and reactive power flow through the grid. SCED based on DC power flow is formulated as an optimization problem to minimize the total generation cost subject to the following security constraints such as generation-demand balance, operation limits of the generators and transmission limits on power flows.

$$\min \frac{1}{2} P_g^T C_2 P_g + C_1^T P_g + C_0 \tag{1}$$

$$S.t \ P_g - P_d = 0 \tag{2}$$

$$P_g \in \left[P_g^{min}, P_g^{max} \right] \tag{3}$$

$$SF \left(P_g - P_d \right) \in [-PL_l, PL_l] \tag{4}$$

In this research, the attacker is assumed to have conducted long term surveillance of the power system. It is assumed that he has obtained crucial operation and control information. The information includes topology of the network, the scheme of connection of devices and electrical parameters of generators/lines and loads, etc. This information will be used by the attacker to conduct an optimal power dispatch study to identify the most vulnerable load which would result in maximum load curtailment and hence maximize damage to equipment/system operation. Thinking along the same lines as the attacker, if the system operator also uses the same information (obtained ethically) to identify the most vulnerable load bus and the amount of load that will be redistributed for maximum load curtailment, then he can better protect the power system and ensure effective methods to be deployed to prevent such attacks. The results from power flow analysis assist the operator in ensuring that all quantities of interest stay within bounds.

4 LR Attack Coordinated with Generator Attack

In a load redistribution attack (LR attack), it is demonstrated that an attacker can inject false data into the power state estimation without being detected [3, 4]. The LR attack is proposed based on the practical conditions where only load and line flow measurements can be subjected to attack. Equations 6–8 of the mathematical model define the LR attack, where Eq. 6 indicates the total demand must be unchanged making the attacker undetectable. Equation 7 is the power balance equation for the total power flow in the system, where SF is the shift-factor matrix, KD is the bus-load incidence matrix and τ is the upper bound of the attack magnitude on the load demand measurements.

The system operating personnel can identify an issue when a certain load/line flow measurement is out of bounds thereby preventing any cascading line outages in the power system. Usually, the generators are physically well-protected ad so it is very difficult to physically attack a generator. When an attacker tries to attack the generators, any change in generation levels can also be detected and mitigated by the operator by remotely controlling the power plant. With these limited options, an attacker can only conduct *cyber attacks on generators*. On the contrary, LR attacks are not easily detectable by the operator. If an LR attack succeeds and the operator misinterpretes the loading conditions at buses, a simultaneous cyber attack on generators will result in extensive damages to the power system. Thus a coordinated attack is one means for the attacker to maximize shutting down of the loads whereas the operator attempts to minimize the same to ensure the reliability of power supply to customers. The attacker needs to strategically decide on which load/generator to attack and with how much R_c to attack. The defender (operator) must use a defensive approach to determine which load/generator will be most vulnerable and use that information to determine and develop protective/corrective control action for the power system. Typically attacking the generator gains higher priority because attacking generator results in larger load curtailment than a LR attack even though attacking a generator is costly. But when the line flow is close to its maximum limit, LR attacks become more effective. If abnormal load changes due to LR attacks are detected quickly by the operator, the reliability of the power system can be increased. LR attacks can be detected effectively by comparing existing load condition with historical measurements. In this paper, we study a LR attack conjugated with an attacking generator scenario to represent a LR coordinated attack.

5   Mathematical Model of LR Attack Coordinated with Generator Attack

The bi-level optimization problem is described in Fig. 1 [3]. The attacker tries to maximize the damage to the power system by enforcing load curtailment coordinating with the compromised generators (coordinated LR attack) to be attacked. This comprises the upper-level formulation of the bi-level optimization problem. The operating personnel in the lower level problem optimally reacts to the compromised power flow estimation which has been successfully manipulated by the attack vector and trips the generators determined from the upper-level optimization problem. Our proposed model features the LR attack model proposed by Yuan et al. [4] along with the generators to be attacked to formulate a coordinated generator - LR attack model. The variables and constants used in the mathematical model are listed in Sect. 9: Appendix A.

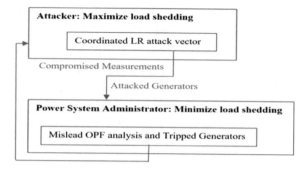

Fig. 1. Illustration of the proposed coordinated LR attack bi-level optimization problem.

$$max_{\Delta D} \sum_{d=1}^{N_d} S_d^* \qquad (5)$$

s.t.

$$\sum_{d=1}^{N_d} \Delta D_d = 0 \qquad (6)$$

$$\Delta P_F = -SF \times KD \times \Delta P_D \qquad (7)$$

$$-\tau P_d \leq \Delta P_d \leq \tau P_d \quad \forall d \qquad (8)$$

$$\Delta P_d = 0 \leftrightarrow \delta_{D,d} = 0 \quad \forall d \qquad (9)$$

$$\begin{aligned}
\Delta P_D + \tau P_{D,d}\delta_{D,d} &\geq 0 \\
\Delta P_D - \tau P_{D,d}\delta_{D,d} &\leq 0 \\
\delta_{D+,d} + \delta_{D-,d} - 2\delta_{D,d} &\leq 0 \\
\Delta P_D + \left(-\tau P_{D,d} - \varepsilon\right)\delta_{D-,d} &\geq -\tau P_{D,d} \\
\Delta P_D + \left(\tau P_{D,d} + \varepsilon\right)\delta_{D+,d} &\leq \tau P_{D,d} \qquad \forall d \\
\delta_{D+,d} + \delta_{D-,d} + \delta_{D,d} &\leq 2 \\
\delta_{D+,d} + \delta_{D-,d} - \delta_{D,d} &\geq 0 \\
\delta_{D+,d}, \delta_{D-,d}, \delta_{D,d} &\in \{0,1\}
\end{aligned} \qquad (9a)$$

$$\Delta PL_l = 0 \leftrightarrow \delta_{PL,l} = 0 \quad \forall l \qquad (10)$$

$$\begin{aligned}
\Delta P_L + M\delta_{L,l} &\geq 0 \\
\Delta P_L - M\delta_{L,l} &\leq 0 \\
\delta_{PL+,l} + \delta_{PL-,l} - 2\delta_{PL,l} &\leq 0 \\
\Delta P_L + (-M - \varepsilon)\delta_{PL-,l} &\geq -M \\
\Delta P_L + (M + \varepsilon)\delta_{PL+,l} &\leq M \qquad \forall l \\
\delta_{PL+,l} + \delta_{PL-,l} + \delta_{PL,l} &\leq 2 \\
\delta_{PL+,l} + \delta_{PL-,l} - \delta_{PL,l} &\geq 0 \\
\delta_{PL+,l}, \delta_{PL-,l}, \delta_{PL,l} &\in \{0,1\}
\end{aligned} \qquad (10a)$$

$$\sum_{d=1}^{N_d} C_{D,d}\delta_{D,d} + 2\sum_{l=1}^{N_l} C_F\delta_{PL,l}$$

$$+ 2\sum_{g=1}^{N_g} C_{G,g}\left(1 - v_{G,g}\right) \leq R_C \qquad \forall v_{G,g} \tag{11}$$

$$S_d^* = arg\left\{\min_{\Delta D} \sum_{d=1}^{N_d} S_d\right\} \tag{12}$$

s.t.

$$\sum_{g=1}^{N_g} P_g = \sum_{d=1}^{N_d}(D_d - S_d) \qquad (\lambda) \tag{13}$$

$$PL = SF.KP.P - SF.KD.(D + \Delta D - S_d) \quad (\mu) \tag{14}$$

$$-PL_l^{max} \leq PL_l \leq PL_l^{max} \quad \forall l \quad (\underline{\alpha}_l, \bar{\alpha}_l) \tag{15}$$

$$v_G P_g^{min} \leq P_g \leq v_G P_g^{max} \quad \forall g \quad \left(\underline{\beta}_g, \bar{\beta}_g\right) \tag{16}$$

$$0 \leq S_d \leq D_d \leq \Delta D_d \quad \forall d \quad \left(\underline{\gamma}_d, \bar{\gamma}_d\right) \tag{17}$$

$$C_G - \lambda + (SF.KP)^T.\mu - \underline{\beta}_g + \bar{\beta}_g = 0 \quad \forall g \tag{18}$$

$$-\mu_l\,\underline{\alpha}_l + \bar{\alpha}_l = 0 \quad \forall l \tag{19}$$

$$C_D - \lambda + (SF.KD)^T.\mu - \underline{\gamma}_d + \bar{\gamma}_d = 0 \quad \forall d \tag{20}$$

$$\underline{\alpha}_l, \bar{\alpha}_l, \underline{\beta}_g, \bar{\beta}_g, \underline{\gamma}_g, \bar{\gamma}_g \geq 0 \quad \forall g, \forall l, \forall d \tag{21}$$

$$\begin{aligned} \underline{\alpha}_l &\leq M\omega_{\underline{\alpha},l} \\ PL_l + PL_l^{max} &\leq M\left(1 - \omega_{\underline{\alpha},l}\right) \\ \bar{\alpha}_l &\leq M\left(\omega_{\bar{\alpha},l}\right) \qquad \forall l \\ PL_l + PL_l^{max} &\leq M\left(1 - \omega_{\bar{\alpha},l}\right) \\ \omega_{\underline{\alpha},l} + \omega_{\bar{\alpha},l} &\leq 1 \end{aligned} \tag{21a}$$

$$\begin{aligned} \underline{\beta}_g &\leq M\omega_{\underline{\beta},g} \\ P_g + P_g^{min} &\leq M\left(1 - \omega_{\underline{\beta},g}\right) \\ \bar{\beta}_l &\leq M\left(\omega_{\bar{\beta},g}\right) \qquad \forall g \\ PL_g^{max} - P_g &\leq M\left(1 - \omega_{\bar{\beta},g}\right) \\ \omega_{\underline{\beta},l} + \omega_{\bar{\beta},l} &\leq 1 \end{aligned} \tag{21b}$$

$$\underline{\gamma}_d \leq M \omega_{\underline{\gamma},d}$$
$$S_d \leq M \left(1 - \omega_{\underline{\gamma},d} \right)$$
$$\overline{\gamma}_d \leq M \left(\omega_{\overline{\gamma},d} \right) \qquad \forall d \qquad (21c)$$
$$D_d + \Delta D_d - S_d \leq M \left(1 - \omega_{\overline{\gamma},d} \right)$$
$$\omega_{\underline{\gamma},d} + \omega_{\overline{\gamma},d} \leq 1$$

$$\omega_{\underline{\alpha},l}, \omega_{\overline{\alpha},l}, \omega_{\underline{\beta},g}, \omega_{\overline{\beta},g}, \omega_{\underline{\gamma},d}, \omega_{\overline{\gamma},d} \in \{0, 1\} \qquad (21d)$$

The attacker's goal is to maximize the total load shedding as described in Eq. 5 under constraints mentioned in Eqs. 6 to 11. Hence, the upper level of the coordinated LR attack bilevel optimization problem is Eqs. 5 through 11. Equation 11 follows the traditional LR attack model mentioned in [4] along with the binary variable, $v_{G,g}$ indicating that the k-th generator is attacked when zero and not attacked otherwise [2]. Constraints 6 through 8 ensure that the attack is an LR attack and the attack magnitude for a load measurement does not exceed a limit determined by τ, which is the upper-bound of $\Delta D_d / D_d$ for each load d. Equation 11 with a binary variable $v_{G,g}$ ensures that the information regarding the generators to be attacked is passed along with the compromised line and load measurements to the lower level optimization for the generator coordinated LR attack and guarantees that the attack satisfies the cyber-attack resource limitation.

The perspective of the system administrator is represented as the lower level problem in Eqs. 12 through 17 by a SCED model, which are parameterized in terms of the upper-level decision variable ΔD. The system administrator aims to minimize the load shedding as shown in Eq. 8 considering the SCED constraints from Eqs. 13 to 17. The administrator is misled by the generator-coordinated LR attack vector from the upper level. The generators are tripped, and he performs an optimal power re-dispatch based on the problem formulation as in Eqs. 12 through 17.

Non-linear Eqs. 8 and 9 in the upper level can be modeled by mixed integer linear programming (MILP) formulation by introducing additional binary variables as in Eqs. 9a and 10a. The entire bilevel problem can be transformed into an equivalent MILP problem [3, 4] by employing the Karush- Kuhn-Tucker (KKT) optimality conditions. Using the KKT optimality conditions, the original bilevel problem from 5 through 17 can be transformed into an equivalent single level MILP model as follows. Equations 6 to 10a are the linear equations pertaining to the upper level model. The constraints 13 to 17 are the SCED primal feasibility constraints in which the non-linear constraint, Eq. 21 is linearized through the expression of complementary slackness conditions as in Eqs. 21a through 21d.

6 Implementation of Bi-level Model

A modified IEEE 14-bus system is shown in Fig. 2. The corresponding optimal power flow analysis of the system is obtained from the MATPOWER[1] package. There are 20 lines, 6 generators and 11 loads in the system. Hence, there are 51 attackable measurements. The maximum capacity of a line, PL from bus 1 to bus 2 is 160 MVA, from bus 2 to bus 3 is 100 MVA. The capacity of the other lines is 60 MVA. The attack magnitude for a load measurement is limited at $\tau = \pm 50$ of its true load value. The attack cost for generators, C_G are assigned in proportion to the generated capacities, P_g^{max} as shown in Table 1.

Table 1. Generator parameters

Generator	1	2	3	4	5	6
Bus	1	1	2	3	6	8
$P_{g,min}$	0	0	0	0	0	0
$P_{g,max}$	100	100	50	30	50	20
C_G	50	50	25	15	25	10

7 Results and Discussion

The proposed bi-level model was implemented on a modified IEEE 14-bus system. The equations for the model were coded using MATLAB and the computational experiments were executed on a PC with Intel Core i5 CPU running at 2.17 GHz. The optimal solution for the bilevel optimization problem was verified using IBM ILOG CPLEX Optimization Studio version 12.8. In Table 2, the effect of varying cyberattack resource on load curtailment is illustrated. As R_C increases, the load curtailment increases proving that with respect to increase in attack magnitude, the magnitude of load shedding increases.

Simultaneously, the impact of generator binary variable $v_{G,g}$ is switched between 0 and 1 to indicate whether the generator is working or tripped respectively. The combination of generators committed or tripped is determined corresponding to maximum load curtailment as in Eq. 11.

In the process of determining maximum load curtailment, the attacker attempts to target those generators that have low Defense Cost Multiplier (DCM). The DCM is defined as the factor that represents the cost of attacking a generator depending on the security measures integrated into the generating stations' control and protection facilities. R_C in Eq. 11 is defined as the cyber-attack resource is in the right-hand side equation. The objective of the attacker is to set the line flow of the system to exceed the transmission capacity with minimal R_C. To maintain a minimal value of R_C, Eq. 11 is set up such that when one of the lines are compromised as the first element in the

---

[1] MATPOWER V6.1 was used: MATPOWER is a free program for numerical computation with strong MATLAB compatibility.

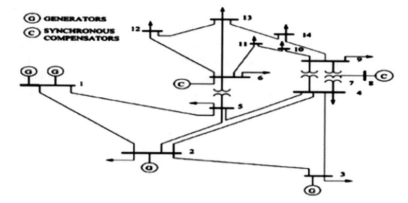

Fig. 2. Schematic of a modified IEEE 14 Bus System

Table 2. Cyber attack resource parameters

Attack resource Rc	Load shedding
10	3.30E−13
15	7.06E−08
20	2.03E−08
25	4.30
30	4.30
35	7.52
40	10.46
45	22.09
50	26.40
55	28.56
60	30.16
65	39.30
70	39.30
75	59.30
80	59.30
85	79.30
90	89.30
95	89.30
100	109.30

equation, the bus which has the heavier load at the two ends of the line is compromised as set up in element 2 of the equation. Further, the remaining part of the equation represent the coordinated attacks from the attacking generators. The level of defence measures in loads, lines and generating stations is accounted for by multiplying each associated cost in Eq. 11 by DCM. Figures 3, 4, 5 and 6 demonstrate the effect of varying DCM on load shedding for different cyber-attack resources. Figures 3, 4, 5 and 6 demonstrate the effect of varying DCM on load shedding for different cyber-attack resources.

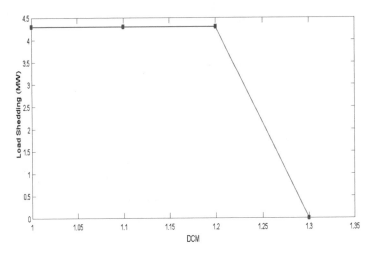

Fig. 3. Load shedding with cyber-attack resource, $R_C = 30$

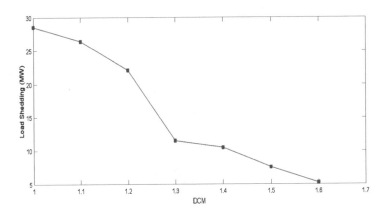

Fig. 4. Load shedding with cyber-attack resource, $R_C = 55$

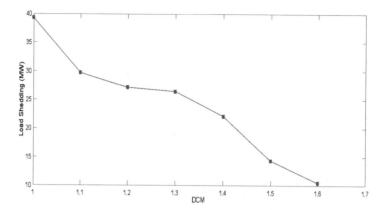

Fig. 5. Load shedding with cyber-attack resource, $R_C = 65$

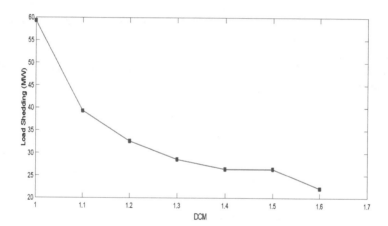

Fig. 6. Load shedding with cyber-attack resource, $R_C = 75$

From the figures, it is observed that for all values of R_C, load shedding eventually decreases to 0. When R_C is 30, load shedding is maximum starting from DCM = 1 till DCM = 1.2 and then reduces to 0 when DCM = 1.3. When R_C is increased to 45, load shedding consistently decreases with increase in DCM and reaches near 0 when R_C is 2.5. Likewise, a similar trend is observed for other values of $R_C = 55$ and $R_C = 65$. With an increase in cyber attack resource, it takes a larger DCM to achieve near 0 load shedding. This proves our hypothesis that, with the attacker bringing in more resources, if the loads, generators and lines are secured with relatively high defense measures, the attacker will not be able to achieve the desired maximum load curtailment.

8 Conclusions

In this research, a bi-level solution was utilized to solve the coordination between LR attack and attack on generators. The approach was implemented on a sample IEEE 14 bus system by varying several factors that decide the impact of these attacks. One of the factors which were considered is the cyber-attack resource and second was the defense cost multiplier. With the increase in cyber-attack resource, load curtailment increased and hence such coordinated attacks caused huge damages to the power system. When the defense cost multiplier increased, providing more security to the generators, resulted in a reduced load curtailment implying that higher the level of protection in generators, harder it will be for the attacker to seriously damage the power system. It is concluded that by proper choice of attacking elements and attack variables, the attacker can mislead the operator to believe that the power system is in secure operating mode, but still cause maximum load curtailment. It is recommended that adequate generation capacity, effective detection, and secured measurements will be critical in preventing coordinated LR attacks. Future work in this area will be to explore other types of coordinated attacks and to investigate other modeling approaches to accurately depict the severity of such attacks. With increased cyber threats and new techniques that the attacker might use to select promising targets, novel advancements in complex network analysis and meta-heuristic techniques will yield a robust electric power grid that is reliable and secure.

Appendix A

Indices	Definitions
d	Index for load demands
g	Index for generators on each bus
l	Index for transmission lines
Constants	
N_d	Number of loads
N_g	Number of generators
N_l	Number of transmission lines
P_g^{max}	Maximum generation output of generator 'g'
P_g^{min}	Minimum generation output of generator 'g'
PL_l^{max}	Maximum capacity of transmission line 'l'
$C_{D,d}$	Cost required to attack load demand measurements
$C_{F,l}$	Cost required to attack line flow measurements
$C_{G,g}$	Cost required to attack the generators
ΔD_d	Attack on the load demand measurements
SF	Shifting factor matrix
KP	Bus-generator incidence matrix

(continued)

(continued)

Indices	Definitions
KD	Bus-load incidence matrix
M	Sufficiently large positive constant
R_C	Cyber-attack resource
ε	Sufficiently small positive constant
τ	Upper bound of the attack magnitude on the load demand measurements
C_0, C_1, C_2	Cost coefficients of generators
Variables	
S_d	Load shedding of load 'd'
ΔD_d	Attack on the load demand measurements
P_g	Generation output of generator 'g'
PL_l	Power flow on line 'l'
ΔPL_l	Attack on power flow measurements
$v_{G,g}$	Binary variable indicating whether the 'g'th generator is attacked
$\left(\delta_{D+,d}, \delta_{D-,d}, \delta_{D,d}\right)$	Binary variables indicating whether the load measurement at 'd'th generator is attacked
$\left(\delta_{PL+,l}, \delta_{PL-,l}, \delta_{PL,l}\right)$	Binary variables indicating whether the load measurement at 'd'th generator is attacked
λ	LaGrange multiplier associated with the power balance equation of the system
μ_l	LaGrange multiplier associated with the power flow equation of line 'l'
$\left(\underline{\gamma}_d, \overline{\gamma}_d\right)$	LaGrange multipliers associated with the lower and upper bounds for the shedding of load 'd'
$\left(\underline{\alpha}_l, \overline{\alpha}_l\right)$	LaGrange multipliers associated with the lower and upper bounds for power flow of line 'l'
$\left(\underline{\beta}_g, \overline{\beta}_g\right)$	LaGrange multipliers associated with the lower and upper bounds for power output of generator 'g'
$\left(\omega_{\underline{\alpha},l}, \omega_{\overline{\alpha},l}\right)$	Binary variables to represent the complementary slackness of the power flow constraints of line 'l'
$\left(\omega_{\underline{\beta},g}, \omega_{\overline{\beta},g}\right)$	Binary variables to represent the complementary slackness of the generation output constraints of generator 'g'
$\left(\omega_{\overline{\gamma},d}, \omega_{\underline{\gamma},d}\right)$	Binary variables to represent the complementary slackness of the load shedding constraints of load 'd'

References

1. Arroyo, J.M., Galiana, F.D.: On the solution of the bi-level programming formulation of the terrorist threat problem. IEEE Trans. Power Syst. **20**(2), 789–797 (2005)
2. Xiang, Y., Wang, L., Liu, N.: Coordinated attacks on electric power systems in a cyber-physical environment. Electr. Power Syst. Res. **149**, 156–168 (2017)

3. Yuan, Y., Li, Z., Ren, K.: Modeling load redistribution attacks in power systems. IEEE Trans. Smart Grid **2**(2), 382–390 (2011)

4. Yuan, Y., Li, Z., Ren, K.: Quantitative analysis of load redistribution attacks in power systems. IEEE Trans. Parallel Distrib. Syst. **23**, 1731–1738 (2012)

5. https://www.cnbc.com/2014/01/03/attacks-on-the-electricity-grid-us-vulnerable-to-physical-and-cyberthreats.html. Accessed 28 March 2019

6. https://www.cfr.org/report/cyberattack-us-power-grid. Accessed 28 March 2019

7. Zhu, Y., Yan, J., Tang, Y., Sun, Y.L., He, H.: Coordinated attacks against substations and transmission lines in power grids. IEEE Global Communications Conference, Austin, TX, pp. 655–661 (2014). https://doi.org/10.1109/glocom.2014.7036882

8. Yan, J., Tang, Y., He, H., Sun, Y.: Cascading failure analysis with DC power flow model and transient stability analysis. IEEE Trans. Power Syst. **30**(1), 285–297 (2014)

9. Albert, R., Albert, I., Nakarado, G.L.: Structural vulnerability of the North American power grid. Phys. Rev. E **69**, 025103 (2004)

10. Crucitti, P., Latora, V., Marchiori, M.: Model for cascading failures in complex networks. Phys. Rev. E **69**, 045104 (2004)

11. Kinney, R., Crucitti, P., Albert, R., Latora, V.: Modeling cascading failures in the North American power grid. Eur. Phys. J. B **46**(1), 101–107 (2005)

12. Wang, J., Rong, L., Zhang, L., Zhang, Z.: Attack vulnerability of scale-free networks due to cascading failures. Phys. A **387**, 6671–6678 (2008)

13. Wang, W., Cai, Q., Sun, Y., He, H.: Risk-aware attacks and catastrophic cascading failures in U.S. power grid. In: Proceedings of IEEE Global Telecommunications Conference, Houston, Texas, USA (2011)

14. Zhu, Y., Sun, Y., He, H.: Load distribution vector-based attack strategies against power grid systems. In: Proceedings of IEEE Global Telecommunications Conference, Anaheim, CA, USA (2012)

15. Hines, P., Cotilla-Sanchez, E., Blumsack, S.: Do topological models provide good information about vulnerability in electric power networks? Chaos 20. arXiv preprint arXiv: 1002.2268 (2010)

16. Yan, J., Zhu, Y., He, H., Sun, Y.: Multi-contingency cascading analysis of smart grid based on self-organizing map. IEEE Trans. Inf. Forensics Secur. **8**(4), 646–656 (2013)

17. Zhu, Y., Yan, J., Sun, Y., He, H.: Risk-aware vulnerability analysis of electric grids from attacker's perspective. In: Proceedings of IEEE Innovative Smart Grid Technologies Conference, Washington, USA (2013)

18. Dobson, I., Carreras, B.A., Newman, D.E.: A loading-dependent model of probabilistic cascading failure. Probab. Eng. Inf. Sci. **19**(1), 15–32 (2005)

19. Nedic, D.P., Dobson, I., Kirschen, D.S., Carreras, B., Lynch, V.E.: Criticality in a cascading failure blackout model. Electr. Power Energy Syst. **28**(1), 627–633 (2006)

20. Mei, S., Ni, Y., Wang, G., Wu, S.: A study of selforganized criticality of power system under cascading failures based on AC-OPF with voltage stability margin. IEEE Trans. Power Syst. **23**(4), 1719–1726 (2008)

21. Chen, Q., McCalley, J.D.: Identifying high risk n-k contingencies for online security assessment. IEEE Trans. Power Syst. **20**, 823–834 (2005)

22. Liu, Y., Ning, P., Reiter, M.: False data injection attacks against state estimation in electric power grids. In: Proceedings of 16th ACM Conference on Computer and Communication Security, pp. 21–32, November 2009

23. Liu, X., Li, Z.: False data attack models, impact analyses and defense strategies in the electricity grid. Electr. J. **30**(4), 35–42 (2017)

24. Yuan, Y., Li, Z., Ren, K.: Quantitative analysis of load redistribution attacks in power systems. IEEE Trans. Parallel Distrib. Syst. **23**(9), 1731–1738 (2012)

A Study on Recent Applications of Blockchain Technology in Vehicular Adhoc Network (VANET)

Subhrajit Majumder, Akshay Mathur, and Ahmad Y. Javaid[✉]

University of Toledo, Toledo, OH 43606, USA
{subhrajit.majumder,akshay.mathur,ahmad.javaid}@utoledo.edu

Abstract. Technology has taken over aspects of our lives today, which were unimaginable a few years ago. Innovations such as high-end transport systems that include connected autonomous vehicles (CAV), and digital currencies, such as Bitcoin and Ethereum, have made our lives more comfortable and convenient. It may not seem at first, but most of these technologies work on similar lines. For example, CAVs use DSRC (Dedicated Short-Range Communication) technology, which has a peer-to-peer aspect. On the other hand, Bitcoin uses a technology called Blockchain, which was founded on its peer-to-peer nature and lack of need for a trusted central authority. Cybersecurity of vehicular ad hoc network (VANET) has become a prime research area for cybersecurity researchers. Minor vulnerabilities in VANET may lead to severe fatalities. This concernment has engaged numerous researchers with the purpose of improving the defense mechanism of these susceptible networks. This paper presents a brief meta-analysis of the field of cybersecurity of autonomous vehicles using Blockchain and discusses how VANETs and CAVs can be protected using the same.

Keywords: Blockchain · Vehicular Adhoc Network · VANET · Security

1 Introduction

In the wireless environment, a connection between devices should be established instantly and should self-configure all the time as the devices may move in random directions all the time since the wireless range is limited. These properties are present in ad hoc wireless network where the network continuously configures itself regardless of the movement of devices, thus eliminating the need for a specific infrastructure for managing the connections. Here the router comes into the picture, which constantly maintains a set of information which is required to forward the data traffic. These routers are either connected to the larger network or can operate by themselves. When these properties of an ad hoc network are used to build an environment for the vehicles, it is called Vehicular Ad-hoc network

K.-K. R. Choo et al. (Eds.): NCS 2019, AISC 1055, pp. 293–308, 2020.
https://doi.org/10.1007/978-3-030-31239-8_22

(VANET). It runs on the principles of the mobile ad-hoc network (MANET) that enables automatic data exchange between vehicles through a wireless network [1]. VANET uses Dedicated Short-Range Communications which are short to medium range communication, specially designed for the automotive environment [2]. It works on the Wireless Access in Vehicular Environment protocol stack which is based on the IEEE 802.11p standard [3].

Autonomous vehicles may communicate to various entities in the network such as other vehicles (known as vehicle-to-vehicle or V2V communication), roadside infrastructure (known as vehicle-to-infrastructure or V2I communication), and other independent, smart devices (known as vehicle-to-everything or V2X communication). The roadside infrastructure may comprise of several components such as traffic signals, traffic cameras, or other roadside units, while smart devices include mobile phones, GPS devices, etc.

The unique characteristics of VANETS require several measures to sustain the privacy of the network and their modules present in it. If a security measure is averting the attacks trying to track the vehicles, it may not be able to prevent the internal modules from sending forged messages. It is coherent that a minor vulnerability can cause significant anomalies in the network as the data which gets transferred spontaneously has to be accurate. There is an adequate number of research works done in the field of VANET to make it more secure from different types of cyber-attacks over the years. It is nearly impossible for a single security measure to guard all the elements in the environment. Hence, if the whole system gets distributed into multiple blocks, the workload gets reduced for each block, and the performance quality of each block will be more accurate.

Simultaneously, the advent of Bitcoin became a sensation in the cyber world. The main reason behind its popularity and success is based on its security measures. The Blockchain is a distributed system which allows us to dismantle an extensive security system into smaller parts that perform simultaneously to reduce the workload from individual components. It offers benefits such as integrity, security, and privacy. Blockchain uses a distributed algorithm to operate, although it contains remarkable concord among its blocks. This notion spurred our paper to provide some valuable contribution to proclaim this promising betterment of VANET's overall performance using Blockchain.

The paper is divided into five sections. Section 2 gives a background about the use of VANET into different levels of autonomous vehicles and Blockchain. Section 3 provides an overview of the work done by researchers in the field of VANETs using Blockchain. Section 4 talks about the various advantages and challenges of using Blockchain in VANET, and Sect. 5 concludes the paper.

2 Background

With vehicles being a part of almost every aspect of our lives, and researchers working hard to develop fully autonomous vehicles, their security is of prime concern. An insecure and vulnerable VANET can lure attackers into taking total control of the vehicle, causing harm to the passengers and the vehicle. With the advent of secure cryptocurrencies such as Bitcoin and Ethereum, blockchain technology has become the talk of the town for its amazing security features. Various researchers and industries are working on introducing blockchain technology in VANET, which is supposed to reduce the complexities and enhance the security of the network.

2.1 Automation in Vehicles

Today, a vehicle is judged by its automation level. At present, based on a vehicle's functionality, there are six levels of automation, as shown in Fig. 1. These levels of automation are described as follows [4,5]:

- **Level 0** autonomous vehicles have no autonomous features in them, i.e., the driver has complete and total control of the vehicle. They refer to popular cars which had no "smart" technology to assist the driver.
- **Level 1** autonomous vehicles aids the driver in controlling a single aspect of driving such as braking and acceleration. Cars with either brake assist or adaptive cruise control fall under this category.
- **Level 2** autonomous vehicles aid the driver with more than one aspect of driving. These include adaptive cruise control, brake assistance, lane-keeping assistance, etc. Tesla's autopilot, Cadillac's Super Cruise, etc. are some of the examples of level 2 systems. But this requires the constant attention of the driver while driving. The system in the car warns about the driving anomalies that need to be corrected, and it is the driver's responsibility to act on them appropriately.
- **Level 3** autonomous vehicles are the ones where we visually get to see automated driving. The use of a wide variety of sensors such as LiDAR, RADAR, and algorithms help the car while driving itself. But these sensors and algorithms are also limited in their functionality and needs driver's intervention as a worst-case scenario or as an emergency measure. As per SAE International's standards, such vehicles must send a "request to intervene" to the driver whenever the need arises. These requests can come up when the speed goes above a certain limit, or when the brakes do not respond, etc.
- **Level 4** autonomous vehicles have high-end automation in them and can manage various dynamic driving tasks, such as steering, accelerating, braking, changing lanes, monitoring the internal and external environment of the vehicles, stopping at traffic signals, etc. But even though it seems automated, there is a need for the driver to intervene when the driving conditions are not appropriate, such as under extreme weather conditions, or when there is snow on the road, and the lanes are not visible. Therefore, they are equipped with a steering wheel and a pedal.
- **Level 5** autonomous vehicles are completely autonomous and do not require any human intervention in any condition. They can be imagined as driving pods that can commute from destination to destination without the driver's intervention. They do not require steering wheels or pedals as the autonomous system manages all the driving tasks on its own.

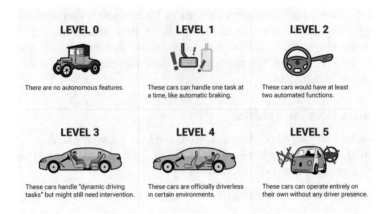

Fig. 1. Levels of automation [6]

2.2 VANET

VANET is similar to MANET as both share many similarities in spite of several dissimilarities. VANET is an ad-hoc network where the channel of communication among nodes present in the network is inconsistent as the nodes represent vehicles in real life which are mobile. In VANET any vehicles may move in any direction with variable speed. Hence channels among them must be built instantly as needed. This allows us to refer VANET as a subspace of MANET because components in it can communicate with each other wirelessly through continuously self-configuring channel [7]. VANET is similar to MANET despite several dissimilarities. The nodes in MANET communicate with each other wirelessly through a self-configuring network. Similarly, the nodes in VANET are moving vehicles connected, making an ad-hoc network where the channel must be built instantly on demand [7]. The nodes in VANET are more mobile than those in MANET and has a highly dynamic and geographically constrained topology. The channel conditions in VANET are unreliable and has to consider the driver behavior along with intermittent connectivity among vehicles, making the real-time deadline more strict, and fragmentation of the network more frequent [8–10].

VANET works with the help of two main components, OBD port-II and Road Side Units (RSUs). OBD port-II is installed in the vehicles whereas RSUs are set on the roadsides. These RSUs work as routers and the OBD port-II gets connected to these RSUs through DSRC (Dedicated Short Range Communication). This network allows the vehicles to communicate through each other with the help of RSUs and are considered to be nodes. Different kinds of information are continuously exchanged between these nodes, such as emergency messages, the speed of the neighboring vehicles, traffic-related information, etc. If this data is tampered with, or an unauthorized user gets access to this network, they can disclose the IDS of vehicles, inject forged data, implement a Denial of Service Attack and much more, which could have catastrophic consequences [11].

2.3 Blockchain

Let's say a block is distributed into smaller blocks; for example, a huge stack of data is stored in multiple connected blocks instead of a huge block. This way, the immense workload on a single block can be avoided as multiple smaller blocks will share the load. There are three main properties of Blockchain which make it applicable to numerous domains. Firstly, vast amounts of data can be taken care of by a set of security measures, each focusing on different blocks. Each block is connected to another block by cryptography; every block contains the cryptographic hash of the previous block along with the timestamp of that block and the transaction data [12]. A unique characteristic make cryptographic hash very valued, it maps any arbitrary data into a predetermined data, but the process is one way. It means that once the data is mapped, there is no turning back [13]. Second, the trusted timestamp, a method used for keeping track of time for any creation or modification of a file, may be used. This is done securely such that not even the owner is allowed to change it after it's done unless the security has been compromised [14]. These properties, as mentioned above, allows Blockchain to exchange data among its blocks in a well-secured way. Hence, dividing the whole security structure into smaller parts not only enable it to operate conveniently but also makes it exhausting for a hacker to breach into multi-layered security levels. Initially, Blockchain used just a limited set of scripting languages, such as UTXO. The next generation, on the other hand, enabled Turing-complete programming languages on the protocol layer level, allowing smart contract capabilities such as Ethereum [15], Qtum [16], and Tezos [17]. The next section discusses a few important related works done in this field, which contribute to this new idea of combining Blockchain technology with cyber-security of autonomous vehicles.

3 Blockchain and VANETs

In VANET, the fundamental requirement is continuously sustained communication among all units in the network. As VANET is blending with our modern world of the connected Internet of Things (IoT), it is tempting the cyber-criminal minds to exploit this enormous amount of information which gets shared among different nodes. One of the most valuable information is the position of a vehicle. Many researchers have modeled such type of attack where the location of a vehicle in the network gets forged. This may lead to road accidents as the wrong information will be shared by the vehicles. Grover et al. had implemented such type of attack where data gets forged in the network. They had included a malicious node in the network which sends out forged information to other nodes. This malicious node used multiple non-existing virtual identities with different locations, which creates an anomaly in the network. Authors have analyzed the effect of this attack on the speed of the vehicle, the number of packets got delivered, and several collisions happened due to this deviation [18]. Another good probe was done by Leinmuller et al. [11]. They modeled the position forging attacker behavior based on four different possible situations, i.e., forge single location, forge multiple positions with varying IDs of a node, forge movement path of a single node, and forge multiple movement paths with different IDs.

Kumar et al. analyzed and implemented a wormhole attack on a VANET. This wormhole attack is quite feasible to launch in an ad-hoc network since the attacker does not need to compromise any node in the network. Despite all packets verifying authentication and other security procedures, the wormhole attack gets executed successfully. This attack requires an attacker to record the packets from one end in the network, tunnel them to another location, and then re-transmit them into the network [19]. Hu et al. introduced the notion of a leash. They came up with two kinds of leashes. First, a geographical leash that makes sure that each packet stays under the maximum allowance for the distance traveled. Second, a temporal leash that ensures each packet has an upper bound on its lifetime, restricting the maximum distance. This mechanism allowed them to detect whether the packets traveled more than the distance than they are supposed to cover or not. This can successfully detect a wormhole attack [20].

Yan et al. implemented a screening process to detect forged data which carry the position of neighboring vehicles [21]. The screening process was divided into two categories, local and global. In local, they used vehicle A to verify the data sent by another vehicle B in the same perimeter. If the data matches, they would approve, otherwise they won't. They combined these local solutions to make it suitable for the global scenario. In another case, Guyette et al. developed a Trusted Platform Module containing various cryptographic keys which authenticate the packets exchanged. This TPM is a piece of hardware which can be attached to a vehicle in the presence of a software environment; it can store and protect the data within its shield. However, these products can be breached if the owner's credentials or the encryption keys get compromised [22].

Yan et al. implemented some security measures using Blockchain technology which can be more effective as the encryption level will be divided into multiple layers making it difficult for the attacker to break in and avoid events such as "black swan" [23]. Raya et al. proposed some ideas on how vehicles will write and aggregate their signatures in a block of a Blockchain network, but it may raise questions on its feasibility [24]. Mishra et al. analyzed the end to end time cycle of the following process and demonstrated that it is feasible to consider Blockchain technology in Vehicular Network. There are some specific areas of VANET in which Blockchain can be more useful, such as:

1. Collect sensor data
2. Generate keys
3. Sign data
4. Transmit data to edge cloud
5. Broadcast signatures to surrounding vehicles
6. Verify incoming Signatures

7. Access Edge Cloud
8. Verify Surrounding vehicle data with own sensor data
9. Aggregate signatures into a block [25]

Several components can be exploited in VANET by attackers to make a cyber-attack successful. Researchers have focused on several vulnerabilities on which security measures can be implemented to prevent attacks. Some examples are given below, where researchers have used Blockchain technology to develop such measures.

3.1 Preventing Forged Data

Forged data is the easiest attack that can be embedded in the vehicular environment. If any data gets manipulated, the automation of the intelligent vehicle gets disrupted.

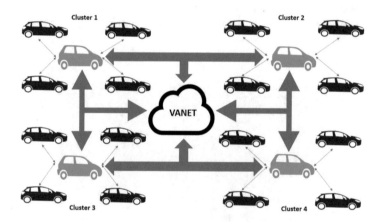

Fig. 2. Communication through cluster heads

Some researchers proposed the idea of cluster heads. The cluster heads are elected from a cluster of vehicles. Instead of all vehicle communicating among each other, one vehicle is elected from a cluster of vehicles as the head of the cluster which will communicate with the other cluster heads. These heads will communicate on behalf of all the other vehicles in its cluster. As shown in Fig. 2, the black cars are communicating with the yellow cars in their respective cluster, which is taking the responsibility of exchanging the data among the other cluster heads of different clusters. The authors came up with Trust Management Scheme (TCMV) which is capable of testing the credibility of the messages getting exchanged among the vehicles, as well as roadside units [26] However, they realized that checking the credibility of the messages sent by the cluster head is not enough to check if the messages being exchanged are genuine or not. This motivated them to come up with a Distributed Trust Management Scheme,

300 S. Majumder et al.

which could take care of this task more efficiently. For this Blockchain scheme, they have used the RSUs as the miners who will be fixed. These miners can be used in various ways such as storing the same messages, creating the list of messages getting exchanged, and can take responsibility for exchanging the data among other miners [27,28]. They distributed the whole authentication process into three levels:

1. **Transmission of VERIF_MSG message:** A vehicle sends an urgent message (ex: road accident) to its cluster head. The cluster head verifies the sender and appeals for its credibility score from the TCMV. The closest RSUs (miners) get notified if the message is correct or not and takes a decision on it using fuzzy logic.
2. **Block creation:** A new block gets generated as the VERIF_MSG is selected by each miner from the local data structure. It consists of several VERIF_MSG which are linked to the previous block. A structure of the block is shown in Fig. 3. It consists of a block header and a block body, each of which is divided into subcategories.
3. **Block Validation:** The first miner validates all other miners by solving their Proof of Work.

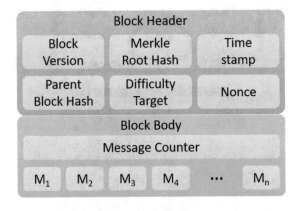

Fig. 3. Structure of a block [29]

3.2 Revocation of Users in the Network

Malik et al. delivered a Blockchain based security mechanism, which not only improves the authentication strength of the overall system but also strengthens the revocation of all the users in the VANET. They decentralized the VANET into subcategories to share the load, which makes the security mechanism of the system inexhaustible. The authors have subdivided the problem statement into the following categories:

1. **Mutual Authentication with Reduced Dependency on Certificate Authority:** Unlike previous schemes, onboard and roadside units can be authenticated at the same time which reduces the number of communications with Certificate Authority (CA) [30].
2. **Scalability:** With scalable attributes, the framework should be reckoned to the vastness of the network.
3. **Privacy Protection:** The authentication of the users should not require them to disclose their identity or perturb their privacy.
4. **Message Confidentiality, Integrity, and Non-repudiation:** The security mechanism must authenticate and verify all the components to prevent any unauthorized intruders at all times.
5. **Speedy Revocation Without Additional Overhead:** Authentication is not the only requirement from a robust and efficient framework. Revoking the malicious components from the network quickly is equally important. Moreover, the system shouldn't revise the entire Certificate Revocation List (CRL) to identify such vehicles and avoid this overhead.

The authors used a private blockchain, which provides complete authority to the RA and CA, limited access to the ledger, RSUs gets read-only access and none to the OBUs. This framework gives a low computational cost since it doesn't require POW (Proof-of-work) [31,32].

3.3   Vehicular Announcement Network

In a network of free vehicles, the vehicular announcement plays a crucial role. Every vehicle sends out an announcement with the information about its surroundings, preserving the identity of the sender as the message contains private information, such as the license plate number of the sender, etc. However, there are two significant issues with this proposition. Firstly, if the identity of the sender stays private, it may affect the credibility of the message sent. Secondly, since most users lack the enthusiasm for forwarding messages, they may stop forwarding them once they realize that their privacy is at risk. Singh et al. came up with a method which consists of a local dynamic Blockchain and a global Blockchain. This system is secured using a unique crypto ID called Intelligent Vehicle Trust Point (IVTP). While communicating, vehicles use and verify this IVTP with the Local Dynamic Blockchain (LDB) to check if a legitimate sender sends it or not [32]. For motivating the users in the network to forward all announcements, Li et al. proposed a privacy-preserving Blockchain based Incentive Announcement Network called Creditcoin. Figure 4 demonstrates the blueprint of Creditcoin. Their incentive mechanism based on Blockchain not only allows the user to earn points by forwarding messages but also keep their identities private to motivate them in forwarding messages in a Blockchain based secure VANET [33].

Sharma et al. gave importance to the architecture of a VANET in their work. They proposed a Blockchain based Distributed Vehicular Network in which all controller nodes are connected in a distributed manner to provide robust services.

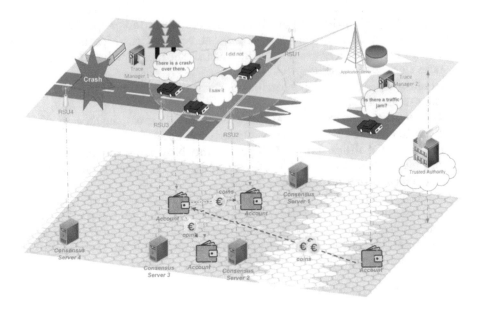

Fig. 4. Overview of credit coin [33]

In their experiment, they considered a random neighboring vehicle as the miner and the others as ordinary nodes. The miner node handles all the requests and responses from the other nodes. This way, they demonstrated the possibility of a highly available and scalable vehicular network [34].

3.4 Ethereum

Ethereum provides a decentralized platform for VANET, based on the blockchain concept. *"Ethereum intends to provide a blockchain with a built-in fully-fledged Turing-complete programming language that can be used for creating 'contracts"'* [35]. It runs on a contract-based Blockchain mechanism which consists of program code, storage, and an account balance. These contracts are posted in the Blockchain along with their respective codes. They execute when a message is received, or any other user/contract tries to reach the contract. For every execution, the code either reads or writes to the storage. To avoid an accidental loop execution, it runs on the account balance, which is known as 'gas' Whenever the code gets executed, it consumes a certain amount of gas. Thus, any accidental loop of execution gets terminated with the term of balance. Leiding et al. used this concept to come up with a novel approach to making VANET self-manageable. In contrast to traditional blockchain architecture, they used the RSUs as a provision of Ethereum based applications which have been deployed into the Ethereum Blockchain instead of using them just like a relay station. Using Ethereum's smart contract system, any application can be deployed and run on the Blockchain network such as traffic regulation, vehicle tax, and vehicle insurance. If any other applications, such as updates on traffic jams or weather forecasts are required, they can be subscribed by the users as well [36].

3.5 Transportation as a Service (TaaS)

Communication among the different types of nodes in VANET does not follow a default interaction standard. Leiding et al. proposed a blockchain based system which allows all the participants in a VANET to exchange any service such as renting a parking lot, paying for a battery, using toll roads, or simple car service check-ups [37]. Sterling et al. used an Agent-Oriented Modeling method which focuses on goal models and named their architecture Chorus. They experimented their prototype in a way such that car insurance companies can provide incentives to their customers based on their good driving behavior [38].

4 Advantages and Challenges

Beside the decentralization properties of blockchain technology, there are many others as described by Sharma et al. which could be utilized for the betterment of VANET [34]. Some of the benefits are:

1. Transparency: All exchanges in the blockchain network are transparent, ensuring the integrity of the records in the network.
2. No fraud: Even the sender cannot cancel the transaction after it is done.
3. Low cost: Exchanges on the Blockchain network can be done at a low cost.
4. Instantaneous Transaction: All exchanges in the Blockchain network are registered instantaneously, and thus the affirmation and compensation can occur in minutes, unlike traditional procedures.
5. Security: Cryptographic and decentralized Blockchain conventions assure good quality security of the network. With peer-to-peer communication, vehicles can have ubiquitous data access in a trusted and secure environment.
6. Financial: Sensitive and individual data does not have to be revealed to the beneficiary.
7. Authentication: One benefit of traditional blockchain implementations is that they often allow peers to join a network without verifying their identity from a trusted party, which expedites the process.

The dynamic environment of detecting, registering, sorting, and conveying, along with continuously adapting with the surrounding, have a significant effect on the security challenges of VANET. Some of the challenges which come with the influence of Blockchain in a VANET are:

1. Adequate Nodes: For a blockchain network, there must be a sufficient number of nodes at all times that share the total workload among themselves and shares data simultaneously.

2. Storage: Every node requires its storage to verify transactions among them. Along with a high number of nodes, large memory storage is also required. In about nine years, the number of blocks in Bitcoin has become 5×10^5 with a total storage of 150 GB [39]. Also, to generate new transaction data, the sender needs access to historical data which must be stored as well.

3. Incentive: The required nodes in the Blockchain based Vehicular network may ask for some benefits for staying active all the time to pass information from one to another, which may cost more.

4. Mobile Architecture: Vehicular Ad hoc Network consists of highly mobile nodes. Even the quantity of steady nodes is not constant. Thus, to provide asset availability progressing and fluctuating application prerequisites must be created by organizing a vehicular network with related convention engineering [40].

5. Soft and Hard Forks: Since all nodes are connected in a Blockchain network, sometimes a software update on a few selected nodes can cause a problem in the overall network. At some point of time, a blockchain network needs to get an update to function at the next level. The nodes that accept this update start using the updated software, whereas the ones that do not accept it keep working with the old version. This change is known as a fork.

6. Anonymity: While it has some benefits of its own in certain implementations, in VANET the legitimate nodes could be exposed to incorrect data or a DDOS attack if some malicious nodes enter the network whose identities are unknown. To counteract this problem, some governing authority (such as USDOT) could distribute digital certificates to the automobile manufacturers to validate a node. Although this would not be "pure" blockchain implementation, this would prove detrimental in preventing illegitimate nodes from disrupting the network [41].

With all the advantages of Blockchain, there are some disadvantages for implementing Blockchain in VANET such as increased number nodes in the network, giving extra incentives for the additional nodes, high-speed data transactions, requires massive computational power and decision making for high mobility of the network's architecture. Despite these hurdles, the advantages are much more appealing. Blockchain might be non-affordable for lightweight IoT devices, but for large systems, such as VANETs, the cost for Blockchain can be justified. As there is an enormous potential for catastrophic outcomes due to constant cyber attacks on VANET, considering Blockchain based architecture for its privacy and security measures can be a significant breakthrough.

To resolve the challenges, several explorations have been done in the recent past. Some researchers have proposed Vehicular Cloud Computing (VCC) for faster access to resources such as data storage, computing, and decision making [42]. To reduce the overwhelming storage requirement [43], came up with the idea of making every vehicle miner-equipped, which will share data with the RSUs for the anonymous authentication. Their experiment setup has impressively shown a storage overhead of only 1809 bytes used for 10 million vehicles. For counteracting the challenges related to Anonymity, *Distributed Trust* can be an effective solution where all the cluster heads generate trust ratings for other cluster heads [44]. This way, it will be challenging for a malicious node to remain undetected in the network and provide illegitimate information about others.

Imagine a scenario where multiple malicious nodes in a cluster have elected a malicious Cluster Head of their own. This head will be detected as an ill-node due to non-satisfactory trust rating from other cluster heads. However, what if numerous malicious vehicles are deployed in different clusters. They elect their respective cluster heads. Eventually, these malignant cluster heads may provide bad ratings for other genuine ones and cast them out. To prevent this possibility, any new nodes must be verified by the original cluster head before entering the network. Therefore, although the present challenges can be resolved with new solutions contributed by brilliant researchers, several loopholes require the attention of researchers.

5   Conclusion

This paper presents an overview of the reasoning behind the idea of using Blockchain in a VANET. There are specific existing vulnerabilities in VANET, which can lead to massive catastrophe as this comprises vehicles on the road with people in them. We have described why Blockchain has the ability to officiate the void present in the current status of VANET's security mechanism. It's "distributed" properties have the strength to manage the immenseness of VANETs. The blend of VANET and Blockchain has an auspicious probability of abating existing VANET vulnerabilities. Recent contributions towards this vision are discussed in the article. In spite of the existence of an adequate number of reasons in favor of this, there are circumstances which may hinder the immaculate growth of this technology. Nevertheless, finding a better way to improve the consolidated performance of VANET while exploiting the advantages of Blockchain technology is relatively unfeasible under the current circumstances.

Acknowledgement. The authors thank the Paul A. Hotmer Family Cybersecurity and Teaming Research (CSTAR) laboratory and the Electrical Engineering & Computer Science Department for their support. We also thank Sukriti Sharada for her assistance in proofreading and editing this work.

References

1. Yousefi, S., Mousavi, M.S., Fathy, M.: Vehicular ad hoc networks (VANETs): challenges and perspectives. In: 2006 6th International Conference on ITS Telecommunications Proceedings, pp. 761–766. IEEE (2006)
2. Kenney, J.B.: Dedicated short-range communications (DSRC) standards in the United States. Proc. IEEE **99**(7), 1162–1182 (2011)
3. Hartenstein, H., Laberteaux, K.: VANET: Vehicular Applications and Internetworking Technologies, vol. 1. Wiley, Hoboken (2009)
4. Hughes, J.: Car automony levels explained (2017). http://www.thedrive.com/sheetmetal/15724/what-are-these-levels-of-autonomy-anyway

5. Chaturvedi, A.: Did you know the 5 levels of autonomous cars (2018). https://www.geospatialworld.net/blogs/five-levels-of-autonomous-cars
6. Dragan Radovanovic, D.M.: This is what the evolution of self-driving cars looks like (2016). https://www.businessinsider.com/what-are-the-different-levels-of-driverless-cars-2016-10
7. Chadha, D., Reena: Vehicular ad hoc network (VANETs): a review. Int. J. Innov. Res. Comput. Commun. Eng. **3**(3), 2339–2346 (2015)
8. Engoulou, R.G., Bellaïche, M., Pierre, S., Quintero, A.: VANET security surveys. Comput. Commun. **44**, 1–13 (2014)
9. Pathan, A.S.K.: Security of self-organizing networks: MANET, WSN, WMN, VANET. CRC Press, Boca Raton (2016)
10. Wang, Y., Li, F.: Vehicular ad hoc networks. In: Guide to Wireless Ad Hoc Networks, pp. 503–525. Springer (2009)
11. Leinmuller, T., Schmidt, R.K., Schoch, E., Held, A., Schafer, G.: Modeling roadside attacker behavior in VANETs. In: 2008 IEEE GLOBECOM Workshops, pp. 1–10. IEEE (2008)
12. Narayanan, A., Bonneau, J., Felten, E., Miller, A., Goldfeder, S.: Bitcoin and Cryptocurrency Technologies: A Comprehensive Introduction. Princeton University Press, Princeton (2016)
13. Halevi, S., Krawczyk, H.: Strengthening digital signatures via randomized hashing. In: Annual International Cryptology Conference, pp. 41–59. Springer (2006)
14. Gipp, B., Meuschke, N., Gernandt, A.: Decentralized trusted timestamping using the crypto currency bitcoin. arXiv preprint arXiv:1502.04015 (2015)
15. Wood, G.: Ethereum: a secure decentralised generalised transaction ledger. Ethereum Proj. Yellow Pap. **151**, 1–32 (2014)
16. Dai, P., Mahi, N., Earls, J., Norta, A.: Smart-contract value-transfer protocols on a distributed mobile application platform (2017). https://qtum.org/uploads/files/cf6d69348ca50dd985b60425ccf282f3.pdf
17. Goodman, L.: Tezos–a self-amending crypto-ledger white paper (2014). https://www.tezos.com/static/papers/white_paper.pdf
18. Grover, J., Gaur, M.S., Laxmi, V.: Position forging attacks in vehicular ad hoc networks: implementation, impact and detection. In: 2011 7th International Wireless Communications and Mobile Computing Conference, pp. 701–706. IEEE (2011)
19. Kumar, P., Student, M.T., Verma, S., Batth, R.S.: Implementation and analysis of detection of wormhole attack in VANET. J. Netw. Commun. Emerging Technol. (JNCET) **8**(3) (2018). http://www.jncet.org
20. Hu, Y.C., Perrig, A., Johnson, D.B.: Packet leashes: a defense against wormhole attacks in wireless networks. In: Twenty-Second Annual Joint Conference of the IEEE Computer and Communications, INFOCOM 2003, IEEE Societies, vol. 3, pp. 1976–1986. IEEE (2003)
21. Yan, G., Olariu, S., Weigle, M.C.: Providing VANET security through active position detection. Comput. Commun. **31**(12), 2883–2897 (2008)
22. Guette, G., Bryce, C.: Using TPMS to secure vehicular ad-hoc networks (VANETs). In: IFIP International Workshop on Information Security Theory and Practices, pp. 106–116. Springer (2008)
23. Yan, G., Olariu, S.: A probabilistic analysis of link duration in vehicular ad hoc networks. IEEE Trans. Intell. Transp. Syst. **12**(4), 1227–1236 (2011)
24. Raya, M., Hubaux, J.P.: Securing vehicular ad hoc networks. J. Comput. Secur. **15**(1), 39–68 (2007)

25. Mishra, B., Nayak, P., Behera, S., Jena, D.: Security in vehicular adhoc networks: a survey. In: Proceedings of the 2011 International Conference on Communication, Computing & Security, pp. 590–595. ACM (2011)

26. Kchaou, A., Abassi, R., Guemara, S.: Towards a secured clustering mechanism for messages exchange in VANET. In: 2018 32nd International Conference on Advanced Information Networking and Applications Workshops (WAINA), pp. 88–93. IEEE (2018)

27. Bahga, A., Madisetti, V.K.: Blockchain platform for industrial internet of things. J. Softw. Eng. Appl. **9**(10), 533 (2016)

28. Lin, J., Shen, Z., Miao, C.: Using blockchain technology to build trust in sharing LoRaWAN IoT. In: Proceedings of the 2nd International Conference on Crowd Science and Engineering, pp. 38–43. ACM (2017)

29. Lei, A., Ogah, C., Asuquo, P., Cruickshank, H., Sun, Z.: A secure key management scheme for heterogeneous secure vehicular communication systems. ZTE Commun. **21**, 1 (2016)

30. Li, C.T., Hwang, M.S., Chu, Y.P.: A secure and efficient communication scheme with authenticated key establishment and privacy preserving for vehicular ad hoc networks. Comput. Commun. **31**(12), 2803–2814 (2008)

31. Malik, N., Nanda, P., Arora, A., He, X., Puthal, D.: Blockchain based secured identity authentication and expeditious revocation framework for vehicular networks. In: 2018 17th IEEE International Conference on Trust, Security and Privacy in Computing and Communications. 12th IEEE International Conference On Big Data Science And Engineering (TrustCom/BigDataSE), pp. 674–679. IEEE (2018)

32. Singh, M., Kim, S.: Branch based blockchain technology in intelligent vehicle. Comput. Netw. **145**, 219–231 (2018)

33. Li, L., Liu, J., Cheng, L., Qiu, S., Wang, W., Zhang, X., Zhang, Z.: CreditCoin: a privacy-preserving blockchain-based incentive announcement network for communications of smart vehicles (2018)

34. Sharma, P.K., Moon, S.Y., Park, J.H.: Block-VN: a distributed blockchain based vehicular network architecture in smart city. J. Inf. Process. Syst. **13**(1), 84 (2017)

35. Buterin, V., et al.: A next-generation smart contract and decentralized application platform. White Paper (2014)

36. Leiding, B., Memarmoshrefi, P., Hogrefe, D.: Self-managed and blockchain-based vehicular ad-hoc networks. In: Proceedings of the 2016 ACM International Joint Conference on Pervasive and Ubiquitous Computing: Adjunct, pp. 137–140. ACM (2016)

37. Leiding, B., Vorobev, W.V.: Enabling the vehicle economy using a blockchain-based value transaction layer protocol for vehicular ad-hoc networks (2018)

38. Sterling, L., Taveter, K.: The Art of Agent-Oriented Modeling. MIT Press, Cambridge (2009)

39. Wang, X., Zha, X., Ni, W., Liu, R.P., Guo, Y.J., Niu, X., Zheng, K.: Survey on blockchain for internet of things. Comput. Commun. **136**, 10–29 (2019)

40. Sung, Y., Sharma, P.K., Lopez, E.M., Park, J.H.: FS-OpenSecurity: a taxonomic modeling of security threats in sdn for future sustainable computing. Sustainability **8**(9), 919 (2016)

41. Barber, R.: Autonomous vehicle communication using blockchain. Ph.D. thesis, The University of Mississippi (2018)

42. Singh, M., Kim, S.: Trust Bit: reward-based intelligent vehicle commination using blockchain paper. In: 2018 IEEE 4th World Forum on Internet of Things (WF-IoT), pp. 62–67. IEEE (2018)

43. Lu, Z., Liu, W., Wang, Q., Qu, G., Liu, Z.: A privacy-preserving trust model based on blockchain for VANETs. IEEE Access **6**, 45655–45664 (2018)
44. Dorri, A., Kanhere, S.S., Jurdak, R.: Blockchain in internet of things: challenges and solutions. arXiv preprint arXiv:1608.05187 (2016)

Author Index

© Springer Nature Switzerland AG 2020
K.-K. R. Choo et al. (Eds.): NCS 2019, AISC 1055, pp. 309–310, 2020.
https://doi.org/10.1007/978-3-030-31239-8

Printed in the United States
By Bookmasters